A Hubterranean View of Syntax

A Hubterranean View of Syntax: An Analysis of Linguistic Form through Network Theory

By

Julie Louise Steele

CAMBRIDGE SCHOLARS

PUBLISHING

A Hubterranean View of Syntax:
An Analysis of Linguistic Form through Network Theory,
by Julie Louise Steele

This book first published 2012

Cambridge Scholars Publishing

12 Back Chapman Street, Newcastle upon Tyne, NE6 2XX, UK

British Library Cataloguing in Publication Data
A catalogue record for this book is available from the British Library

ISBN (10): 1-4438-3577-3, ISBN (13): 978-1-4438-3577-0

~This study is dedicated~

~to Robert and Becky~

TABLE OF CONTENTS

ACKNOWLEDGEMENTS

First and foremost, I'd like to express my deepest thanks to John Ingram my principal supervisor. Immensely patient and always supportive, his encouragement can be found in every aspect of my time at UQ. John constantly provided me with valuable feedback; always there to drive my understanding and offer insightful and thought provoking questions. Thank you John for everything but especially for the most important gift you could give: thank you for believing in me. Also, my warmest thanks to Mary Laughren, David Lee and especially Lynn Wales. Lynn was my first port of call at UQ and I am very thankful for her supportive role.

Some very patient and clever people saved my computer bacon by writing programs to assist my research and I am eternally grateful to them. Many thanks to Andrew Smith for tutoring me in the Leximancer software as well as tirelessly writing modifications especially for my research. I would also like to express my immense gratitude to Hans Jørgen Klarskov Mortensen for writing the program that calculated the frequencies and probabilities, for his patience and tenacity in the development of the software, for recommending literature to read and for enduring with such good humour my first tries and mistakes at writing the Danish language. Thank you so much Hans! Warmest thanks to Sean Fowler who wrote a very handy program that saved me days of tedious work and who never fails to make me laugh with his endearing humour. I'm also grateful to David Lee who shared his macro wizardry and made my computer life so much easier – thank you! Cheers also to Mike Proctor for his help with Perl.

I'd like to thank Mikael Bodén and Rob Pensalfini for reading through my candidature documents. Thanks also to Rob for his comments in the initial stages of my research. I'd also like to honour the contribution that John Ingram, Lynn Wales, Michael Harrington, Mary Laughren and Rob Pensalfini made in establishing a supportive environment for all the students in the Linguistic program at UQ by hosting parties and get-togethers. Their effort makes studying linguistics at UQ the pleasure it is. A big round of applause goes to those attendants of the UQ Linguistics Seminar Series, for their thought-provoking questions. Special thanks go to John Ingram, Lynn Wales, Mary Laughren, Michael Harrington and Henry Brighton for their helpful comments. Many thanks to the EMSAH administration team, especially Annette Henderson, for their support.

I'd also like to give my fellow members of the Phonetics Lab a hearty pat on the back, especially Sacha DeVelle (an honorary member!), Thu Nguyen, Abdel el Hankari, Diana Guillemin, Tina Pentland and Sacha Rixon for their company. Thanks also to Nik Geard, Ben Skellet and Penny Drennan. To Lorraine and Jon Creedon, thank you for your friendship all these years. I also wish to express my appreciation for the support given by Misi Brody, Yves Le Clézio, Stefani Anyadi, Eleni Gregoromichelaki, Ann Warunwatcharin, Marsha Hill and Crispin Lane in those early days. My thanks to Ricard V. Solé for his permission to use the pretty graphic on the cover of this manuscript. And thank goodness for PhD comics!

I would especially like to convey my immense gratitude to the organisations that have helped me financially through my study. Firstly to the Victim support "Criminal Injuries Compensation" group in the UK whose compensation payment helped me finance my first year at UQ. I am immensely appreciative of the generosity of CRLPL at UQ who gave financial support for two years of my study and am indebted to the EMSAH department and the UQ Graduate School for the postgraduate research stipend which made studying at UQ financially possible. Also, I would like to express my appreciation and immense thanks for the support I received from Richard Fotheringham as the head of school.

In these closing words, I could not pass up the opportunity to tip my golden straw hat to my little friends in the land of grain. To my parents Mick and Carole, my brother David and my Nan Mona, much love and thanks for everything. In memory of absent family - especially beloved Nan Brenda and my cousin Linda - thinking of you all. And finally, to Robert Feyrer, whose supportive and wonderfully wise words continue to have such a profound impact on my life. Thank you with all my heart.

ABSTRACT

Language is part of nature, and as such, certain general principles that generate the form of natural systems, will also create the patterns found within linguistic form. Since network theory is one of the best theoretical frameworks for extracting general principles from diverse systems, this study examines how a network perspective can shed light on the characteristics and the learning of syntax.

It is demonstrated that two word co-occurrence networks constructed from adult and child speech (BNC World Edition 2001; Sachs 1983; MacWhinney 2000a) exhibit three non-atomic syntactic primitives namely, the truncated power law distributions of frequency, degree and the link length between two nodes (the link representing a precedence relation). Since a power law distribution of link lengths characterises a hubterranean structure (Kasturirangan 1999) i.e. a structure that has a few highly connected nodes and many poorly connected nodes, both the adult and the child word co-occurrence networks exhibit hubterranean structure.

This structure is formed by an optimisation process that minimises the link length whilst maximising connectivity (Mathias & Gopal 2001 a&b). The link length in a word co-occurrence network is the storage cost of representing two adjacently co-occurring words and is inversely proportion to the transitional probability (TP) of the word pair. Adjacent words that co-occur often together i.e. have a high TP, exhibit a high cohesion and tend to form chunks. These chunks are a cost effective method of storing representations. Thus, on this view, the (multi-) power law of link lengths represents the distribution of storage costs or cohesions within adjacent words. Such cohesions form groupings of linguistic form known as syntactic constituents. Thus, syntactic constituency is not specific to language and is a property derived from the optimisation of the network.

In keeping with other systems generated by a cost constraint on the link length, it is demonstrated that both the child and adult word co-occurrence networks are not hierarchically organised in terms of degree distribution (Ravasz and Barabási 2003:1). Furthermore, both networks are disassortative, and in line with other disassortative networks, there is a correlation between degree and betweenness centrality (BC) values (Goh, Kahng and Kim 2003). In agreement with scale free networks (Goh, Oh, Jeong, Kahng and Kim 2002), the BC values in both networks follow a power law distribution.

In this study, a motif analysis of the two word co-occurrence networks is a richly detailed (non-functional) distributional analysis and reveals that the adult and child significance profiles for triad subgraphs correlate closely. Furthermore, the most significant 4-node motifs in the adult network are also the most significant in the child network. Utilising this non-functional distributional analysis in a word co-occurrence network, it is argued that the notion of a general syntactic category is not evidenced and as such is inadmissible. Thus, non- general or construction-specific categories are preferred (in line with Croft 2001).

Function words tend to be the hub words of the network (see Ferrer i Cancho and Solé 2001a), being defined and therefore identified by their high type and token frequency. These properties are useful for identifying syntactic categories since function words are traditionally associated with particular syntactic categories (see Cann 2000). Consequently, a function word and thus a syntactic category may be identified by the interception of the frequency and degree power laws with their truncated tails. As a given syntactic category captures the type of words that may co-occur with the function word, the category then encourages consistency within the functional patterns in the network and re-enforces the network's (near-) optimised state. Syntax then, on this view, is both a navigator, manoeuvring through the ever varying sea of linguistic form and a guide, forging an uncharted course through novel expression.

There is also evidence suggesting that the hubterranean structure is not only found in the word co-occurrence network, but within other theoretical syntactic levels. Factors affecting the choice of a verb that is generalised early relate to the formation and the characteristics of hubs. In that, the property of a high (token) frequency in combination with either a high degree (type frequency) or a low storage cost, point to certain verbs within the network and these highly 'visible' verbs tend to be generalised early (in line with Boyd and Goldberg forthcoming). Furthermore, the optimisation process that creates hubterranean structure is implicated in the verb-construction subpart network of the adult's linguistic knowledge, the mapping of the constructions' form-to-meaning pairings, the construction inventory size as well as certain strategies aiding first language learning and adult artificial language learning.

Keywords: syntax, network theory, child language, power laws, frequency

Australian and New Zealand Standard Research Classifications (ANZSRC): 200408 Linguistic Structures (incl. Grammar, Phonology, Lexicon, Semantics) 65%, 010299 Applied Mathematics not elsewhere classified 35%.

LIST OF FIGURES

LIST OF TABLES

CHAPTER ONE

INTRODUCING THE HUBTERRANEAN WORLD OF LINGUISTIC FORM

This study explores the idea that patterns in nature may be realised in the linguistic form of our own conversations; that our words dance to the same tune that is played out in our world. It takes little effort to observe in nature how certain patterns reappear in quite diverse environments. For example, the branch configuration of a tree and its leaf structure echoed in the distributary arrangement in a river delta and the blood vessels of a kidney. Recall the spiral of a shell, its shape reflected in the wind currents of a tornado, the florets of a sunflower head and the curl of a ram's horn.

Interestingly, these pervasive patterns have quite simple factors underpinning their shape. Spiral structures may be captured by the mathematical equiangular spiral configuration, a structure which retains its shape as its size increases. The biological branching structure is captured by simple principles related to hierarchical branching, size invariance of 'terminal units', and the "minimization of the energy and time required to distribute resources" (Brown, West and Enquist 2005: 735; West, Brown and Enquist 1999). The enquiry of this study is whether language too reflects omnipresent patterns in nature and in the world at large; patterns that are created by very simple and general principles. If so, these principles may explain why language has certain properties once thought of as specific to language and/or to cognition.

Studies have shown that when one views linguistic form as a word network, language does share specific properties with other systems (Ferrer i Cancho and Solé 2001a), such as social networks (Granovetter 1973, 1983; Amaral, Scala, Barthélémy and Stanley 2000), genetic transcription networks in certain mammals (Potapov, Voss, Sasse and Wingender 2005), ecological food webs (Montoya and Solé 2002), to name but a few. These properties are known as the small world and the hubterranean structure. Before we start this discussion, it would be pertinent to consider how such eclectic systems are compared.

1.1 Network approach

When comparing systems from diverse domains, the network approach is one of the most versatile ways of embodying such systems, as the theoretical machinery behind the network model is very simple. The entities being observed are represented by *nodes* or *vertices* (the circles in Figure 1.1) and the relation between nodes is denoted by a link. A link can be either undirected i.e. an edge or directed i.e. an *arc*. An arc is shown as the arrow in Figure 1.1. Any relation between entities can theoretically be modelled by a network.[1]

For example, in a food web, node A in Figure 1.1 may represent a predator and node B the prey; the link denoting that entity A feeds on entity B. In a language network, the nodes may represent words (Ferrer i Cancho and Solé 2001a; Steyvers and Tenenbaum 2005), phonological word-forms (Vitevitch 2008), sets of synonyms (Miller 1990) or word meanings (Steyvers and Tenenbaum 2005), etc. Links between words would then denote syntactic/collocation relations, association relations or the relation of phonological neighbour, respectively. Whereas sets of synonyms are linked by "conceptual-semantic and lexical relations" (Wordnet 2008), and word meanings by hypernymic and meronymic relations.

Figure 1.1. A network link (arc). Nodes represent entities whilst links represent relations

---

[1] Note that this thesis utilizes the network approach as a tool for representing relations between entities. On this view, the network is not a shorthand notation for a generative algorithm that generates grammatical sentences. If the network were such an algorithm, then the precedence relation represented by the network connections would seem overly generative. However, since the network is a means of representing distributional patterns of relations, the network facilitates the recognition of patterns within the domain of study and allows us to compare these patterns with those found in other systems in the world. (I'd like to thank Michael Harrington for discussing this issue).

In this study, the focus of attention is on how syntactic relations in adult and child speech ay reflect other general properties of systems found in our world. Basic syntax is concerned with word combinations and as such, basic syntactic relations will be identified by distributional analysis. Adopting a purely (non-functional) distributional analysis approach means that language may be represented as a network in which the nodes symbolise words and the links indicate the words that occur adjacently within a sentence.

An example of what a word co-occurrence network would look like is given below. The network is formed from converting word co-occurrences in the text (box a) to links (c). The direction of the link indicating the linear temporal order (precedency) of the words i.e. the preceding word points to the following word.

In representing linguistic form as a word co-occurrence network, this study suggests that (at least) two broad characteristics of language, namely basic constituency and the functional- lexical distinction may be explained by appealing to a certain property that is ubiquitous in the world of patterns: the small world effect. It was observed that many systems exhibit a small average shortest path and a high clustering co-efficient (Watts and Strogatz 1998; see below for definitions). Since these properties make it easier, that is, quicker to traverse from one node to any other in a given network, they were termed small world properties.

A path length between two nodes equates to the number of connections traversed across the network. For example, the path length between the word *about* and the word *and* in Figure 1.2 is two, since two connections are traversed between one node and the other. When the shortest path between all the nodes in the network is calculated and then averaged, in a small world network this value[2] is small compared to a random network[3] and "and offers a measure of a network's overall navigability" (Barabási and Oltvai 2004: 102).

---

[2] This value is also known as the network diameter (Albert, Jeong and Barabási 1999).

[3] A network in which the nodes have roughly the same number of links is known as a random network (see Erdös and Rényi 1960).

Figure 1.2. Language as a network of word co-occurrence links. Sourced from Solé, Corominas-Murta, Valverde and Steels (2005: 2, Figure 2A&C).

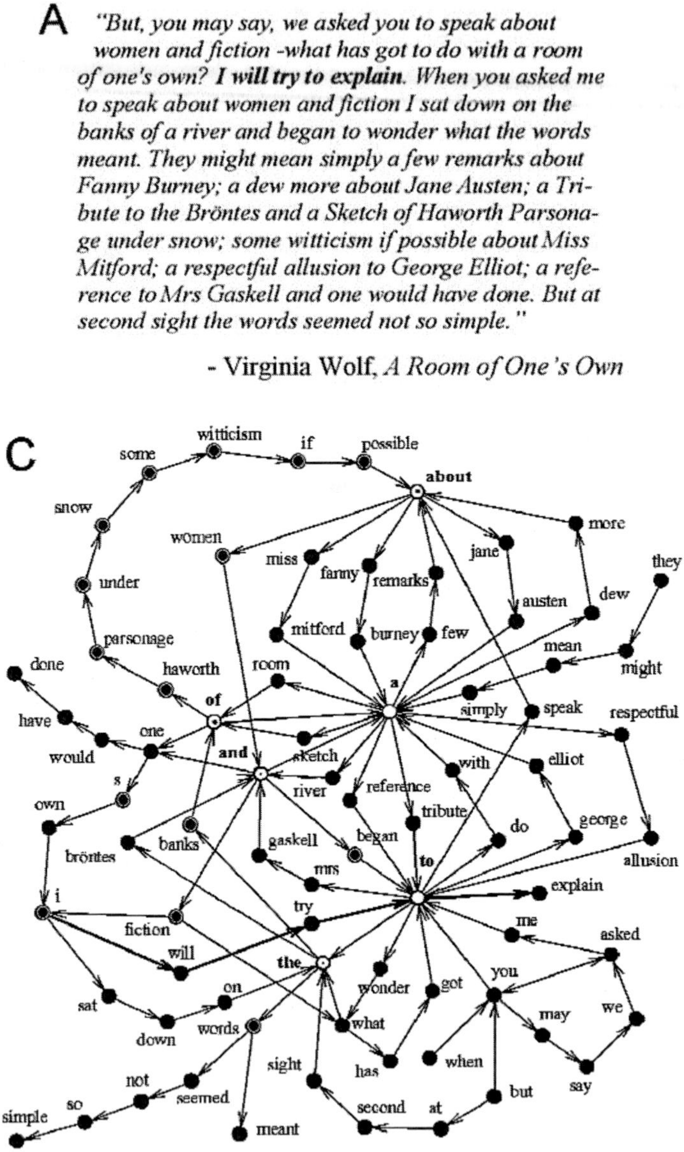

"The average clustering coefficient,<C>, characterizes the overall tendency of nodes to form clusters or groups" (Barabási and Oltvai 2004: 102) in the sense that if node A connects to node B, and node B connects to node C, then how likely is it that node A is connected to node C. A small world has a high propensity to form such groups, in that "the average clustering coefficient of all nodes with *k* links" is high (Barabási and Oltvai 2004:102).

1.2 Hub structure

One way of generating a small average shortest path length - as well as a high clustering co- efficient - is to introduce hubs into the network. Hubs are highly connected nodes, relative to the connectivity of all the nodes (Jeong, Tombor, Albert, Oltvai and Barabási 2000). Hubs increase the overall connectivity of the network by providing short-cuts between two nodes. This idea can be easily visualised by a transportation network. In this network, the nodes are population bases e.g. villages, towns, cities whilst the connections represent the road system servicing and connecting these places. Major cities are the hubs of the network since many highways and roads converge in these cities. If one wanted to travel from a village in one state to a village in another state, the quickest way is to aim for largest city closest to the target village, perhaps passing through other major cities on the way i.e. hubs in the network, rather than travelling on the back roads. Thus, it is easy to see how hubs help navigation through a network and so tend to create a small world.

You may not be surprised that language exhibits these short-cuts, that is, hubs, when language is represented as a word co-occurrence network. It has been shown that the connectivity i.e. degree distribution of nodes in a linguistic network may be represented by a power law (scale- free) function (Ferrer i Cancho and Solé 2001a). Formally, the function represents "the probability [P] that a selected node has exactly *k* links" which is inversely proportional to the power y of k i.e. $P(k) \sim k^{-y}$ (Barabási and Oltvai 2004:102). This distribution equates to a network with many sparsely connected nodes with a few nodes that are highly connected i.e. hubs. These hub nodes tend to represent the 'function' words (Ferrer i Cancho and Solé 2001a).

If the hubs tend to denote words that are construed as function words, then perhaps the hub nodes within a word co-occurrence network are implicated in the definition and the explanation of the theoretical concept. This proposition hints at the possibility that the hubterranean structure may account for very basic syntactic characteristics of language. The functional-lexical (f-l) distinction is the obvious place to start investigating but also since the network encodes precedence relations between words, it would be worthwhile exploring if constituency (i.e. how words combine) might also be related to this hubterranean structure.

Thus, this study examines the possibility of the f-l distinction and constituency being correlated with the properties of the hubterranean structure in a small world network. Furthermore, when this correlation stands (and it will be argued that it does), it would be interesting and productive to know the mechanisms and processes behind hub formation since this will inform us of why language exhibits these particular properties.

Mathias and Gopal (2001 a&b) propose that a hub structure is formed by optimising two factors related to the links in the network. Firstly, the number of links in the network is maximised and secondly, the cost of wiring i.e. the length of a new link, is minimised. When these two factors are optimised, a hubterranean network arises. In the context of the word co- occurrence network, the connectivity factor relates to the shortcuts within the network. That is, shortcuts are words that form many connections and as such maximise the number of word combinations that occur adjacently within a sentence.

An ideal candidate for the cost factor in the linguistic network is the storage cost of linguistic form. It is well known that when two elements occur frequently together, a form of chunking occurs which is very often accompanied (in a speech context) by the reduction of phonetic form as well as a loss of internal structure (Bybee 2002a). The chunk also gains autonomy and each element within the chunk primes the other (Bybee 2002a; see section 2.7.1). Thus in storage terms, the cost of storing two elements is reduced when it is stored in a chunk. Since chunking occurs when the two elements occur frequently together, then it is proposed that frequency of co-occurrence or transitional probability relates to the network cost factor.

In this study, the transitional probability[4] that $word_x$ precedes $word_y$ is calculated by observing how many times $word_x$ precedes $word_y$ and then dividing by the total number of co- occurrences of $word_x$ in the data set i.e. $word_x$'s token frequency.

Equation 1.1 The probability *(P)* of $word_b$ given $word_a$ is calculated by the number of times $word_x$ co-occurs with $word_y$ divided by the number of times $word_a$ ($T(f_x)$) occurs in the text. $T(f_x)$ and $T(f_y)$ are the token frequency of $word_x$ and $word_y$, respectively.

$$P(Wordx, y) = \frac{T\,(fy) \mid T\,(fx)}{T\,(fx)}$$

When the two words occur often together, their transitional probability is high and so chunking is more likely to occur, lowering the storage cost of the two elements. In short, transitional probability is inversely proportional to storage cost. Thus we now have an equivalent linguistic measure for the second factor in the formation of hubs: the factor of storage cost.

In regard to the formation of hubs, Kasturirangan (1999: 1) noted how in small worlds, the links between nodes have "diverse length scales" which may be represented as a power law. Since the length of the link between two nodes in our word co-occurrence network relates inversely to transitional probability of two words co-occurring together, then it is predicted that the distribution of the inverse transitional probability values of the network may be represented as a power law function. This distribution is indeed observed (see chapter five and six) for both a network constructed from adult speech (British National Corpus World 2001) and for one constructed from early child speech (Sachs 1983; MacWhinney 2000a).

It will be argued that the overall distribution of transitional probability values in a corpus relates to cohesion or syntactic constituency, in that "sequentiality is basic to language and constituent structure emerges from sequentiality because elements that are frequently used together bind

[4] Note that the term 'transitional probability' does not denote that there is a transition from one state to another (in contrast to *transition* probability). Note that transitional probability is also known as 'conditional probability' (Bell, Jurafsky, Fosler- Lussier, Girand, Gregory, and Gildea 2003: 1015).

together into constituents" Bybee (2002a: 109). Thus, the higher the probability value the tighter the syntactic constituency. On this view, this notion of syntactic constituency (defined in chapter two) relates to a general property of a hubterranean network.

1.3 Universal properties

Earlier on in this introduction, I enquired whether ubiquitous patterns in natural systems would be able to explain certain properties once thought of as specific to language and/or cognition. Further to basic syntactic constituency (as defined by Bybee 2002a: 109 above) being a linguistic manifestation of a hubterranean property, it will be argued that the properties of hubs in the network correlate directly with the *syntactic* properties of a function word i.e. a high token frequency and a high type frequency.

Indeed, these two properties aid the identification of function words within a network. Function words are useful to locate as they tend to co-occur with certain syntactic categories e.g. the word *a* is associated with an English Count Noun. Since extending our linguistic knowledge to a new construal requires us to ascertain regularities within this knowledge, identifying syntactic categories is an important task. On this view, hubs aid the discovery of such categories.

Consequently, the generation of abstract syntactic elements are facilitated by general systemic properties of the small world hubterranean effect. Both constituency and the f-l distinction are observed in both the child and the adult network and as such these properties are potentially *universal* properties of language (as traditionally assumed). Naturally, one can not reasonably posit a *universal* when only one language has been analysed (see Croft in press). The term 'universal' is used, in this study, to express a hypostudy encouraging further cross-linguistic research into possible exceptions, in the spirit of Dryer (1998).

Notice too how such *universals* do not need to be specified genetically since the process of creating a small world - through the formation of hubs in a network - is driven by the optimisation of two very general factors: storage cost and connectivity. As such, the idea that a property found cross-linguistically is evidence of a genetic endowment or even a cognitive property is not water-tight. Indeed, this type of network investigation helps clarify and rule out which characteristics of language could possibly be genetically pre-described, by finding out first which properties are specific to a kind of system such as a small world.

In this study, it is argued that the *universal* properties of constituency and the f-l distinction are compatible with the functionalist notion of universal (if the properties are proven to be consistent cross-linguistically), since functionalist universals represent cross-linguistic similarities, driven by economy and/or iconicity "with functional, cognitive and semantic explanations" (Dryer 1997: 116). On the approach adopted in this study, there is an additional explanation, one that involves general system properties that are not unique to cognition but are employed in cognition and relate to how the mind stores linguistic representations, specifically linguistic form.

In line with the functionalist approach (Dryer 1997), those syntactic universals with an explanation that is "functional, cognitive and semantic" as well as from a general system, obviate the need for the formal explanation i.e. a prescribed statement referring to a universal syntactic relation (see Dryer 1997: 116). The non-formal explanation is sufficient in capturing the cross-linguistic similarities.

That the network properties discussed above i.e. the hubterranean structure and the power law distribution of storage cost is found in a network composed of child language suggests that children analyse linguistic form similarly to adults. In fact it will be proposed that the factors that are optimised in the formation of the hubterranean structure relate to the learning strategies children adopt when segmenting linguistic form. Thus, the way linguistic form is stored is the same way in which we capture knowledge about linguistic form. In fact, hub properties also relate to extending our linguistic knowledge beyond our previous linguistic experiences (more about this proposition below). On this view, learning and storage processes are one and the same.

That the *universals* pertain to frequency values emphasises that language knowledge and language learning is all about experience. Thus, this study adopts a usage-based constructionist approach where individual speech events are retained. Abstract knowledge emerges through similarities in the semantics and the distribution of these events (Langacker 1988). The word co-occurrence network is able to show the relations between these simple constructions i.e. words, as a very detailed (non-functional) distributional analysis.

The next section of the introduction gives a brief, chapter by chapter summary of the study content, starting with chapter two which details the syntactic perspectives of the study.

1.4 Content

1.4.1 Chapter two

Chapter two introduces the theoretical linguistic perspective adopted in this study - Construction Grammar, in which language is viewed as meaning to form pairings (see Langacker 1987; Goldberg 1995, 2003, 2006). Although this study focuses on the form part of this pairing, certain basic tenets underpinning usage-based approaches (Kemmer and Barlow 2000) and Construction Grammar (Goldberg 1995, 2006), are captured by the representation of language as a word co-occurrence network[5] and are briefly discussed.

Syntactic categories specific to a given construction are adopted (Croft 2001).This study diverges from Croft (2001) in that it will be argued that syntactic relations do exist and are characterised by frequency metrics e.g. syntactic constituency i.e. the grouping of linguistic form and the f-l distinction. Furthermore, the strategies for adult's learning an artificial language are proposed to derive from the optimal storage of linguistic form.

1.4.2 Chapter three

In fact, I will further argue in chapter three[6] that the process of learning how to segment the linguistic form is driven by the optimal storage of this linguistic form.[7] Indeed, children as young as seven months show sensitivity to transitional probabilities (Thiessen and Saffran 2003). This is

---

[5] Symbolic relations are not discussed in this study (see Ferrer i Cancho and Solé 2003a). See also Steels (2005) for a practical experiment in grounding meaning in robotics and for computer simulations of embodied construction grammar, see Bergen and Chang (2005) and (Chang and Mok 2006) as well as Smith (2003) for the software to represent semantic concepts.

[6] This chapter reviews usage-based language learning literature. Note that most researchers use the traditional categories of noun and verb rather than construction-specific categories (Croft 2001).

[7] Naturally, the successful mapping of the semantics to the linguistic form is another important and relevant skill that will not be pursued in this study.

not a species-specific skill since monkeys exhibit such sensitivity too (Hauser, Newport and Aslin 2001).

To become fluent in a language, we generalise beyond our linguistic experience. To do this children adopt general cognitive skills such as intention reading, analogy, schematisation and distributional analysis (Tomasello 2006). This study focuses on distributional analysis in the form of a word co-occurrence network. The network properties of the word co-occurrence network helps shed light on why certain verbs are construed to new syntactic environments by children (see Sethuraman and Goodman 2004a). Five relevant factors were observed – all confounded by frequency (Sethuraman and Goodman 2004a) and are presented. All these factors although seemingly eclectic will be related to the characteristics of a hubterranean structure in chapter nine. Thus, in chapter three we begin to see how the syntactic properties of the linguistic input could be derived from the internalisation of linguistic form.

1.4.3 Chapter four

Network theory provides the theoretical scaffolding to support such a hypostudy. In chapter four, network theory is introduced and the discussion focuses on word co-occurrence networks and small worlds. Studies in this area are discussed although none deal with the concepts of constituency and f-l distinction. This study will address this issue in that these basic properties of linguistic form are resultant of general system properties.

Studies on child language and network theory are summarised in section 4.4 (Ninio 2006a; Corominas-Murtra, Valverde and Solé 2007 a&b; Ke and Yao 2008). In section 4.5, our discussion focuses on Mathias and Gopal's (2001 a&b) explanation as to why and how small world networks form. In a simulation, it was shown that small worlds occur when the *cost* factor in connecting two nodes is optimised by the path length (corresponding to the number of connections between nodes). As noted by Kasturirangan (1999), multiple link lengths that are distributed by a power law function characterise a small world, hubterranean network (see Mathias and Gopal 2001 a&b). Thus, a distribution of cost values that form a power law would be indicative of a small world network as well as hubterranean structure.

1.4.4 Chapters five and six

Chapters five and six will show that both the adult and child word co-occurrence networks do indeed possess the properties of a small world: a small average shortest path between a pair of nodes and a high clustering co-efficient. In addition, the distribution of word frequencies and word degree i.e. connectivity, is captured by a power law in line with other small world networks with hubs. In agreement with a syntactic dependency network (Corominas-Murtra, Valverde and Solé 2007 a&b), the two word co-occurrence networks were found to exhibit disassortative behaviour, in that highly connected nodes resist connecting with other hub nodes.

Furthermore, the distribution of link lengths is captured by a power law, in agreement with Kasturirangan (1999) and Mathias and Gopal's (2001 a&b) observation that hubterranean structure is accompanied by this power law distribution. In linguistics terms, the link length between two nodes correlates inversely with the transitional probability of the respective word pair; that is, the degree of chunking. Chunking produces "short coding" (in the sense of Haspelmath (2008a: 2); see section 2.7), reducing the storage cost of the linguistic forms (see chapter two). Since constituents are formed from chunking elements by how frequently they co-occur (see Bybee 2002a), then the power law distribution of (inverse) transitional probabilities represents the simplest form of syntactic constituency. Consequently, this fundamental linguistic phenomenon originates from a hubterranean structure and is not a property specific to linguistic form.

This power law function is found in the child word co-occurrence network and so this simplest form of syntactic constituency is found in early child language as well as adult language. This finding may establish syntactic constituency – defined by the power law distribution of cost - as a general property. Naturally it is an empirical question as to whether this relation is cross-linguistic. This *universal* hypostudy is easy to refute since it only requires one language not to adhere to the power law cost function. If such an exception is found then the power law may represent a *statistical* universal (Dryer 1998).

1.4.5 Chapter seven

This chapter introduces the network perspectives that will help identify abstract syntactic properties and structure within the network e.g. syntactic categories. Since syntactic categories are specific to a construction, it would be relevant to analyse the local properties of the network, such as the link patterns. This chapter focuses on significant link patterns in the network known as motifs: their definition and their identification. The motifs of the child and adult language are presented.

1.4.6 Chapter eight

The specific instances and extensions of motifs in the adult and child word co-occurrence network are discussed. For example, the motif 6 link pattern tends to reiterate around (i.e. is connected to) a specific function word, to which the traditional verb and noun classes are anchored. It is argued, however, that from a network approach such general syntactic categories are inadmissible. Rather, construction-specific syntactic categories are preferred in line with Croft (2001). One may also propose that function words occur in the child language using the criteria established in this chapter, since the child network has a hubterranean structure. Thus hubs represent function words in the child language even though the set of function words differs from that of an adult.

1.4.7 Chapter nine

Chapter nine discusses how the hubterranean structure is implicated at many different theoretical levels. Firstly, it is argued that the optimisation process of storing linguistic form facilitates the segmentation of the caregiver's input, in that the probabilities between words are constrained by a power law distribution i.e. a few words tend to be highly connected (and frequent). This approach provides an explanation as to how adjacent monosyllabic words are segmented in a probabilistic model (since transitional probabilities between part words are usually employed to identify words). These words are generally highly frequent and represent hub words that are created when linguistic form is stored optimally i.e. in a hubterranean configuration.

Since function words tend to be represented as hub words, being defined and identified by a high type and token frequency, the interception of the truncated power laws of degree and frequency will identify a function word. This interception facilitates the discovery of (construction-specific) syntactic categories both in the adult and the child network, as specific syntactic categories are associated with certain function words. The process of automorphic duplication creates the category which itself establishes a new level of syntactic analysis i.e. abstract syntax. On this view, 'tinkering' may be implicated in the (initial) formation of abstract syntax (for tinkering see Jacob 1977; Solé and Valverde 2006).

The hubterranean structure is also implicated in the five factors that identify which verbs are construed early in a new syntactic environment (see Sethuraman and Goodman 2004a). Being either hub words or linking to them, these verbs are easy to identify (see Boyd and Goldberg forthcoming). Since the hub structure originates from optimally minimising the storage cost of word representations whilst maximising word combinations, then early verb generalisations are facilitated by this optimisation process.

Furthermore, the hubterranean structure is implicated in the verb-construction subpart network, the construction form-to-meaning pairings and the construction inventory, as well as certain general principles guiding first language and artificial language learning.

1.4.8 Chapter ten

In the concluding chapter of this study, I discuss the limitations and issues falling out from this analysis. The message of the study is very clear: not only can we view language as consisting entirely of constructions **"it's constructions all the way down"** (Goldberg 2006: 18), but these constructions and their relations are stored together in a way that form hubterranean structures.[8] To misquote Goldberg (2006), **it's hubs all the way up!**

[8] At least for the storage aspect of syntax i.e. the knowledge of language, hubterranean structure is prevalent. It is more likely that syntactic computation, i.e. the utilisation of this knowledge, will be represented by a taxonomic/branching structure. However, this issue is beyond the scope of this study.

CHAPTER TWO

SYNTACTIC CONSTRUCTIONS

2.1 Overview

This chapter will discuss the theoretical linguistic perspectives adopted in this study – notably construction grammar with a usage-based approach - with particular emphasis on the property of frequency and how it relates to two broad, coarse grained properties of syntax: namely syntactic constituency and the functional-lexical (f-l) distinction.

The key insights of construction grammar are summarised in section 2.2. Constructions, i.e. the meaning-to-form pairings, are the syntactic primitives of language (Croft 2001). Syntactic categories are not cross-linguistic entities but are construction-specific (Croft 2001). Construction representations are stored as a taxonomic network (see Goldberg 1995; section 2.3) and psychological principles effect their organisation (Goldberg 1995; section 2.4).

A usage-based approach is also emphasised in this study in that language experience underlies much of our linguistic knowledge (section 2.5). As such, grammaticality is related to the familiarity of a given construction and not to the adherence of a set of rules. That is, repetition i.e. frequency, is implicated in grammaticality. It will be shown that certain syntactic relations exist and that these are potentially cross-linguistic although not innate i.e. the relations are characteristic of a particular general system – specifically, a small world hubterranean network.

In respect to cross-linguistic similarities, frequency has an important role to play in the characterisation of linguistic form. One such characterisation is constituency. Croft (2001) convincingly demonstrates that constituency tests are inconsistent and not useful in identifying constituents (section 2.6.1). In fact, there are different formal groupings, each creating chunks of linguistic form that do not overlap. However, (non-)linguistic elements that co-occur highly frequently form chunks (section 2.7). Effects known to occur with elements frequently co-occurring together are also observed in the chunking of linguistic form: loss of internal structure, priming, autonomy and form reduction (section 2.7.1).

In fact, Bybee (2002a) has shown that constituent structure can be identified by appealing to frequency co-occurrence values (section 2.7.2). One problem for this account is discontinuous dependencies (section 2.7.3). This is unsurprising since the traditional notion of constituency incorporates the symbolic relation that "conceptual grouping motivates syntactic contiguity" (Croft pc. 2009; see also the psychological principle II given in section 2.4). Thus, the traditional concept of constituency conflates two phenomena: a symbolic relation and a relation that appeals to the chunking of linguistic form, conditioned by frequency and linear order.[1] This study focuses on the latter phenomenon, the constituency of linguistic form, which will subsequently be referred to as syntactic constituency.

This study argues that syntactic constituency may be represented by the distribution of frequency co-occurrence values, revealing a potential primitive of linguistic form - one that is not specific to language but characteristic of a particular general system i.e. a small world (see chapter 8). To capture this frequency distribution, formal accounts would use the term 'constituency' to 'explain' the associated cross-linguistic properties (see Dryer 1997). There is however no need to stipulate a formal cross-linguistic primitive, since the explanation for the primitive derives from an independent factor, functionally motivated by the storage cost of linguistic form (see Dryer 1997).

Another coarse-grained property of linguistic form under examination is the functional-lexical distinction (section 2.8). The syntactic properties of function words are discussed in respect to their frequency properties and their local relations within a given construction (2.8.1). Studies in how we learn artificial languages highlight the non-gradient nature of the function word (section 2.8.1.1).

Other properties of functional expressions discussed are: "restricted syntactic contexts" (Cann 2000: 41; section 2.8.1); closed class expression (Cann 2000; section 2.8.2); association with a general category (Cann 2000; section 2.8.3); consistent occurrence with all expressions that have a

---

[1] Thanks to William Croft for discussion on this issue.

particular property (Cann 2000; section 2.8.4); modification of a lexical element's syntactic category (Cann 2000; section 2.8.5); for functional expressions "there are no processes that alter their selectional properties as there are for lexical expressions" (Cann 2000: 42; section 2.8.6).

It will be argued that the properties of a function word fall out from a frequency-based approach attending to both token and type frequency. Likewise with constituency, these properties relate to optimising the storage of linguistic form (see chapter five). As such, there is an independent explanation for these properties and so the f-l distinction is redundant as a cross-linguistic 'universal' relation in a formal sense (Dryer 1997). The next section will give a brief outline of the fundamental tenets adopted by Construction Grammar.

2.2 Constructions

Language is composed of constructions: conventionalized and "LEARNED PAIRINGS OF FORM WITH SEMANTIC OR DISCOURSE FUNCTION" (Goldberg 2006: 5, caps are authors; Goldberg 1995; Fillmore, Kay, and O'Connor 1988; Langacker 1987). Constructions are found at "ALL LEVELS OF GRAMMATICAL ANALYSIS" (Goldberg 2006: 5, caps are authors), in various sizes (Wade and Möbius 2008) and complexity, ranging from single morphemes such as *–ed* to abstract linguistic structures such as the transitive construction (see Table 2.1).

There is no distinction between the core grammar and the peripheral constructions in a language; the latter being considered just as important in gaining insights into language structure. Construction Grammar, which can explain non-core cases, can be equally applied to core cases (Goldberg 1995: 6; for a similar account of the continuity between the idiosyncratic elements and generalisations using rules and (fragments of) phrase structure, see Culicover and Jackendoff 2006).

Table 2.1 "Examples of constructions, varying in size and complexity" (Goldberg 2006: 5, Table 1.1).

Morpheme	e.g. *pre-, -ing*
Word	e.g. *avocado, anaconda, and*
Complex word	e.g. *daredevil, shoo-in*
Complex word (partially filled)	e.g. [N-s] (for regular plurals)
Idiom (filled)	e.g. *going great guns, give the devil his due*
Idiom (partially filled)	e.g. *jog* <someone's> *memory, send* <someone> *to the cleaners*
Covariational Conditional	The Xer the Yer (e.g. *the more you think about it, the less you understand*
Ditransitive (double object)	Subj V Obj1 Obj2 (e.g. *he gave her a fish taco, he baked her a muffin*)
Passive	Subj aux VPpp (PP_{by}) (e.g. *the armadillo was hit by a car*)

Contructions do not have a derivational component i.e. an underlying syntactic structure; they are 'monostratal' (Goldberg 1995:7). Furthermore, since all linguistic structures are constructions, there is no "strict division …between the lexicon and syntax" (Goldberg 1995: 7). A lack of strict division is also pertinent to semantics and pragmatics. Both types of information are represented in constructions. A schematization of a construction's pairing is given in Figure 2.1. Meaning consists of a semantic, pragmatic and discourse functional component, whilst the construction's form represents the surface form i.e. the syntactic, the morphological and the phonological component.

Figure 2.1 "The symbolic structure of a construction", sourced from Croft (2001: 18, Figure 1.2)

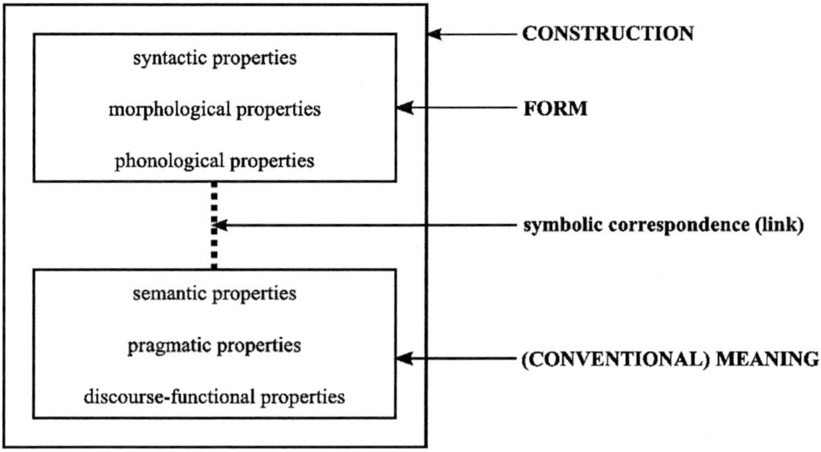

Since a semantic/pragmatic component is inherent in constructions, each construction has a slightly different interpretation. Consequently, the construction and the verb within "are interrelated but independent" (Goldberg 1995: 9). This relationship can be seen clearly in the below examples (Example 2.1 a-e).

Example 2.1 Sourced from Goldberg (2006: 7, Examples 8-12)

a) He sliced the bread (transitive)

b) Pat sliced the carrots into the salad (caused motion)

c) Pat sliced Chris a piece of pie (ditransitive)

d) Emeril sliced and diced his way to stardom (way construction)

e) Pat sliced the box open (resultative)

In each of the examples given, the verb's meaning of 'a cutting action with a sharp object' is retained. The different interpretation attributed to each example is resultant of the form-meaning pairing of the individual construction.[2] For example, the transitive construction (Example 2.1a) imparts a general meaning that there is "something acting on something else", the caused motion construction (Example 2.1 b) "something causing something else to move", in the way construction (Example 2.1 d) "someone moving somewhere despite obstacles", etc (Goldberg 2006: 7). Consequently, any pattern in language that is not predictable from its function or form components (or from another construction) may be represented as a construction. In fact, "patterns are stored as constructions even if they are fully predictable as long as they occur with sufficient frequency" (Goldberg 2006: 5).

Furthermore, this study emphasises the non-reductionist nature of language: constructions are *primitive* grammatical units i.e. independent units (Croft 2001). "Complex constructions, made up of multiple elements, may have either schematic or substantive elements, or a combination of the two. But atomic "constructions" can only be substantive, that is, they can only be specific words" (Croft 2001: 362-3).

2.2.1 Syntactic relations

Although Croft (2001) assumes no syntactic relations between elements in a construction (only knowledge of syntactic roles) – see Figure 2.2, this study maintains that frequency of co-occurrence between words is a primitive syntactic relation (albeit a probabilistic and graded one). The primitive is generated by representational storage concerns of words and is hypostudyed to be general to language. I will be arguing that the functional-lexical distinction makes use of syntactic relations within a given construction as shown by the dotted line in the diagram; the distinction is identified by primitives generated by the optimisation of the storage and connectivity of words.

Since a formal syntactic universal is merely a stipulation that 'explains' cross-linguistic similarities, note that no such universal is required for an explanation of the characteristics ascribed to the f-l distinction (see Dryer 1997). In that, the optimal storage of a system with maximal connections accounts for the linguistic characteristics of function and lexical words and so the formal stipulation of the f-l distinction is redundant (see Dryer 1997).

---

[2] The alternative approach is to associate multiple meanings to the verb (e.g. Pinker 1989). For advantages of constructionist approach i.e. not having to posit null verbs, see Goldberg (2006: 6-9).

Figure 2.2 Sourced from Croft (2001: 176, Figure 5.1) "The internal structure of a construction (exploded diagram)". Bolded lines in this diagram represent the analysis of a construction by Radical Construction Grammar (excluding the dotted syntactic relation).

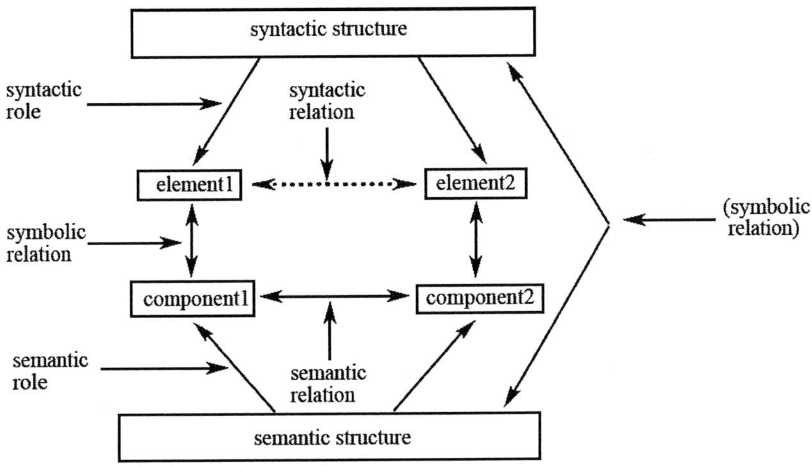

2.3 A network of constructions

Constructions are not just representational islands in a sea of syntactic knowledge - they form a taxonomic network particular to the given language. In this network, constructions represent nodes (Croft 2001: 25) and the links between the nodes represent inheritance relations. As Goldberg (1995: 67) points out, "The inheritance network lets us capture generalizations across constructions while at the same time allowing for subregularities and exceptions". Figure 2.3 below from Croft (2001: 56) demonstrates the argument structure links between the English transitive and intransitive construction. Note that the doted lines represent taxonomic links between construction components whilst the solid lines indicate taxonomic links between the constructions as a whole.

Both the transitive and intransitive constructions contain an argument of the verb i.e. SbjArg, meaning the 'Subject as an argument' and a predicate i.e. Pred (Croft 2001). The transitive construction captures facts about how the verbs *kick* and *kiss* are associated with two arguments: the transitive subject (TrSbj) and the transitive object (TrObj; Croft 2001). Idioms such as *kick the bucket, kick the habit* are also constructions since their meanings are not strictly predictable from their constituent parts but nevertheless have similar surface form to the *kick* construction i.e. two arguments relating to the verb (Croft 2001).

Figure 2.3 Radical Construction Grammar representation of construction taxonomy" Sourced from Croft (2001: 56, Figure 1.15).

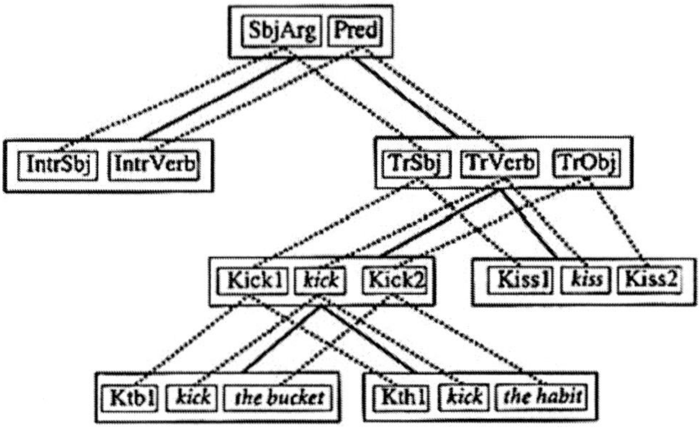

In line with Goldberg (1995) construction networks also have subpart links, "when one construction is a *proper subpart* of another construction and exists independently" (Goldberg 1995: 78). Goldberg (1995) gives an example of the intransitive motion construction being a subpart of the caused-motion construction (see Example 2.2).

Example 2.2 (a) caused-motion construction and (b) the intransitive motion construction. The theme argument is bolded and the goal argument is italicized.

a) She wheeled **the settee** into *the wall.*
b) **The settee** rolled into *the wall.*

Both constructions have two syntactic arguments relating to a theme and a goal. The theme is represented by the syntactic object in the caused-motion construction (a) and by the subject in the intransitive motion construction (b), bolded in the example. The goal is an oblique object in both constructions (italicized in the example). Consequently, "the syntactic and semantic specifications of the intransitive motion construction are a subpart of the syntactic and semantic specifications of the caused-motion construction" (Goldberg 1995: 78).

2.4 Principles of organisation

Certain psychological principles underpin the theory behind construction networks (Goldberg 1995). These principles are given below, having been quoted from Goldberg (1995: 67-68). The ensuing discussion is a summary of Goldberg (1995: 67-72).

Principle I: "The Principle of Maximized Motivation" (Haiman 1985:20; Lakoff 1987)
"If construction A is related to construction B syntactically, then the system of construction A is *motivated* to the degree that it is related to construction B semantically… Such motivation is maximized" (Goldberg 1995:67)

Principle II: "The Principle of No Synonymy"
"If two constructions are syntactically distinct, they must be semantically or pragmatically distinct (Bolinger 1977; Haiman 1985:24, Clark 1987; MacWhinney 1989)."

"Pragmatic aspects of constructions involve particulars of information structure, including topic and focus, and additionally stylistic aspects of the construction such as register.

Corollary A: If two constructions are syntactically distinct and S(emantically)-synonymous, then they must not be P(ragmatically)-synonymous.

Corollary B: If two constructions are syntactically distinct and P-synonymous, then they must not be S- synonymous." (Goldberg 1995:67)

Principle III: "The Principle of Maximized Expressive Power" (Goldberg 1995: 67).
"The inventory of constructions is maximized for communicative purposes"

Principle IV: "The Principle of Maximized Economy" (Haiman 1985:24-25)
"The number of distinct constructions is minimized as much as possible, given principle III" (Goldberg 1995: 67-68)

Principles II and III relate to the notion of isomorphism in language. Principle II captures the idea that variance in the surface form equates to variance in the semantics or pragmatics. Differences in semantics or pragmatics would then produce differences in the surface form, principle III. Divergence from isomorphism is observed in language as evidenced by polysemy and homonymy which minimise the amount of distinct constructions, in line with principle IV. As Goldberg (1995: 69) notes, principle III and IV oppose each other since the former principle encourages distinct forms whilst the latter principle restrains them. Together "these two principles mutually constrain each other" (Goldberg 1995: 69).

The principle of Maximised Motivation (I) captures the balance between "predictability and arbitrariness" found in language (Goldberg 1995: 69-70). Goldberg cites Langacker's (1987: 47) example of nouns that take the plural form: *scissors, pants, glasses* and *binoculars* "as a matter of convention". Even though one cannot predict this behaviour from their designated sign, the pluralisation does predict a duality in character of the designated object. Thus, the dual character of the object motivates the plural form.

In regard to constructions, a construction is more motivated if it shares some (non-conflicting) relation with another construction. There is motivation for one construction to inherit properties from another i.e. to have a degree of predictability. The effect of maximizing 'motivation' in respect to constructions is that they inherit more properties and consequently introduce more redundancy into the taxonomic network, creating a network that coheres more.[3] Consequently, the inheritance links between constructions are maximised, with principle II setting a lower limit to the drive for motivation, curtailing synonymic closeness between the semantics or pragmatics of constructions and their syntactic form.

It will be argued in chapter nine that these principles relate to the same optimisation process that forms hubterranean structure in word co-occurrence networks.

---

[3] However, the designations of the constructions would still be arbitrary. For example, the use of the copula and the *–ed* morpheme in the passive construction are not motivated but arbitrary.

2.5 A Usage-based approach

The usage of language - both production and comprehension – forms the basis of the linguistic system (Kemmer and Barlow 2000); a system that is shaped by "mechanisms that are not limited to the linguistic domain" (Abbot-Smith and Tomasello 2006: 276). Language is "maximalist" and "non-reductive" since specific details of language usage are retained and stored in varying degrees of size and complexity (Langacker 1988). For example, a study has shown that three word sentences are stored as complete units with high frequency strings being recognised faster as grammatical sentences (Bod 2001; see Jurafsky 2003 and Ellis 2002 for a review of frequency studies). Furthermore, whole sentences are able to be both recognised and recalled "at above chance rates" after just one exposure to a 300 word text with pictures (Gurevich and Goldberg forthcoming: 2).

On this approach, knowledge of language consists of very specific knowledge of constructions combined with abstract knowledge i.e. a generalisation or *schema* (Kemmer and Barlow 2000: xxiii). The generalisation or schema develops from the repeated usage of items perceived to be similar i.e. sharing "a set of co-occurring properties" (Kemmer and Barlow 2000: xxiii; Abbot-Smith and Tomasello 2006). Language is also driven by a "bottom-up" process since it is these commonalities between specific linguistic items or knowledge that generate more general knowledge of language (Langacker 1988).

Since the linguistic system is based on exemplars of language use, usage events are not only "products of the speaker's linguistic system", they continue to shape, and reshape the linguistic system "in a kind of feedback loop" throughout the life of the speaker (Kemmer and Barlow 2000: ix; Bod 2006). Thus, the linguistic system is dynamic (see Culicover and Nowak 2003a) in that it is "ever coming into being rather than static, categorical, and fixed" Bybee (2006:4).

Furthermore, "learning is the result of unsupervised interactions with exemplars" made simpler by the fact that the ubiquitous examples in the learning environment reflect the general properties of the linguistic system (Hershberg and Ninio 2004). For example, the use of high frequency verbs by caregivers. Such verbs are usually semantically light and their general semantics reflect certain argument constructions (Goldberg, Casenhiser and Sethuraman 2004) or verb class (Hershberg and Ninio 2004; Ninio

1999a). Such typical or useful examples (Hershberg and Ninio 2004) facilitate the representation of the linguistic input as linguistic knowledge.

The knowledge of grammaticality and its judgement is very closely tied to the usage of language. As Bybee (2007: 695) notes the "lack of grammaticality is lack of familiarity". A study that rated English sentences for grammaticality supports this view (Luka and Barsalou 2005). In this study, it was found that sentences read earlier in the experiment received a higher grammaticality rating (Luka and Barsalou 2005). Thus, attention to the frequency of the exemplars is very important in characterising language (Kemmer and Barlow 2000; Bybee 1995, 1988; Haiman 1991).

Indeed, frequency is heavily implicated in grammaticalisation; being not just "a result of grammaticalization, it is also a primary contributor to the process" (Bybee 2003: 602). Specifically, Rostila (2004:1) emphasises the "role of type frequency in the spread of pas into the context of all verbs" in the grammaticalisation of *pas* in French. Asymmetries in frequency of co-occurrence have been proposed to account for linguistic phenonema (Haspelmath 2008 a&b) that previously had been explained by iconic relations in regard to complexity, quantity and cohesion (Croft 2008; Haiman 2008). Frequency asymmetries also account for certain *universal* contrasts in reflexive marking (see Haspelmath to appear). In respect to morphological patterns, type frequency is related to the pattern's productivity (Bybee 1995).

Furthermore, frequency is implicated in the internal structure of a construction, and the binding of elements within an utterance (Bybee 2002a). Bound elements form linguistic groupings known as constituents. There are however different types of linguistic groupings (Croft 2001). For example, the traditional notion of a constituent utilises "conceptual grouping" (Croft 2009 pc). However, as just noted, constituents are derivable from attending to the frequency of sequential elements – motivating the reduction of their linguistic form (Croft 2009 p.c.).

The next section will briefly discuss the traditional notion of constituency. This study will argue that in agreement with Croft (2001), "constituency is not an abstract global structure for the syntax of constructions" in respect to the traditional sense of constituency. Furthermore, it will be argued that a frequency-based approach to the syntactic grouping of linguistic form will reveal a cross-linguistic functional primitive that captures basic constituent formation.

2.6 Groupings

Identifying categories and constructions assumes that language is made up of parts that bind together to form larger – more complex - units. The mainstream view of this binding mechanism is the notion of constituency. The Wikipedia (2008) entry on 'constituent (linguistic)' defines a constituent as "a word or a group of words that functions as a single unit within a hierarchical structure". For example in the below phrase structure tree (Figure 2.4), the verb joins with a noun phrase to form a verb phrase and this verb phrase joins together to form a sentence with another noun phrase (IP). On this view, there is a hierarchy of phrases, with the sentence phrase (IP) dominating the verb phrase (VP). In Generative grammar, hierarchical structure is generally seen as innate and primitive.[4]

Figure 2.4 An example of a simplified Generative tree structure representing hierarchical phrase structure.

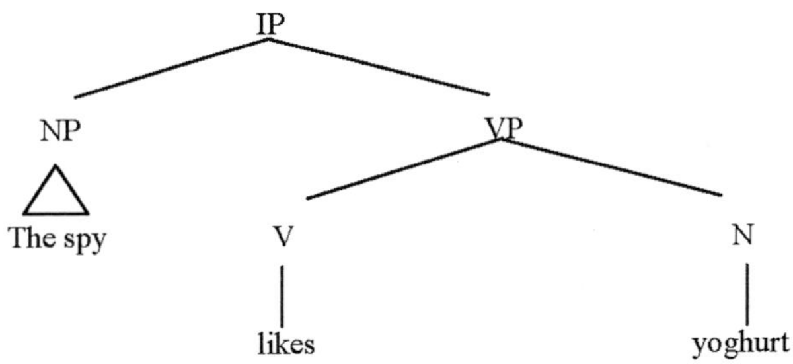

Thus, mainstream linguistics assumes that syntactic units known as constituents form a hierarchy that is primitive. In short, constituency is not derived from linear order.

2.6.1 Constituency tests

Traditionally, certain tests have been used to identify constituents. For example, a constituent may be substituted by a proform e.g. ***a duck*** *is diving* versus ***it*** *is diving* where *a duck* is a nominal constituent. In addition, the constituent may be used in isolation e.g. in answer to the question: *What just dove? A:* ***a duck***. Also, constituents tend to be able to

[4] However, see Culicover (2001:1) for the view that hierarchy is "a reflex of the linear order and the corresponding conceptual structure".

occur in various places in utterances. For example, the constituent *the duck* may occur as the subject or as an object: ***a duck*** *ate the bread; the hunter shot* ***a duck***. Furthermore, it is said that only constituents may be co-ordinated e.g. the noun phrase *a duck* in the utterance ***a duck*** *and a swamp hen were fighting*, whilst ellipsis constructions reveal constituent structure e.g. *the duck will* ***eat the grain*** *and the swamp hen will* ***Ø*** *too* , where *eat the grain* is the ellipsed verb phrase constituent.

It is important to note that these constituency tests do not conclusively identify constituents since not all elements that are regarded as constituents in one test necessarily obey all the other tests. Consequently, the tests do not constitute proof of a constituent; instead they help form an *argument* that a group of elements may or may not be classed as a constituent[5] (see Adger 2003:66, cited in Croft 2006). For example (see Croft 2001), the elements *60 kilos* is a constituent since it can stand alone as an entire utterance e.g. in response to the question *how much do you weigh?,* and it can be substituted by a proform with a supporting point gesture e.g. *I weigh that* – the speaker pointing to where the weight is recorded. However, the element

60 kilos is not found in all positions in an utterance as Example 2.3 illustrates. Even so, *60 kilos* is still classed as a constituent, "It is even more important to notice that when one of these tests fails, we cannot conclude that the sequence is not a constituent, we can only conclude that there is no evidence that it is" (Adger 2003:66, cited in Croft 2006).

Example 2.3 Passivisation (see Croft 2001: 35)

a) I weigh 60 kilos
b) *60 kilos is weighed by me

This claim contradicts the practice of identifying specific syntactic units: "failing to form a Passive is normally taken as evidence against the Direct Object status of the argument in the counterpart Active" (Croft 2006: 3). As shown by the *60 kilos* example, whilst it is not identified as a direct object it still has the status of a constituent - even though it has failed the same test. If failing a test can sometimes disqualify a group of elements from being a certain type of constituent, and yet at the same time the

---

[5] As Croft (2006) points out, an element failing a test has spurred linguists to argue against its constituency.

failure does not prove fatal against the constituency judgment (Croft 2006: 3), how is this inconsistent reasoning ironed out?

It seems that ad hoc accounts must be introduced, as to why in particular the grouping *60 kilos* does not undergo passivisation. "And almost all of the time, distributional tests do not match if you examine the full range of units that are supposed to fill the role in the construction used for the distributional test" (Croft 2006: 3). As a result, many ad hoc measures would be required. Furthermore, as Croft (2006) points out, since not all tests occur in all languages, then different tests will need to be used to identify constituents cross-linguistically. This point leads onto the further question of how one can be sure that it is the same syntactic unit being tested for across the different languages (Croft 2001). Consequently, "cross-linguistically valid criteria" is required for evidencing general abstract constituent structure (Croft 2001: 191).

Croft (2001; section 5.4.3) proposes three criteria for formal groupings: Congruity/separability, intonation units and the unit of first-position self-repair. Congruity/separability of elements is generally based on the element's semantic closeness e.g. ***look up*** as in *We will **look up** the info*. Mismatches commonly occur between the semantics and the form e.g. *We will **look** the info **up***. Thus, contiguity does not provide evidence for the syntactic relation of constituency, but rather "in most cases, syntactic contiguity reflects semantic relations" (Croft 2001:186) and is reflective of a symbolic relation. Likewise, the comparison of intonation units with grammatical units reveals that these two types of groupings are not always isomorphic (Croft 1995, 2001:192).

The final grouping Croft discusses is the unit of first-position self-repair. When a speaker errs "before the point of projected closure of the utterance" (Croft 2001: 194), a repair strategy repeats (i.e. recycles) the segment or grammatical unit again with the correction. The segments that are repeated with the repair may be defined as a "type of formal grouping for the English Noun Phrase, the Prepositional Phrase, and the Clause" (Croft 2001: 195). However, it seems that verb phrases do not occur as a first-position self-repair grouping.[6] Since "the three criteria do not match, and yield different formal groupings" (Croft 2001: 195), Croft (2001: 202) concludes that "constituency is not an abstract global structure for the

[6] However, as Croft (2001) notes this effect may be due to the data having English pronominal subjects which may be interpreted as proclitics. Thus the recycling occurs at the start of the English pronominal rather than after.

syntax of constructions". Indeed, without stipulating a constituent structure, these three criteria for constituency are accessible to the hearer (Croft 2001), when attending to linear order, repetition and prosodic information in a given set of utterances. Croft (2001) goes further and proposes that there are no syntactic relations within constructions: coded dependencies such as constituency, linear order and agreement are actually symbolic relations (see Croft 2001 chapter 6 for a detailed discussion which is beyond the scope of this study).

However, this study will propose that there is another type of formal grouping that is syntactic (rather than symbolic) and may constitute cross-linguistically valid criteria for syntactic constituency i.e. frequency of co-occurrence. This domain of study will concentrate on words – the smallest substantive constructions. Approached from this theoretical angle, the next section will discuss the frequency of co-occurrence criteria for constituency.

The discussion is based on Bybee's (2002a) hypostudy and her argument as to how sequential order is primitive and derives hierarchical structure. The study opens the door to viewing the constituency relation as a relation between word constructions. It will be argued in this study that the distribution of word constructions will be a general property of language. The property is generated by representational storage concerns i.e. optimising the cost of storing words whilst allowing the maximal combination of words. Thus, recourse to a formal (cross- linguistic) syntactic relation which captures constituency is unnecessary.

2.7 Frequency of Occurrence

Frequency of co-occurrence has been implicated in identifying small constituents, such as words, without recourse to knowledge of semantic relations. For example, the initial segmentation of a string of syllables in an artificial language can be achieved by purely attending to transitional probabilities i.e. noting adjacent word co-occurrences (Saffran, Aslin and Newport 1996; Aslin, Saffran and Newport 1998; see chapter three for discussion on transitional probability studies in learning an artificial language). Attention to both low and high transitional probabilities has also been evidenced in a recent experiment to track word-to-object mappings (Vouloumanos 2008). In fact, not only may frequency be implicated in constituent i.e. segment identification, it may be a factor in constituent formation.

It has been known for some time now that the process of repetition increases the fluency of the repeated skill. The *power law of practice* illustrates this relationship: the more repetitions of a skill, the quicker the act is performed (see Newell and Rosenbloom 1981). Repetition increases cohesion within the discrete elements that make up the skill or act. For example, the discrete elements 'going to' which form part of the frequently used phrase 'be going to', often reduce to the single form 'gonna' (Bybee 2002a).

Notice, however, that phonological reduction does not always match conceptual groupings, as the *gonna* example highlights (Bybee 2002a). Reduction, then, may not serve to indicate symbolic relations (such as contiguity). Rather cohesion is produced by the frequent co- occurrence of elements. For example, the phonetic form of the word *of* was more reduced in highly frequent collocations (e.g. *kind of*; *lot of*), than in the less frequent ones (Vogel Sosa 2000).

In fact, an on-line word identification task evidenced the chunking of highly frequent 'X-of' collocations such as *lot of*, *sort of*. For these collocations, adults took longer to identify the word *of* in these units than less frequent collocations such as: *think of*, *any of*, and *example of* (Vogel Sosa and MacFarlane 2002). Furthermore, it is in the most frequently occurring collocations that the negated English auxiliary *don't* reduces the most e.g. *I don't know* (Bybee and Scheibman 1999), where *don't* follows the word *I* and precedes "know, think, have(to), want, like, etc." (see Bybee and Scheibman 1999: 581).

In respect to reduction, Haspelmath (2008b: 59) posits that "frequency of use implies short coding" where short coding equates to "shorter and more cohesive expression types" (Haspelmath 2008a:2). Haspelmath (2008b: 59) cites the reason for this implication as "frequent items are more predictable", in that "the cause of the reduction is not routinization, but the speaker's tendency to save energy when part of the message is predictable". Thus, the predictability of elements allows their form to become more economical.

The view adopted in this study is similar but with the emphasis on the cost factor rather than the predictive factor. The drive for economy encourages many elements to be stored as cost effectively as possible, without damaging or interfering with the interpretation by the hearer. Therefore, it is the form of predictable elements i.e. the most accessible elements, which will be reduced, since the reduction will impact less on the interpretation.

Note that when comparing frequency of usage and phonological reduction or cohesion, many studies use the token frequency of particular elements (Bybee and Slobin 1982; Bybee and Scheibman 1999; Bybee and Thompson 2000; see Bybee 2001).[7] However, it is pertinent to note that when identifying independent units in artificial languages, adults and children attend to *transitional probabilities* i.e. relative frequencies (Saffran, Newport and Aslin 1996; Aslin, Saffran and Newport 1998; see chapter three for discussion).

In fact, it has been shown that transitional probabilities play a role in bounding consonant reduction, namely the process of palatalisation e.g. *band you're* (noun + complementiser), *but you'd* (conjunction + pronoun), *put your* (verb + determiner) (Bush 1999: 268). Bush (2001: 268) asserts that "the non-random distribution of natural speech …develops chunks of stored information out of frequently-used, multi-word strings, consequently forming segments that represent single agglutinated phonological domains".

Bush (2001: 273) shows that palatalisation and therefore the formation of phonological segments, is not triggered by a specific phonological environmental but occurs when multi- words such as *would you* exhibit "a very high transitional probability as a collocated pair". It is this frequency-dependent chunking process that gives rise to 'emergent constituent structure' (Bush 2001: 273) as illustrated below in Figure 2.5

[7] Token frequency is also used in studies when comparing the frequency of usage and typological markedness of individual elements in the language (Croft 2008, Haspelmath 2008b). Nevertheless, relative frequencies have been utilised when comparing "the frequency of one category and the frequency of another category" within a construction (Haspelmath 2008a: 9).

Figure 2.5 The relationship between transitional probability and emergent constituent structure in naturally occurring discourse. Utterances were taken from Carterette and Jones 1974/MacWhinney 1995 corpus (Bush 2001).

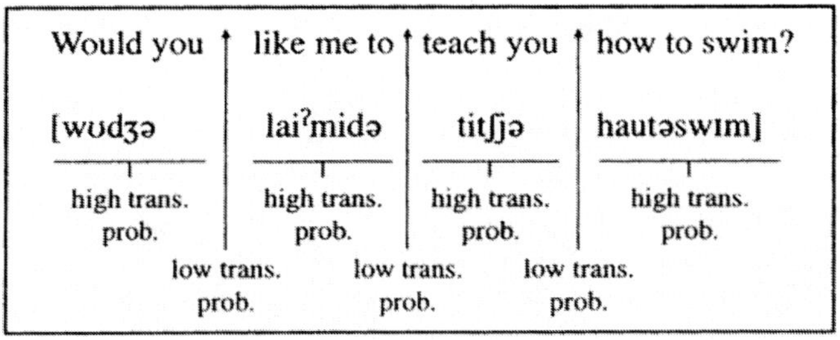

2.7.1 Chunking

When discrete elements are processed as a whole unit, the process of chunking has occurred. The notion of chunking has been used in many scientific domains. For example, when solving problems (De Groot 1978); in memory and perception studies (Miller 1956). See also Gobet, Lane, Croker, Cheng, Jones, Oliver and Pine (2001) and Ellis (1996) for a discussion on chunking. In respect to syntax, when forms become chunked, the chunk represents a *syntactic* grouping. Many effects that have been associated with chunking are also found in linguistic groupings too:

- The loss of internal structure e.g. *gonna* example (see Bybee 2002a; Rostila 2006; Langacker 1987: 58) and Mauritian Creole Nouns (Guillemin 2007)
- Priming in that "one unit of the chunk primes or automates the other unit" (Bybee 2002a: 112) e.g. the non-attributable form of *dry* primes the use of *up* i.e. *dry up* (Hoey 2006)
- Autonomy, in that "the morphemes or words inside a chunk become autonomous from other instances" (Bybee 2002a: 112) e.g. there is a recognition problem with the word *of* in highly frequent phrases such as *sort of X, kind of X* (Sosa and MacFarlane 2002) ;
- Reduction as discussed above.

2.7.2 Frequency of co-occurrence and constituency

It has been shown "that a low type frequency for each slot in a construction coupled with a high token frequency of particular items in a construction creates the tightest constituent structure" (Bybee and Schreibman 1999: 594). Thus, repetition (token frequency) and frequency of co-occurrence are implicated in a type of constituent formation. To evidence if repetition or more specifically, frequency of co-occurrence corresponds to traditional constituency (constituency$_{trad}$) and determines its boundaries, Bybee (2002a) undertook a study on the English noun phrase. The noun phrase was chosen as it generally tests well for a constituent$_{trad}$ since it tends to occur in the same form in various places in an utterance (in a subject/object and oblique role, although certain collocations resist - see example 2.3) and is used in isolation in casual speech (Ono and Thompson 1994). The eleven most highly frequent lexical nouns from Switchboard corpus were used (Bybee 2002a); the speech consisting of telephone conversations with 2.4 million words recorded.

The hypostudy under investigation was that "specific items that are used together frequently will form tighter bonds than items that occur together less often" (Bybee 2002a: 116). The prediction from this hypostudy "is that items that occur within a traditional constituent are more likely to occur together in running discourse than are items that are in different constituents" (Bybee 2002a: 116). The study evidenced this hypostudy in that nouns occurred more frequently with elements in the nominal constituent$_{trad}$ i.e. the noun phrase, than elements occurring outside the noun phrase. For a more detailed discussion, see Bybee (2002a).

Bybee's (2002a) study uses absolute frequency of co-occurrence values rather than relative frequency of co-occurrence (i.e. transitional or conditional probabilities). In this study, the latter is preferred since studies have shown how adults (and children) - when segmenting and learning an artificial language - attend to transitional probabilities and the fixed position of a high type frequency word (Saffran, Newport and Aslin 1996; Saffran, Aslin and Newport 1996; Gómez 2002, 2006; Gómez and Maye 2005) as discussed below. Furthermore, Thompson and Newport (2007) show how transitional probabilities facilitate phrasal groupings of a miniature language. However, as Ellis (2010 p.c.) observes, since type and token frequency are implicated in a construction's productivity, both absolute and relative frequencies may play an important role in explaining the emergence of syntactic structure.

2.7.3 Discontinuous dependencies

Discontinuous dependencies are tricky to analyse using sequential order since the sequential analysis of an utterance would need to be temporally suspended and picked up again at another point down stream in the utterance. It has been noted that other neural motor activities do just that (Bybee 2002a:128). However, in jazz music, Culicover (2005) notes that discontinuous dependencies are not apparent and may be due to the fact that there is no semantic component to the formal system of jazz unlike the mismatch of form to meaning that creates long distance dependencies in language.

Thus, it seems that discontinuous dependences exist only when observing the interaction between semantics and syntax – i.e. when examining the symbolic relation between meaning and form. To avoid discontinuous dependencies in syntax, we can focus on the kind of constituency that pertains purely to linguistic form, being referred to as syntactic constituency. Our objective is to identify form boundaries and syntactic relations, so we are interested in how elements combine together, starting with the smallest primitives in language, the word constructions and in future research, working our way upwards in complexity.

Examining the syntactic relations between word constructions entails examining the word distribution. Since it has been shown that adults (and children – see chapter three) use transitional probabilities for segmentation of language and non-linguistic tasks, it is pertinent to examine the transitional probability (TP) values for words in a given text: both for adult and for child language. If we find a common pattern of TP values, we are on the way to hypostudying a syntactic relation.[8] If we find that such a pattern is characteristic of the type of network language generates when its representations are stored optimally, we may well have found a linguistic syntactic *universal.* If the network, generated by linguistic form, is a general system produced by optimising a generic cost and connectivity function, we then have an explanation of why the universal exists, that is,

[8] In free word order languages e.g. Australian Aboriginal languages, it would be more plausible to identify the transitional probabilities in the morphology, since inflection provides information on the organisation of linguistic elements in these languages (see Evans and Levinson to appear, section 5 for discussion of constituency in free word order languages).

why linguistic form exhibits a certain distribution of forms.

On this view, syntactic constituency would be explained by independent syntactic factors and so the term syntactic constituency as a formal cross-linguistic syntactic stipulation would be redundant. These hypostudyes are evidenced and discussed in chapter five and six. From syntactic constituency, we move to another broad characterization of linguistic form that is often used when describing adult and child language: the functional-lexical (f-l) distinction.

The next section will briefly review the basic syntactic properties ascribed to f(unctional) expressions (for a more detailed discussion see Cann 2000). It will then be argued that the f-l distinction is couched in terms of a syntactic role that is ungraded, local to a given construction and defined by a high token and type frequency. Furthermore, this syntactic role will be shown to occur independently of syntax, through the optimisation process of storing linguistic form.

2.8 The functional-lexical distinction

So intuitive is the f-l distinction, that it has been postulated as universal/innate and psychologically real (Stromswold 1990, 1994). In fact, intuitions behind the distinction are so strong that linguists have extended the distinction from words to categories. So that now one often hears of the *functional* category of inflection or the *lexical* category of nouns. Although there is evidence for a distinction between function words and lexical words, there are inconsistencies in the definition of these elements. The next section will illustrate this problem in regard to defining function words by their syntactic properties. Cann (2000) provides an excellent discussion of the characteristics of functional (f-) and lexical (l-) expressions. The next section is based on his observations of the syntactic properties of f- expressions.

2.8.1 Function words are found in a restricted syntactic environment (Cann 2000: 41)

Even though f-expressions have a high token frequency, they "usually appear in just a few syntactic contexts…and these are definitional of the class they belong to" (Cann 2000: 41 §2.3). Examples are: the function word *the* occurs before an English nominal and its modifiers; an English verb follows the function word *would*. From a constructionist approach, that f-expressions appear in a few syntactic environments is due to the f-expressions defining the syntactic environment, particularly in a partial schema construction in which there is an item-specific (substantive) part i.e. usually the f-expression and an abstract part shown as a syntactic category (a general phrase/category in traditional accounts or a construction-specific category in Radical Construction Grammar).

For example, the copula and the –ed morphology identifies the English passive construction (copula + $\text{Verb}_{\text{passive}}$-ed); the English progressive construction (copula + $\text{Verb}_{\text{progressive}}$-ing) is recognisable by the copula and the –ing morpheme. This type of construction is often referred to as a slot-filler schema (especially in language learning literature) since the abstract part of the schema allows different (but conceptually relevant) elements to specify the slot. Thus, f- expressions are the elements that are specific to the construction, which is why they "usually appear in just a few syntactic contexts" as Cann (2000: 41) points out.

The fact that f-expressions "are definitional of the class they belong to" (Cann 2000), is due to their being linked to different but related elements that form a class/category within and specific to that construction. Take for instance, the word combinations 'an + aardvark, 'an + amoeba', 'an + apricot'. The combinations form a construction 'a(n) + English $\text{Noun}_{\text{indefiniteness}}$', where 'English $\text{Noun}_{\text{indefiniteness}}$' is defined by the objects' conceptual information in that they are construed as "unitary, individuated, countable objects" (Lee 2001: 140). That is, the elements of the category crowd around the f-expressions, furnishing the f- expressions with a high type frequency (relative to the words in the construction).

In fact, the higher the type frequency, the more schematic or abstract the role in the slot is within the construction since "a schematic unit represents a generalization over specific instantiations" with schematicity being "a relation between a schema and its instantiations" that is, "the relation between a type and its tokens". (Clausner and Croft 1997: 251). Schematicity is related to grammaticalisation, in a gradient way (Rostila 2006: 53), in that "the more schematic a construction is, the more grammaticalized it is" (Rostila 2006: 53). Schematicity also correlates in a graded way with a construction's productivity. As the slot elements increase in number, the productivity and schematicity also increase. Thus, as Rostila (2006) observes, a "high type frequency of a string can cause its storage as a productive template". See Rostila (2006) for an account of how storage factors drive grammaticalisation.

Interestingly, the variable elements in the high type frequency slot tend to have a low token frequency in highly productive (morphological) schemas (Bybee 1985: 133) whereas low productivity is instantiated with high token frequency elements occurring with a low type frequency element (Bybee 1985: 133). A high token frequency tends to entrench the element and so the stored form is quite autonomous and resists being regularised e.g. *went* v **go-ed* (Bybee 1985: 119-124).

Furthermore, "the more a form is used, the more its representation is strengthened, making it easier to access the next time. Words that are strong in memory and easy to access are not likely to be replaced by new forms created with the regular pattern" (Bybee and Thompson 2000: 67; Bybee 1985). As a result, high token frequency reinforces irregularity of the stored representations of form (Bybee 1985: 119). In short "high type frequency enhances productivity, while high token frequency inhibits productivity" (Clausner and Croft 1997: 254).

To sum up this part of the discussion, productivity (Bybee 1985, 1995; Clausner and Croft 1997; Goldberg 1995) and grammaticality (Rostila 2006) are directly correlated with type frequency. High token frequency, on the other hand, is proportional to grammaticality but inversely proportional to productivity (Bybee 1985: 119; Bybee 2001: 165). It is no surprise then that f-expressions have a high type and token frequency whilst l-expressions tend to be elements with a low type and token frequency. Consequently, f-expressions will have an entrenched nature and be associated with highly productive constructions whereas l- expressions will be regularised easily and form the productive part of a construction.

Function words, with their high type (and token) frequency are often referred to as grammatical words. As such, the notions of grammaticality and functionality (in respect to the f-l distinction) could be conflated. On this view, since functional elements identify the construction, an element would be classed as more functional, if the construction it identifies with is more grammaticalised. In contrast, lexical words tend to have a lower type frequency (relative to the elements in the construction) and form very close collocations e.g. 'pay heed', or 'addled brains/eggs' (Cann 2000: 42).

Two important points fall out from this discussion. Firstly, it seems that the notion of a functional word is a theoretically graded property. However, the onus of proof is in the psychological realisation of whether one element is functional or lexical. It is more likely, given evidence that will be presented below, that the distinction between function and lexical words is a clear cut distinction. As such, the graded nature of grammaticality – with particular reference to familiarity and repetition - is quite different from the identification of a function word and as such, the terms 'functional' expression and 'grammatical' element refer to very different concepts, and are not to be conflated.

The second point concerns the fact that the terms 'functional' and 'lexical' refer to a local syntactic relation *within* a particular construction. They are not umbrella terms for words used across constructions. Indeed, in artificial language studies, adults categorize words by being sensitive to the distribution of co-occurrence patterns and position of words *within* certain (frequent) frames (Mintz 2002).[9]

In fact, a word may be construed as functional in one construction but lexical in another i.e. it can behave as an f-expression and an l-expression depending on its environment. That the terms functional and lexical refer to relations relative to a construction, explains why there are discrepancies in certain data analyses that propose general functional categories. For example, Newmeyer (1998, 2000, 2004) proposes that 'near' in Example 2.4 acts both like a adjective$_{trad}$ (selects a preposition$_{trad}$ as in Example 2.4 a) and a preposition$_{trad}$ (selects an object$_{trad}$ as in Example 2.4 b).[10]

Example 2.4 Sourced from Ross (1972)

a) The shed is near to the barn
b) The shed is near the barn

On Newmeyer's view (1998, 2000, 2004), in Example 2.4a, *near* is a lexical word but in b, it is a functional word. Notice how the potential type frequency of *near* is much higher in (b) than (a) (considering the referential terms can be used to replace the word *the*). In Example 2.4a,

[9] In respect to 12-month old infants, it has been shown that the infants use their knowledge of distributional patterns to categorise words. And although "there is some indication, although preliminary, that 12-month-olds might be selectively attentive to frames containing Verbs", it remains to be proven whether infants utilize frequent frames generally (Mintz 2006: 42).

[10] Ross (1972) views 'near' as inhabiting a categorial space in between the two categories.

near is strongly collocated with *to,* in the same way that *next to* and *close to* are. The fragment 'near + Slot + the + barn' is not productive whereas 'near + Slot' is. Thus, an f- expression is recognised as such by its high type and token frequency within the construction whereas l-expressions grouping around a high token frequency element generate the element's high type frequency within the given construction.

2.8.1.1 High Type Frequency and Learning

Support for attending to high type frequency in learning abstract relations is found in the learning of causal grammars. With the aim of investigating which type of causal relations are easier to learn, a study used blocks on a computer screen which could then be moved around, with some blocks producing a light and sound when moved together (Tenenbaum and Niyogi 2003). That is, they caused a 'blicket' to be activated. The blicket activation was governed by a causal grammar represented by one of four networks (Figure 2.6). In the networks, nodes represented blocks and the arrow between the nodes, the causal links (the blicket activation). The task of the (adult) participant was to learn a grammar by experimenting with the block interactions.

Causal links i.e. causal relations, learnt the most easily were those represented by the network where the nodes had consistently the highest type frequency (G_1 and G_2). Furthermore, two distinct classes were learnt, in which each object was a member of class A or B and where an object in one class lit up the objects in the other class without lighting up objects in the same class.

Consequently the relations between members of the two classes were asymmetric and automorphic (see chapter seven). Elements are automorphic when each element's relation with the other elements in the network is the same throughout (although Tenenbaum and Niyogi 2003 do not express such relations in automorphic terms). In the grammars most easily learnt, the objects in each of the classes displayed this property.[11]

---

[11] Class 1 grammars are represented in network theory as a 204 subgraph (see chapter seven). Thus, this experiment suggests that the 204 subgraph facilitates the mapping between two levels of representation.

Figure 2.6 Sourced from Griffiths and Tenenbaum (2007: 14, Figure 9a). The four networks representing the causal grammars under investigation by Tenenbaum and Niyogi (2003). G1 and G2 were more easily learnt than G3 and G4.

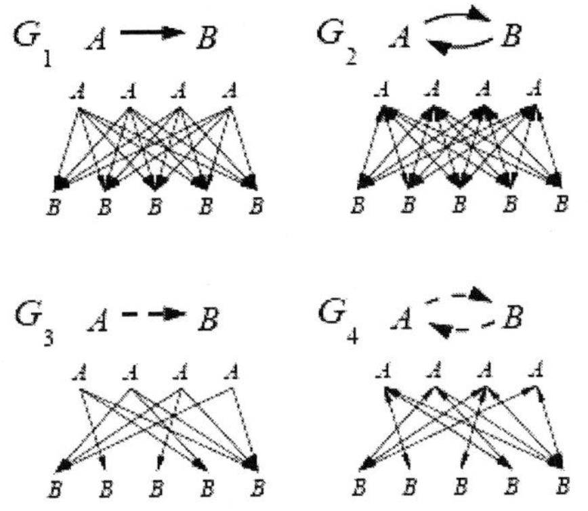

High type frequency also correlates with the notion of anchoring (see Valian and Coulsen, 1988, see section 3.5.1) in that function words are perceived as perceptual anchors within a sea of variation. Learning is sensitive to invariant structure (Gibson, 1969; Gibson 1966) and so a stable form, held constant in a variable environment, highlights its position within the utterance (Gómez 2006). Evidence is found in artificial language studies, where children as young as 18 months have shown sensitivity to the non adjacent structure of two elements intervened by a set of 18 or more items (Gómez 2002; Gómez and Maye 2005 – set sizes covered were 24, 18, 12, 6, 2). With a set of 12 intervening items, adults and children play attention to the adjacent transitional probabilities rather than the non adjacent dependencies. Thus, it seems that there are two types of learning occurring, one that focuses attention on adjacent relations when the values of the transitional probabilities are higher, and the other learning strategy that attends to the position of elements when the transitional probabilities reach a certain low value.

A further experiment evidenced the two learning strategies in adults (Gómez 2006). The position of elements were kept constant in a grammar described by aX and Yb, where X and Y had a set size of either 1 or 24, with one element representing a and one element for b (Gómez 2006). With the large set size, position could be utilised and transitional probabilities could also theoretically aid the learning of the grammar - although it would seem unlikely given the results of previous studies

(Gómez 2002, Gómez and Maye 2005). In the small set size, the learner could note the transitional probabilities between a and x (and y and b) and importantly, the hearer has a chance to utilise the position of a and b to learn the grammar. Thus, the experiment sought to determine "whether learners would be more likely to notice one form of structure over another as a function of the variability of X- and Y-elements" (Gómez 2006: 20). Results confirmed that two distinct learning strategies are employed:

> "information generated by participants in Set-size 1 almost entirely reflected Verbal knowledge of co-occurrence. In contrast, Verbal knowledge of position predominated in the Set-size 24 condition… Furthermore, seven participants in this condition spontaneously reported that the a- and b-words "popped out," whereas none of the participants in Set-size 1 reported such an effect. In Set size 24, the relevant information was position of the a- and b- elements" (Gómez 2006: 23).

Furthermore, the transition between the two types of learning does not seem to be gradual (Gómez 2002), since with 12 items, no sensitivity to position was recorded, whereas with 18 items, position was attended to. Returning to f-expressions, it may be tempting to draw a parallel between the two learning strategies for nonadjacent and adjacent dependencies and the functional-lexical distinction. When transitional probabilities are low i.e. when the type frequency is high, the position of an item is attended to in a way reminiscent of the slot filler schema with its anchor and a variable set of elements filling the slot. On this view the functional expression is the anchor (and the lexical expressions tend to be the variable elements).

In regard to the strategy employed for the adjacent dependencies (frequency of co- occurrence), attention to high transitional probabilities could be construed as the lexical items within a construction i.e. close collocations. However, in many languages inflectional markers are found in 'bundles' in a fixed order (Example 2.5). The bigram transitional probability values of these inflectional 'bundles' would be potentially quite high.

Example 2.5 An example of a Lamalama sentence. Lamalama is spoken in the Princess Charlotte Bay area of Cape York, North Queensland, Australia (Pentland 2009: 2, example 2, BAS-72-35A). NPAST – present/non-past tense; 1SG.0 – 1st person, singular, direct accusative; 3SG.S – 3rd person singular nominative.

Maka-	m- na- la
See-	NPAST-1SG.O-3SG.S
'He sees me'	

As a result, it would be pertinent to view the perception of a functional element as being related to its high type (and token) frequency and focus on the learning of its (fixed) position within the construction (even for languages where the bundles of inflectional material may occur in fixed positions within the utterance). And so, conversely, variable elements in the construction - the low type frequency elements generating the high type frequency of an f- expression - are perceived as lexical. Thus, the asymmetry of the type frequency establishes the two syntactic notions of functional and lexical expressions within the *same* construction.

Notice how the perception of position i.e. the anchor function word, is clear cut and not graded (recall that there was no position sensitivity to set size of 12 elements in Gómez's (2002) study whereas for a set size of 18 elements, position was attended to).[12] Hence the role of a function word is not graded but it is reliant on the relations between co-occurring elements i.e. its role is based on the perception of type frequency. Interestingly, with respect to the strategy for learning adjacent relations within an artificial grammar, the reliance on transitional probabilities is in keeping with the knowledge of the constituent structure discussed above, specifically word constructions. Thus attending to word constituency and f- expressions derive the two learning strategies.

It is also interesting to note that the two learning strategies studied by Gómez and colleagues relate to the connectivity of elements i.e. high type frequency, and to how strong the cohesion is between two elements. As discussed previously, the latter factor relates to how frequently two elements occur together, and the cost of storing this relation. The more familiar the relation i.e. the more the elements occur together (relative to their overall occurrence), the lower the storage cost. Consequently, the two learning strategies relate to the factors of connectivity and cost. These factors will be shown to underlie the distribution of word co-occurrence patterns in the English data for adult and child language. Thus, learning strategies are a product of efficiently storing the maximal number of relations between representations of linguistic form in an optimal way.

---

[12] One might argue that perhaps the gradience is within the 13-17 set size. However, gradience by its very nature does not start nor finish abruptly rather it waxes and wanes. Thus if the learning of position was a graded sensitivity, some recognition of position would be observed in the set sizes leading up to 18 i.e. the set size 12.

2.8.2 Function words belong to a closed class (see Cann 2000)

New function words do not occur as often as new lexical words; the latter appearing in the language anew from borrowings e.g. *zeitgeist*, from morphological generalisations e.g. *do- able* or from coinages e.g. *bling*. New function words are accrued over time through grammaticalisation but this is usually a slow process across generations of speakers, whereas many lexical words may join a language in just one generation i.e. *blog, dongle,* etc. By understanding the role of a function word in respect to its high type frequency, it is easy to understand why such words aren't imported or created so quickly since the word is required to co-occur with a high type frequency before it is perceived as a function word.

In fact, if Gómez's artificial language study (2002) is revealing of how we perceive natural language, and this is quite probable given that linguistic development points to general cognitive abilities matched to a particular domain (see Emergentist accounts such as MacWhinney 2000b, 2005), then a putative function word needs to have a high type frequency and inhabit a fixed position within a schema (as well as a high token frequency) for the role of function word to be established. Note that since 'function word' is a syntactic role, there is no need to elicit semantic criteria[13] and so the critique that artificial languages do not encode meaning is not applicable here. Thus, the potential function word is required to occur in a productive construction. Since grammaticalisation is a slow process, function words tend to develop slowly.[14]

In respect to open and closed classes, Cann (2000: 40) points out, that within open classes, there are closed subclasses "For example, auxiliary

---

[13] This point does not obviate that a "grammatical distinction may be semantically predicable" (Dryer 1997: 127. Only that the distinction in syntax may be made on distributional similarities and differences in linguistic form which is quite separate from the semantics.

[14] If no functional expressions are present, then on this view, new ones with be generated by the (child) speakers. Such evidence is likely to be found in newly created languages such as the Nicaraguan Sign Language. Consequently, the learning strategy of position is not a passive strategy but an active one imposed by the process of storing the representations of linguistic form in the most optimal way. The drive of optimal storage creates the distribution patterns of forms such that words with high type frequency occur. These words allow for maximal combinations at the lowest cost – as discussed previously.

Verbs form a closed subclass of Verbs and days of the week, months of the year, points of the compass, etc., form closed subclasses of Nouns that show idiosyncratic syntactic behaviour". Example 2.5 below illustrates the restrictions on co-occurrences.

Example 2.5 Closed subclasses, Cann (2000: 40 Example 1a-e)

a) I'll see you Tuesday/on Tuesday/the first Tuesday after Easter.
b) I'll see you tomorrow/*on tomorrow/*on the first tomorrow after Easter.
c) I'll see you *breakfast/at breakfast/at the first breakfast after Lent.
d) The exiles went North/to the North/to North London/North of Watford.
e) The exiles went *London/to London/to London town/*London of Ontario

However, if one adopts Croft's (2001) notion of construction-specific syntactic categories such inconsistencies above are smoothed away. In Example 2.5, three constructions can be established each with its own restrictions i.e. on + English Noun_{on}, at + English Noun_{at}, to + English Noun_{to}.

2.8.3 Functional expressions associate with certain general syntactic categories (Cann 2000)

Cann (2000: 43) asserts that "Groups of functional expressions also tend to cluster together around a particular major class (e.g. determiners and quantifiers with Nouns, tense, aspect and agreement with Verbs) and these groupings define syntactic domains of a particular type". In fact, it is more probable that the source of such grouping is related to the underlying conceptual structure of the tokens that form the f-expression's high type frequency that creates a productive construction.

Note that each of the general classes of noun, verb (and adjective) exhibit a "prototypical correlation" with a semantic class and a pragmatic function, i.e. a propositional act (Croft 1990: 248). For example, a word classified as a noun is prototypically an object associated with the pragmatic function of referring whereas for a verb, the word is prototypically an action with the pragmatic function of predicating (Croft 1990).

On this view, function words (as well as non-prototypical members and inflectional morphology) specify three other pragmatic functions: the "MINOR propositional acts of categorizing, situating and selecting" (Croft 1990: 251). For example, quantifiers select individuated objects, whereas determiners (articles) situate in a deictic way an "entity in a mental space" (Croft 1990: 265). Both types of function words act on objects but in two pragmatically different ways. Hence, they both group around words that denote objects.

In respect to tense and aspect, tense deictically situates an event in a temporal dimension, whereas aspect categorises an event (Croft 1990: 258 and 255 respectively). Both function words pragmatically act on events and thus group around events. Furthermore, general non construction-specific syntactic classes are not evidenced by distributional analysis (Croft 2001), so it is not the case that f-expressions group around a general category. Rather f- expressions with similar semantic and pragmatic properties will co-occur with similar semantic items.

2.8.4 Functional expressions occur consistently with all expressions that have a particular property (Cann 2000)

Examples given are that *be* occurs with any transitive verb that has an /–en/ form and *the* occurs with any common noun (Cann 2000: 42). Given that constructions may have their own meaning, independent of the elements within the construction (see Example 2.1 from Goldberg 2006), then each expression used in the construction will be interpreted within the remit of the construction's meaning. Thus, any expression with compatible properties[15] may occur in the construction. Since f-expressions identify the partial schema constructions[16] then it follows that any expression compatible with the semantics and pragmatics of the construction will

---

[15] Compatibility in the sense that the element is in "semantic proximity to a cluster of attested instances" (Goldberg 2006: 101), although limited by statistical pre-emption: "more specific knowledge always pre-empts general knowledge in production, *as long as either satisfy the functional demands of the context equally well*" e.g. *went* is preferred to *go-ed*, the former pre-empting the latter (Goldberg 2006: 94-95).

[16] Lexical expressions on the whole do not form item specific parts of a construction, but they do provide semantics for prototypical meanings of more abstract constructions e.g. *give* in respect to the ditransitive construction, *make* in respect to the resultative construction (Goldberg 1995, 2006).

appear with the f-expression. Consequently, Cann's (2000) observation reflects the independent meaning of a construction that interacts with the meaning of the expression that occurs in the construction. Thus, symbolic and semantic relations underpin this f-expression property.

2.8.5 The modification of a lexical expression's syntactic category (Cann 2000)

For an example of this property, Cann (2000: 44) points out that "articles are always nominal whatever the lexical expression they appear with" whilst lexical words appear in many different syntactic constructions. As noted previously, the categorial status of l-expressions is dependent on which construction the l-expression is occurring in, since each construction has its own unique syntactic categories (Croft 2001). In partial schema constructions, these categories arise from the high type frequency of the f-expression, a constant element in the construction. Consequently, it is not that an f-expression 'alters' the categorical status of an l- expression. An expression has a category status by virtue of the construction in which it occurs. In other words, an l-expression does not have an inherent *syntactic* category – it is assigned one by the construction it occurs in.

Furthermore, a lexical expression by definition does not form an item specific part of a partial schema construction (although they do for argument constructions see footnote 16) as by definition the l-expression does not have a high type frequency from which a syntactic category could be construed. That is, the l-expression would not be responsible for creating a syntactic group with an f-expression as a member, since the definition of such an l-expression is the variable part of a partial schema that occurs with a high type frequency expression (the f-expression). Thus, Cann's (2000) observation serves to highlight the asymmetry in type frequency between the f-expression and the l-expression in a given partial schema construction. By definition, the former specifies the syntactic class within the construction, whilst the latter is a felicitous member of the class. Furthermore, the high token frequency of an f-expression curtails the potential for the element to occur in a productive schema, whereas l-expression by definition has a low token frequency and consequently felicitously reinforces a schema.

2.8.6 For functional expressions "there are no processes that alter their selectional properties" (Cann 2000: 42)

Cann (2000: 42) notes that "there are no processes that apply to functional expressions that alter the status of their semantic arguments". Passivisation alters the selection of semantic arguments in that the agent of the verb is demoted and occurs after the verb within *by* (*I washed the car* versus *The car was washed by me*). The selectional properties of the verb are altered since the verb in the active transitive sentence obligatorily selects an agent role whereas the verb that has undergone passivisation does not. Consequently, the verb‘s selectional properties have been modified by the passivisation 'process'.

These selectional processes such as passivisation are internal theoretic operations that derive one syntactic structure from another. As noted earlier in this chapter, Construction Grammar is monostratal. As such, a syntactic structure with a different semantic/pragmatic function is treated as a distinct construction. Thus, there are two different constructions that capture the active and passive sentences: the English transitive construction and the English passive constructions. If we want to demote agentivity, we use the English passive constructions identified by the use of particular f-expressions (copula + $V_{passive}$-ed + by).

Selectional requirements are not applicable on this view, rather the items need to fulfil a semantic requirement to occur in the construction and that requirement is based on the semantics and pragmatics of pre-occurring expressions within the construction. As Goldberg (2006: 99) notes, "Constructions that have been heard used with a wide *variety* of types are more likely to be extended broadly than constructions that have been heard used with a semantically circumscribed set of types". For example, if the pre-occurring expressions shared similar semantic and pragmatic properties, then for a new expression to occur in the construction, it would need to share these properties too. Or if the expressions were eclectic then it is more likely that the other potential expressions would be equally eclectic (see Goldberg 2006).

Since the selectional property of an expression is really the influence of the construction's independent meaning and partial schema constructions are identified by the f-expression, then altering a f-expression's selectional property would mean changing the construction's independent meaning that the f-expression identifies. Such process occurs through grammaticalisation but it is a diachronic change rather than synchronic one since modifying the independent meaning of a construction creates a new construction. Rather, l-expressions which occur in the variable part of the construction are construed with the independent meaning of the construction and thus "change" their "selectional properties". In the passivisation example the verb *wash* may be construed as having a demoted agent since the construction specifies it as such.

We can now understand why in regard to functional expressions "there are no processes that alter their selectional properties" (Cann 2000: 42), as selectional properties of l-expressions are not altered. Rather the l-expression by definition occurs in the variable part of a partial schema construction in that it occurs as a token of the exemplars that make up the high frequency type i.e. the f-expression. For an f-expression to have its selectional properties altered, it would mean in constructionist terms that the f-expression would form a variable part in another construction. However, an f-expression, by definition will have a high type and token frequency. Having a high token frequency, as previously noted, curtails productivity - that is, schematisation. Hence, the entrenched forms of f-expressions would not reinforce a schema (Clausner and Croft 1997) and therefore would not have their selectional properties altered.

2.9 Summary

The chapter started off with a review of the theoretical linguistic perspectives adopted in this study namely Construction Grammar, and the very compatible usage-based approach. Constituency and the functional-lexical distinction - two broad characteristics of linguistic form - are discussed, both of which are potentially domain-general properties when frequency (type, token and co-occurrence) values are observed.

Croft (2001) argued that constituency is not evidenced since formal groupings such as contiguity, intonation units, etc do not coincide. Rather, constituency is a coded dependency in that it encodes symbolic relations (Croft 2001, chapter six). However, high frequency co- occurrences do provide groupings if one looks at syntactic relations rather than symbolic relations. It will be shown (in chapter five and six) that in adult and child language the frequency of word co-occurrences form a power law distribution (corresponding to the multiple link lengths characteristically found in small worlds). Thus, constituency is a property of a general system and so it is predicted that such a distribution will be found cross-linguistically.

With respect to the second broad property of language discussed, the f-l distinction, it was argued that some of the f-expression's syntactic properties related to symbolic and semantic relations within the particular construction. Other properties were attributed to syntactic relations and related specifically to the token and/or type frequency of the f-expression within a productive construction. It will be shown that in a network sense, such f-expressions are the hubs of a word co-occurrence network (both adult and child) around which certain productive link patterns appear. Since hub structure is a key characteristic of a small world then the notion of an f-expression may be a *universal* phenomena, although not innate but rather a property of a general system. On this view, the formal label of the f-l distinction as a way of explaining the potential cross-linguistic similarities is redundant, as this study offers an independent explanation.

Table 2.2 summarises the different kinds of frequencies discussed. The cause of a given high frequency is shown, as well as its linguistic effect.

Table 2.2 Summary of different kinds of frequencies and their associated linguistic effect

type of high frequency	the cause	the effect
token	repetition	entrenchment
type	schematisation	productivity
transitional probability between words	Automisation (low storage cost)	reduction of predictable elements (short coding)

Furthermore, in the discussion on the properties of f-expressions, it was observed that in a study of how artificial languages are learnt, two learning strategies were adopted. The first attended to transitional probabilities (relative pairing frequencies) which are related (inversely) to the cost of storing form representations. The second learning strategy attended to the type frequency of fixed-position words in a chunk of linguistic form. This second strategy relates to the connectivity of words. It will be evidenced in chapter five and six that the two factors of cost and connectivity that produce the linguistic patterns in a word co- occurrence network relate to the storage of the linguistic representations. Thus, not only is the optimisation of connectivity and cost the most optimal way to store linguistic forms but this optimisation is also behind the strategies for learning a language. Thus, the same strategies for optimal storage are the same for optimal learning. Chapter four will introduce the theory behind this optimisation. However, the next chapter will present the perspectives on *first* language learning

CHAPTER THREE

PERSPECTIVES ON LANGUAGE LEARNING

3.1 Overview[1]

When learning a language, children are constructing "structured conventional linguistic units" i.e. a construction inventory (Langacker 1987, 1991) rather than acquiring grammar. Children learn these constructions from their caregiver's speech (Theakston, Lieven, Pine and Rowland 2001; Cameron-Faulkner, Lieven and Tomasello 2003; Abbot-Smith, Lieven and Tomasello 2008); the initial stages of child directed speech having a high degree of specificity (Cameron- Faulkner et al 2003: 862). Since in constructionist grammar there is no distinction between the lexicon and syntax, it is predicted that the development of a vocabulary is positively related to the child's development of constructions.

Indeed, it has been found that for English speaking two-year olds, the lexical and syntactic richness in the input is related to vocabulary development (Hoff and Naigles 2002; for French and Austrian German, see Bassano, Laaha, Maillochon and Dressler 2004) rather than age (for American English, see McGregor, Sheng and Smith 2005; for Italian, see Devescovi, Caselli, Marchione, Pasqualetti, Reilly and Bates 2005). The correlation may be linear or non-linear depending on the language examined (Caselli, Casadio and Bates 1999; Devescovi, Caselli, Marchione, Reilly and Bates 2003; Devescovi, Caselli, Marchione, Pasqualetti, Reilly and Bates 2005; Szagun, Steinbrink, Franik and Stumper 2006)

General cognitive abilities facilitate learning (Tomasello 2006: 9) as does constraints imposed by the linguistic processing system and cognition (see Christiansen and Chater 1999; Morgan and Newport, 1981; Newport and

---

[1] Syntactic roles and parts of speech in child language are only specific to the construction in which they occur (Croft 2001). However, almost all literature (for an exception see Tomasello 2006) assumes general verb and noun categories and so this review will be including these studies in this discussion.

Aslin, 2000; Saffran, Hauser, Seibel, Kapfhamer, Tsao and Cushman 2008). Indeed, "Infant learning abilities may serve both to facilitate natural language acquisition and to impose constraints on the structure of human languages" (Saffran, Hauser, Seibel, Kapfhamer, Tsao and Cushman 2008: 479-480). Another important source of constraints on which very little work in linguistics has been done, is the general systemic constraint i.e. a constraint imposed by virtue of the system being of a certain type. The constraint may represent an optimised solution to the task at hand. In this case, the linguistic task is the storage and representation of utterances.

This study aims to address this shortfall and to examine the evidence that language has the functional-lexical (f-l) distinction and a certain type of syntactic constituent structure by virtue of language being represented by a certain type of system. Thus, properties attributable to this general system will form the "Prerequisites for Natural Language" (Goldberg 2008: 522). Indeed, the very system that creates the f-l distinction and constituency may aid the segmentation process (see chapter nine).

To learn their native language, children build up and accumulate their own repertoire of constructions in their constructicon by segmenting and linearly sequencing the input from caregivers (section 3.4). These processes rely heavily on the detection and analysis of multiple cues in the input language i.e. prosody (Cutler and Carter 1987; Mattys, Jusczyk, Luce and Morgan 1999), phonotactic sensitivity (Cairns, Shillcock, Chater and Levy 1997), semantic information (Pinker 1989), sequential probabilities including repetition and familiar names (Bortfeld, Morgan, Golinkoff and Rathbun 2005). All these cues assist in the segmentation process which in turn aids linear sequencing (see Newport and Aslin, 2000 for the correlation of syntactic cues). Furthermore, cue integration itself seems to be an important source of segmental knowledge (Christiansen, Allen and Seidenberg 1998:45; Christiansen, Conway and Curtin 2005; Mattys, Jusczyk, Luce and Morgan 1999: 485; for computer simulations and connectionist studies see Brent and Cartwright 1996; Christiansen, Allen and Seidenberg 1998).

This study will focus on one of those cues: word co-occurrence patterns of constructions, particularly transitional probabilities and frequencies. Tracking transitional probabilities is not a species-specific skill (Hauser, Newport and Aslin 2001) and such probabilities are attended to at an early age - as young as seven months (Thiessen and Saffron 2003; section 3.4). Sensitivity to linear sequencing has been established in an experimental setting as young as nine months old (Mintz 1996).

Once children have segmented the input, they must then extract patterns from these linguistic chunks to form constructions (section 3.5). These early constructions form a network, linked together by the highly frequent linguistic chunks found by the segmentation process. Previous analyses of child language have recognised that high frequency linguistic elements form anchor points in the distributional analysis (section 3.5.1). Experimental evidence suggests that the anchors can be familiar names (section 3.5.2), pronouns (section 3.5.3) or even repetitious linguistic material (section 3.5.4).

Furthermore, it has been demonstrated that children are able to establish certain continuous and discontinuous dependencies through their sensitivity to word co-occurrence relations (section 3.5.5 and 3.5.6). Whilst it has been shown that a frame with two discontinuous anchors i.e. A _ B, aids the child's categorisation process (Mintz 2003), this facilitatory effect may only be language- specific (Erkelens submitted, section 3.5.6). Better categorisation results obtain from attending to the co-occurrences of each anchor separately (Monaghan and Christiansen, 2004).

The frequency of linguistic elements also bears on the issue of how a child goes beyond the input received and generalises to a new construal. The focus of attention is on verbs (section 3.6). In the input speech, there are certain factors related to which verb gets extended early (see Sethuraman and Goodman 2004a): frequency (section 3.6 and 3.6.1), a verb being heard in a syntactically consistent environment (section 3.6.2), a verb being semantically light (section 3.6.3), a verb being heard in multiple constructions (section 3.6.4) and the use of pronominal frames (as discussed in section 3.5.3). These factors are seemingly eclectic and all confounded by frequency, with two factors being seeming contradictory ('syntactic diversity' and 'syntactic consistency').

In chapter nine these factors will be shown to be consistent, when contextualised within the network approach and will be related to the formation of a hub structure. Thus, the storage, the analysis and the generalisation of language constructions are three faces of the same process i.e. the small world hubterranean structure of a network created by optimising the storage of a maximal amount of word combinations. This theoretical line of reasoning resonates with Hershberg and Ninio's (in press; see also 2004) notion of optimal exemplar learning: "in natural cognitive systems the inside, the outside, and the process of internalization are tightly interconnected" (section 3.7).

3.2 Usage-based approach

This study adopts a usage-based approach to language learning. The theory emphasises that language is learnt from the input of the caregiver. The actual usage of the language – the individual instances and exemplars, the word co-occurrence patterns and frequencies - are (identified and) stored, forming the basis of lexically-specific constructions and also the more traditional abstract linguistic generalisations i.e. the transitive construction. On this view, it is not that the child is identifying a grammar, rather the child is learning how to use language (Bates and MacWhinney 1982; see Seidenberg, 1997). For example, the number of multi-clausal sentences used by children aged 54-60 months old is sensitive to the input the child has received at a given time in the language learning process (Huttenlocher, Vasilyeva, Cymerman and Levine 2002).

Children's constructions are very much like adult constructions, only that they are "simpler and more concrete, with fewer and weaker abstractions - because they are based on less linguistic experience" (Tomasello 2003: 97). Furthermore, on this view, the dual mechanism of learning regular and irregular patterns (Pinker 1991; Pinker and Prince 1991) is not required logically since if a child by necessity must learn irregular patterns, then the child could also learn regular patterns with the same mechanisms (Goldberg 2000). Even basic linguistic properties such as structure dependency may be learnt, as structural dependency errors have been observed (Ambridge, Rowland and Pine 2008) contra a previous study (Crain and Nakayama 1987).

Indeed, if computational models and simulations have the capacity to show how language learning could occur, then such models have arguably shown that it is possible for learning to not necessarily require direct evidence (Lewis and Elman 2002; Reali and Christiansen 2003, 2004, 2005; see Ingram 2007 for a review of the connectionist research). 'Missing evidence' in the input is also not a problem for other computational models of indirect learning that bring a probabilistic approach to the language learning task i.e. a "simple Bayesian learning procedure" (Regier and Gahl 2004) .

3.3 Distributional analysis

Since children start initially with concrete item-based experiences of language, Tomasello (2006: 9) suggests four "processes (or classes of processes)" that facilitate the forming of abstract constructions "(i) intention-reading/cultural learning, (ii) schematization /analogy, (iii) constraint, and (iv) distributional analysis". This study will focus on the fourth process.

Although linguists themselves have extensively utilized distributional analysis to form grammars of languages, it has been questioned whether it is possible for children to make use of such knowledge (Pinker 1984). The following discussion sumarises these critiques with their respective convincing counter-arguments (Redington, Chater and Finch 1998: 431).

The first criticism to be addressed here is the contention that the amount of different relations possible in the data would be so high that a mechanism using distribution analysis would be overcome with all the different possibilities (Pinker 1984: 49). However, recent distributional models look at specific properties such as transitional probabilities. As this study will evidence, transitional probabilities between words form a particular pattern within speech i.e. a power law. Consequently, distributions of words are constrained by the type of network they occur in i.e. a small world hubterranean network. This network represents how words and constructions are stored. Thus, the learner has certain constraints on the distributional analysis imposed by the linguistic storage system and so the distributional analysis is not an open ended affair.

It is also argued that "linguistically relevant properties are abstract" (Pinker 1984:49). Pinker (1984: 50) assumes that "serial positions and adjacency and co-occurrence relations among words-are in general linguistically irrelevant". Clearly this value judgment is theory internal. Generative grammarians choose to posit abstract underlying structure such as relations within phrase structure, general syntactic categories and abstract grammatical relations (see also comment by Redington, Chater and Finch 1998: 431). As discussed in chapter two, observing "serial positions and adjacency and co-occurrence relations" provides very useful syntactic information.

The endeavour of "testing for correlations among linguistically relevant properties" is questioned and considered "unnecessarily wasteful, for not only do languages use only certain properties and not others, they sanction only certain types of correlations among those properties" Pinker (1984: 50). Given that the input generally determines the type of constructions included in the child's inventory, then the child will only be concerned with the linguistic properties relevant to the language at hand. Syntactic theories that account for the properties found in all and possible languages i.e. Generative Grammar, do have to solve the problem of what constitutes a relevant property to the given language being learnt and the wastefulness of descriptions that include properties not found in all languages (Redington, Chater and Finch 1998). In respect to syntactic usage-based theories, this is not a relevant issue.

Finally, it is claimed that distributional analyses give rise to "spurious correlations". The argument goes that if the child heard the first three sentences given in example 3.1 (a-c), then the child might erroneously conclude that sentence (3.1d) would be admissible.

Example 3.1 Sourced from Pinker (1984: 49).

a) John eats fish
b) John eats quickly
c) The fish is good
d) *The quickly is good

Contemporary distributional analyses do not rely on singular sentence examples as those above, but take into account the knowledge of many different constructions (Redington, Chater and Finch 1998). Indeed, Tomasello (2003) offers a functional distributional analysis, whereby functional patterns are correlated with surface patterns: "children form paradigmatic categories of linguistic items - either words or phrases - that play similar communicative roles in the utterances they hear around them".

Furthermore, Culicover (1999) observes how children are both *conservative* and *attentive* to the usage of language. Children do not overly generalise beyond their experience of language. They are attentive to generalisations that correlate with their language knowledge. Israel's (2002) proposal for dealing with generalisations i.e. creativity, in child language is also in the same vein. Children's generalisations remain locally consistent with the knowledge of language in that "linguistic units activated online in usage events," would "be as consistent as possible with entrenched utterance types" (Israel 2002: 123). In addition, children's generalisations are globally consistent with the "the repertoire of constructions as a whole", and seeks "units [to] be represented in ways which maximize their consistency with each other" (Israel 2002: 123). These skills go a long way in preventing "spurious correlations".

The next section will review the evidence of how distributional information facilitates the process of segmentation and linear sequencing of the input by the child.

3.4 Segmentation and linear sequencing

As this study is concerned with the role of transitional probabilities in a word co-occurrence network, the review will focus of sequential probabilities and frequencies of words (type and token) as cues for segmentation and linear sequencing and also on the role of distributional word patterns in the input and what that can tell us about syntactic structure. The next section will briefly examine how children use transitional probabilities to segment the input heard into words, categories and constituents.

3.4.1 Attending to transitional probabilities

As early as seven months, infants are utilising transitional probability information to help develop a more language-specific strategy for segmentation. Since two phonemes occurring across a word boundary generally have a lower transitional probability than word-internal phonemes, infants make use of these statistics to segment speech. For example, seven month old infants attended less to stress cues and more to transitional probabilities than nine month olds when segmenting words from part words (Thiessen and Saffran 2003). Experiments indicated that eight month old infants successfully segmented word boundaries by tracking transitional probabilities of one phoneme occurring adjacent to another (Saffran, Aslin and Newport 1996; Aslin, Saffran and Newport 1998). Infants were exposed to two minutes of continuous synstudyed speech with no word boundary cues other than transitional probabilities.

Indeed, the higher the transitional probability the better the recognition of the words, in the case of adults (Saffran, Newport and Aslin 1996) and seven month old infants (Goodsitt, Morgan and Kuhl 1993). Furthermore, in a follow-up study of eight month old infants, the frequency of words and part words in an artificial language were matched, demonstrating that discrimination of word boundaries relied specifically on transitional probabilities (Aslin, Saffran and Newport 1998) rather than on relative frequency.

It seems that the children aren't just picking up on phoneme sequences either but are learning what constitutes possible lexical items (Saffran 2001) since eight month olds prefer artificial words over artificial part words set in English frames e.g. **I like my** pabiku. But in nonsense frames e.g. **Zy fike ny** pabiku or tone sequences, there was no preference. This result emphasises that "infants prefer to listen to materials that are segmented according to the structure of their native language...[and that] they are using the statistical properties of sound sequences to acquire linguistic knowledge: possible words in their native language" (Saffran 2001:165).

Infants (aged nine months) are able to differentiate between phonotactic patterns that are legal and illegal in their native language (Friederici and Wessels 1993; Jusczyk, Friederici, Wessels, Svenkerud and Jusczyk 1993), with infants preferring patterns that are more frequent in the native language (Jusczyk, Luce and Charles-Luce 1994). Indeed, it has been demonstrated that nine month infants are "sensitive to how phonotactic sequences typically align with word boundaries" (Mattys, Jusczyk, Luce and Morgan 1999: 465) Thus, infants aged nine months can segment speech using phonotactic probabilities, in that a consonant cluster with a high probability when straddling words and a low probability within words provides good cues to word boundaries (Mattys and Jusczyk 2001).

Unlike six month old infants which lacked the skill of integrating cues, nine month old infants are sensitive to sequential and rhythmic correlations (Morgan and Saffran 1995) as well as the "increasing familiarity with the prosodic features of the native language" (Jusczyk, Cutler and Redanz 1993: 675). By the age of seventeen months, infants are able to identify minimal word pairs in an experimental environment (Swingley and Aslin 2000; Werker, Fennell, Corcoran and Stager 2002).

Attention to sequential probabilities is not just specific to language stimuli. Children use this cue to trace the boundaries of non-linguistic tonal sequences (Saffran, Johnson, Aslin and Newport 1999) as well as in the visual domain (Fiser and Aslin 2002; Kirkham, Slemmer and Johnson 2002). Nor is being able to keep track of probabilities a species-specific skill, since Tamarin monkeys are able to compute basic sequential patterns

(Hauser, Newport and Aslin 2001) and rats also detect relative frequencies of co-occurring sounds (Toro and Trobalón 2005). In addition, it seems that tracking sequential probabilities does not require explicit direction to this task as shown by experiments in human infants (Saffran, Aslin and Newport, 1996), cotton-top tamarin monkeys (Hauser, Newport and Aslin, 2001), and rats (Toro and Trobalón 2005).[2]

3.4.2 Linear sequencing

Linear sequencing assists infants in identifying word boundaries (Saffran, Aslin and Newport 1996; Saffran, Johnson, Aslin and Newport 1999; Saffran 2001; see section 3.4.1) as well as with combining elements in multi word or gesture utterances.[3] Indeed, one of the learning mechanisms found by observing Nicaraguan Sign Language showed "a predisposition for linear sequencing". Experimental work demonstrates that infants, in attending to sequential information, prefer familiar orders (Mintz 1996). After nine month old infants were introduced to three random word strings containing four distinct words for a total of 120s, the infants attended more to word strings in which words had a familiar order rather than word strings with words that occur in a variety of positions (Mintz 1996: 81).

Knowledge of transitional probabilities provides a basis for identifying sequential ordering (Saffran and Wilson 2003). Infants aged twelve months distinguished illegal and legal word strings from an artificial language. Words in the training strings were only characterised by transitional probabilities (1 across words and 0.25 between words). So to discriminate legal strings from illegal strings the infants would first have

---

[2] Whilst it has been shown that tasks such as colouring do not detract from detecting sequential probabilities (Saffran, Newport, Aslin, Tunick and Barrueco 1997), more distracting tasks have shown that tracking is quite possibly dependent on the attentional resources of the tracker (Toro, Sinnett and Soto-Faraco 2005). At test, attentional distractions did disrupt the computation of serial patterns (Toro, Sinnett and Soto-Faraco 2005). The distractions used were tasks to detect pitch changes or sound repetitions, such as a car engine running, overlapped into (but not correlated with) the artificial language - the delay between these sounds and the artificial language being between 650-750 ms- or repeated pictures shown at a 30°angle for 500ms

[3] Multi gesture combinations may be sequenced linearly, even those relating to events that are simultaneously represented (see Nicaraguan Sign Language , Senghas, Kita and Özyürek 2004).

used transitional probabilities to identify legal words.[4]

Having segmented the input, the task of the learner is to identify the distributional patterns. How the learner does this and in what form the patterns are represented, will be the topic of the next part of the literature review with the focus on the knowledge of (input) frequency.

3.5 Extracting patterns

It has been proposed that word distributions provide information on syntactic categories such as a (general) verb and noun category (see Maratsos and Chalkley 1980 for semantic-distributional patterns; Mintz, Newport and Bever 1995; Mintz 1996; Cartwright and Brent 1997: 121, for sentences that were minimal pairs; Redington, Chater and Finch 1998; Mintz, Newport and Bever 2002). Recall that this study supports construction-specific categories rather than general categories. A proposal indirectly supporting this view suggests that early constructions are verb-specific i.e. a verb has its own syntactic or semantic relations associated with it, so that there does not exist a general syntactic category of verb, abstracted across constructions (Tomasello 2003).

Although some studies evidence that constructions act as verb-specific islands in the child's constructicon (Tomasello, Akhtar, Dodson and Rekau 1997; see Tomasello 2003 for discussion), there is evidence to the contrary. In a study of English speaking children aged 2.5, the transitive construction was generalised to novel verbs in experimental conditions (Childers and Tomasello 2001). Importantly, the effect was facilitated by pronominal frames. Thus it seems that the child's early syntactic knowledge does not just consist of an archipelago of verbal islands, but rather there are interconnections between individual constructions through pronominal frames (and perhaps other high frequency items, see sections 3.5.1 to 3.5.4). In other words, right from the start children are working with a "system" of knowledge (Ninio 2003).

The next section reviews evidence that children's early constructions tend to be organized around highly frequent items that act as anchors, forming constructional schemas (Pine, Lieven and Rowland 1998). The anchors are fixed items that reoccur in the schema whilst the filler item is variable

---

[4] An alternative explanation is that the infant may have been attending to four or more syllable sequences (see Gómez 2007).

(with the possibility of a set of variables representing a syntactic category). This type of schema has been known as pivot schemas (Braine 1963), slot-filler schemas (Pine and Lieven 1997), lexically-specific patterns and positional patterns (Lieven, Pine and Baldwin 1997). Anchors can consist of familiar names e.g. mommy (Bortfeld et al 2005), pronouns (Lieven, Pine and Baldwin 1997), verbs (Tomasello 1992) and function words in multi-phrases e.g. 'there's a x' (Lieven, Pine and Baldwin 1997).

For example, of the utterances representing the first twenty five lexical patterns produced in early speech of eleven children (1;8 to 2;8 year olds language; MLU 1.41 to 3.75), a third of the total were positional patterns (Lieven, Pine and Baldwin 1997:195) whilst 60% of all utterances fitted into a pattern. This figure included frozen utterances[5] (Lieven, Pine and Baldwin 1997).

3.5.1 Anchors and Frames

Since the eighties it has been known that learning an artificial language is aided by so-called anchor points in the input (Valian and Coulson 1988; see also Green 1979). These anchors are highly frequent words, words that are "reliably associated with a single structure"[6] (Valian and Coulson 1988: 72) much like the highly connected function words in natural language. The artificial language was constructed by the phrasal schema aA bB, where a and b are the more frequent markers and A and B are the less frequent content (lexical) words (Valian and Coulson 1988). Phrasal learning was the fastest when the anchor token was six times as frequent as the content word token. Learners failed totally to extract two different phrases when the most frequent marker occurred only 1.5 times more often. Thus, a high token frequency of a word provides a way of tracking dependencies, in this case the "marker-content pairing" (Valian and Coulson 1988: 82).

Even though the study consisted of a tiny artificial language, the results predict that markers will be utilised by children to identify categories and their structure, and that these markers will occur very frequently in the

---

[5] "Utterances which contain two or more words which have not previously occurred alone in the child's vocabulary or which contain one such word, provided it has not occurred in the same position in a previous multiword utterance" are classified as frozen (Lieven, Pine and Baldwin 1997: 195).

[6] This single structure will be captured by the automorphic motif as discussed in chapters eight and nine.

child's input (Valian and Coulson 1988: 83). Valian and Coulson (1988) cites Kucera and Francis's (1967) analysis of written text as evidence – the twenty most frequent words are all function words.

3.5.2 Familiar names

It seems that anchors or more specifically, frequent lexical information – may play an important role in learning language. For example, in respect to frequently used familiar names such as the infants own and *mommy*, infants as young as six months old recognise the word after the familiar name but not after unfamiliar names (Bortfeld, Morgan, Golinkoff and Rathbun 2005). Altering the sound of the familiar name i.e. *mommy* to *mommy's* does not produce any recognition by the infant. Since the input is such that the child could use transitional probabilities to recognise the word following *mommy's*, the infant appears to be relying on the frequency of the familiar word to anchor h(is/er) attention.

3.5.3 Pronouns[7]

Recent studies have also shown that lexically-based patterns in early child language often consist of pronouns that act as a fixed part of a slot filler schema (Lieven, Pine and Baldwin 1997; Pine, Lieven and Rowland 1998). In their study of early speech of eleven children (1;8 to 2;8 year olds language; MLU 1.41 to 3.75), the pronoun *it* was used so frequently after the verb, that Lieven, Pine and Baldwin (1997: 210) contemplated that "we may be seeing the development of an emergent category of verb".

In fact, since the object pronoun *it* is found so frequently after transitive verbs, the pronoun may be viewed as a clitic and represent the variable part of the transitive predicate schema since it allows the direct object complement of the verb to be expressed overtly i.e. "it marks the predicate's general or partially incomplete character and the type of argument it should combine with" (Ninio 1994: 20).

When children (aged between 2;1 and 4;4) corrected novel word orders to the canonical SVO, around half of the corrections included pronouns whereas full noun phrases were used for non- canonical word orders. In fact, "whenever children used a pronoun with the non-SVO verbs, they

[7] Pronoun case errors will not be discussed as this topic is beyond the scope of this study.

used SVO order" (Akhtar 1999:351). These results suggest that pronouns form the anchor part of positional patterns in the child's language (Akhtar 1999). The results were replicated by Abbot- Smith, Lieven and Tomasello (2001:688). Pronouns were produced in the SVO word order corrections by "2-year-olds almost five times as often, 3-year-olds more than twice as often" as nouns.

In line with this study, children (approximately 2.5 years old) were more adept at producing transitive utterances with nonce verbs when trained on utterances using nouns and a pronoun schema e.g. *he's verb-ing it*, rather than with just nouns (Childers and Tomasello 2001). It is not clear whether the effect is solely attributable to the pronoun schema or the fact that the nouns alternated with the pronominal schema providing variation in the slot (Childers and Tomasello 2001). Indeed, as Childers and Tomasello (2001: 747) point out, "children need both some "anchor point" in terms of some specific, high-frequency morphosyntactic pattern (Valian and Coulson 1988), as well as diversity in terms of type variation in some places in this pattern, including verbs (Bybee 1995)".

Thus, a highly frequent word with an equally high type frequency is helpful for learning constructions. In network terms, these kinds of words are hub words (see chapter 4). In addition, pronouns are frequently found in child-directed speech (Cameron-Faulkner, Lieven and Tomasello 2003). Indeed, as Dodson and Tomasello (1998) point out, children under 2;6 years old, when learning novel verbs that appeared in sentences without arguments, produce transitive verbs with subjects *I* and *me* instead of nouns.

In a follow up study examining how helpful pronominal schemas are in the comprehension of the agent/patient roles within a novel transitive verb (Childers and Tomasello 2001), it seems that one particular pronominal pattern was helpful *he's [verb]-ing it.* This finding is perhaps due to *it* being the more frequent pronominal sequence, compared to say *it's [verb]-ing her*, since *it* is more frequently used as an object pronoun rather than a subject (Childers and Tomasello 2001).

Another study used a computational model (MOSAIC) to simulate child speech by analysing her mother's speech (see Jones, Gobet and Pine 2000 for technical details). MOSAIC's analysis of the mother's speech closely matched the child's output. Furthermore, in the child's speech (aged from 1;10.7 to 2;9.10), her mother's speech and the MOSAIC analysis all strongly showed pronominal frames.

Indeed, Sethuraman and Goodman (2004a) suggest that pronominal frames assist in the early generalisation of verbs, in that verbs frequently heard with pronouns in the input are generalised early (although there are exceptions – see Sethuraman and Goodman 2004a: 85). These studies point to pronominal schemas facilitating a learning path from specific to general, more abstract knowledge - at least from the perspective of the English language. The fact that pronouns are so frequent in the input makes them suitable candidates for schema anchors.

3.5.4 Repetition

Infants may well be attending to patterns based on repetitious input. In one study, infants aged seven month old were trained for two minutes on one of two artificial languages based on one of the following templates: ABA 'ga ti ga' and ABB, 'ga ti ti' each comprising 16 three word sentences that were repeated three times. At test, infants attended longer to sentences based on the untrained grammar in contrast to the trained grammar with new words (Marcus, Vijayan, Bandi Rao and Vishton 1999). Recourse to transitional probabilities was not possible (all words at test were novel). Thus, Marcus *et al* (1999) suggest that infants had learnt a variable placeholder, much like an algebraic term. On this view, the infants differentiated between the two grammars, because infants may have formed algebraic-like rules (Marcus *et al* 1999).

Since reduplication in the 'grammar' was present, the infant's may have been expressing sensitivity to the repetition of sounds rather than forming algebraic-like rules (Gómez and Gerken 1998). This approach has been evidenced in that the learners transferred pattern knowledge which consisted of repetitious elements but failed when the input lacked reduplication (see Gómez and Gerken 1998 for infants and Gómez, Gerken and Schvaneveldt 2000 for adults), even though infants were still sensitive to sequential dependencies (Gómez and Gerken 1998). Moreover, attention to repetition in pictorial form i.e. pictures of cats and dogs, has been demonstrated suggesting that this sensitivity is not limited to linguistic stimuli but maybe a domain general property (Saffran, Pollak, Seibel and Shkolnik 2007).

3.5.5 Adjacent/Continuous dependencies

There is evidence to suggest that twelve month old infants can associate words that mimic a prosodic characteristic of function words, with their adjacent categories (Gómez and LaKusta 2004). The infants were trained on one of two artificial languages, constructed with forms: *aX* and *bY*; *aY* and *bX*, where disyllabic words were substituted for the *X*'s and monosyllabic for the *Y*'s, since the length of function words is usually very short. The infants were tested on their training language but with novel *X* and *Y* words and demonstrated that "12-month-olds are capable of forming categories of X- and Y-elements based on a shared feature [of syllable structure] and, furthermore, form associations between particular a- and b-elements and these categories" (Gómez and LaKusta 2004: 567). Importantly, the infants were not just attending to the syllable structure of the *X* and *Y*'s but also to the co-occurring elements *a* or *b*.

Furthermore, Gómez (2006) extended this study (Gómez and LaKusta 2004, experiment 1) to include a variable set size for the function-like words to examine the effect of variability in word set size on the infants ability to associate function-like words with categories. One group of infants were trained on a language with 6X's and 6Y's (see Gómez and LaKusta 2004, experiment 1) whilst the other group's training language had 3X's and 3Y's. The infants were trained on their language. Both languages with a set size of 6 consisted of twenty-four randomly presented phrases and twelve phrases for set size 3 presented twice as often to keep frequencies of *a* and *b* words constant across the study. In Gómez and LaKusta (2004) study, novel words replaced the trained X and Y words whereas the a and b words were respectively retained.

For set size 6, of the twenty-four infants, eighteen "listened longer to strings from their training language than to strings from the other language" (Gómez 2006), but for the set size 3 there was no preference for the training set. It seems that even though the syllabic structure of training was preserved at test, the attention to syllables was not enough in the smaller size set to generalise to novel X and Y elements. Infants with set size 6, showed that most could discriminate between their training language, suggesting "sensitivity to the co-occurrence relations between markers and X- and Y-categories based on their distinguishing features" (Gómez and LaKusta 2004: 572). Furthermore, "learning was to some degree abstract (involving grouping of the X- and Y- elements according to syllable number)", being based on the variability of members in the category and thus implies categorisation of the elements (Gómez and LaKusta 2004: 572).

3.5.6 Non-adjacent/discontinuous dependencies

Certain studies have shown that a particular discontinuous frame is effective at capturing syntactic categories (Gómez 2002; Mintz 2003). The pronominal frame is an example i.e. *he's [verb]-ing it.* There are two anchor points and a medial variable slot. Since the items in the frame are non-adjacent, such frames show non-adjacent (or discontinuous) dependencies. English speaking infants as young as eighteen months attend to discontinuous dependencies e.g. the copula gerund construction 'Y is VERB-ing' (Santelmann and Jusczyk 1998). The sensitivity to the dependency is only across a restricted window of one-three syllables.[8]

Interestingly, a study has examined the relation between the types of elements intervening the discontinuous dependency 'hat-ge-X-t' e.g. 'Grossmutter hat immer gelacht' and the effect on dependency recognition by nineteen-month German-speaking infants (Höhle, Schmitz, Santelmann and Weissenborn 2006). When the intervening element was an adverb, the dependency wasn't tracked, whereas with a noun phrase of the same syllabic length, the dependency was recognized. Höhle et al (2006) argue that a general familiarity with the intervening element is not a probable explanation given that two year old children's comprehension of a sentence was disrupted with a familiar but incorrectly positioned element (Gerken and McIntosh 1993). However, familiarity of an element *is* closely tied to its sequential position in an utterance, particularly with so-called function words i.e. *the* precedes but rarely follows a noun.[9] Thus, the incorrectly positioned element may have seemed unfamiliar due to an unusual and unexpected placement.

In a study on the learning of non-adjacent dependencies in an artificial language, eighteen month old infants were tested on their ability to detect

---

[8] However, as Santelmann and Jusczyk (1998) note not all languages have such short dependencies, e.g. in the German sentence 'Wir haben diese Bilder schon gesehen', 'we have already seen these pictures', haben and gesehen have five intervening syllables. Consequently, the restriction on the processing window may be language-specific if German 18 month old infants are sensitive to this construction.

[9] In most cases, there is a preceding pause as in list-type utterances 'the **one**, **the** only Mohammed Ali' where ***one*** occurs before ***the.***

discontinuous over adjacent dependencies using a "preferential listening procedure" (Gómez 2002). Two languages were used and consisted of "three-element strings" of the type "*aXd* and *bXe"* or *"aXe* and *bXd"*. The medial variable element has varied from a set size of {3, 12, 24} elements. So that "infants exposed to set size 24 heard each of the 48 strings (2 dependencies x 24 X elements) produced by the training language one time during training" (Gómez 2002:433). Once the three minute training session was completed, the infants were tested on the trained language strings and the alternative language strings.

Results showed that "learners acquired nonadjacent dependencies only when the middle element was most variable, when adjacent dependencies were least predictable in set-size = 24" (Gómez 2006: 18). Before this set size, the learners still seemed to be focussing on the adjacent dependencies, since there was no correlation between the incremental size increases in the medial set and sensitivity to non-adjacent dependencies (Gómez 2002; Gómez 2006). Thus learners attend to adjacent transitional probabilities even after these probabilities are not informative. Only when the type frequency increases to a certain point between dependencies, does the learner discern the discontinuous dependencies.

Similar results were found for adults after an eighteen minute training session with a middle set size of {2,6,12, 24} elements with three dependencies "*aXb, cXd,* and *eXf"* or *"aXd, bXf* and *cXb"* (Gómez 2002). A subsequent study of seventeen month old infants used a maximum middle set size of eighteen variable items and acquired the same result - although twelve month olds failed at this task: "by 15 months of age infants begin to show sensitivity to this structure" (Gómez and Maye 2005:199). It seems that when transitional probabilities are low between adjacent dependencies, then the learner attends to type frequency between non-adjacent elements. Just how much variability is required "presumably varies as a function of infant age and other factors contributing to task difficulty" (Gómez and Maye 2005).

On to natural language and language input (CDS). In a distributional analysis of CDS from four CHILDES corpora (MacWhinney 2000a), it was found that by observing the word co-occurrences surrounding a word i.e. following Y, preceding X, it is possible to identify the syntactic categories of verb and noun (Mintz, Newport and Bever 2002; Redington et al. 1998; see Cartwright and Brent 1997 for a slightly different approach).

Since this adjacency environment around a word may be used for category identification, it would be pertinent to see how informative this type of "distributional unit" is to language learners (Mintz 2003). Mintz (2003) extended previous studies to include both preceding and following adjacent environments to form a frame. Only frames that were frequent would identify categories. As Mintz (2003) points out, Mintz et al. (2002) and Redington et al. (1998), studies differed since selection was based on the frequency of the actual target word not the words adjacent. A frequent frame, as a unit, is defined as "ordered pairs of words that frequently co-occur with exactly one word position intervening (occupied by any word). Any words that occur as the intervening word inside a given frequent frame are categorized together" (Mintz 2003: 93).

Distributional analysis of adult speech was preformed on six corpora from CHILDES collected from conversations of children up to the age of 2;6 years (MacWhinney 2000a). All frames were extracted - as per the above definition (see Mintz 2003: 96 for a full account of the criteria). Table 3.1 is from Mintz (2003) study, showing frequent frames that appeared in two or more corpora. The numbers at the top indicate how many of the corpora the frames occurred in.

Whilst most of the elements that compose the frames are function words e.g. "to__it", "the__in", "what__that", some frames do incorporate lexical words e.g. "do__want"; "put__on", "I__know", "do__think". Furthermore, "the tokens of the categorized types making up the 6% contained in frequent frames constituted half the tokens in a given corpus" (Mintz 2003: 103). Consequently, on the view that there is a general noun and verb category, the learner need only attend to a relatively small number of contexts to make in roads into categorising the input (Mintz 2003).

Table 3.1 "Frames that were frequent frames in at least two corpora, organized by number of corpora in which each occurred" Sourced from Mintz (2003: 102, Table 5).

6	5	4	3	2
do__want	a__of	I__think	I__know	I__it
put__on	put__in	I__you	are__doing	a__on
the__in	to__the	are__going	did__do	can__it
the__on		is__a	go__the	do__think
to__it		it__the	the__of	don't__it
want__to		to__a	the__one	have__got
what__you		would__like	there__is	have__look
you__a		you__that	to__to	here__are
you__it			what__it	in__box
you__me			why__you	put__back
you__the			you__with	shall__put
you__to				the__and
				the__is
				there__are
				to__on
				to__with
				we__a
				we__to
				what're__doing
				what__is
				what__that
				what__we
				you're__to
				you__have
				you__some
				you__what
				you__your

Mintz (2003) argues that frequent frames provide a "potentially more informative context than the bigram contexts investigated in prior studies, and also provides a cross-linguistically viable framework for focusing on the level of analysis (e.g. sub-lexical morpheme, word) that might be most informative for a particular language" (Mintz 2003: 115). Training a feedforward neural network on child directed speech (Monaghan and Christiansen 2004: 5), it was observed that bigram information, specifically the schema *aX* i.e. an anchor word preceding the slot, provided "slightly lower accuracy [in syntactic categorisation] but much higher coverage of the language" than the *aXb* frame proposed by Mintz (2003). The simulations suggest that whilst "learning is most effective when information about the preceding word and the succeeding word is available…this is only the case when the AxB frame is not computed as a whole. Learning must also be based in part on the relationship between A and x and between x and B."

Whilst it seems that frames are effective in categorising French language input (Chemla, Mintz, Bernal and Christophe 2009), it seems that this effectiveness may not be universal. For example, when novel words were presented verbally in frequent frames, twelve and sixteen month old Dutch infants failed to categorise these novel words (Erkelens submitted). Erkelens (submitted) notes that "the frame-based categories for English are much more accurate than the Dutch frame- based categories. This fact makes the frequent frame cues less reliable for Dutch children to use in initial categorization".

Frames are interesting in a Radical Construction approach since the frame and slot can equate to an actual construction e.g. I __it (taken from Mintz 2003: 102, Table 3.1 above) with the variable part represented by the construction-specific category. From a usage approach, language is built "piecemeal" and so various fragments combine to produce longer more complex constructions. Consequently, frames may combine together to produce more adult-like constructions and provide a higher level of chunking compared to just bigram probabilities. As will be shown, by analysing link patterns in a word co-occurrence network, such frames do occur more often than by chance in the adult and child texts. However, there are other patterns in the links that also occur significantly more often than by chance, so this type of analysis is one of many patterns that exist in language and would be open for inspection by the child through form and functional regularities.

From a network approach, since certain syntactic structures are a product of storing linguistic representations in a small world hubterranean network, token frequency as well as type frequency has a bearing on the structure formed i.e. through the transitional probabilities, word connectivity (Ferrer i Cancho and Solé 2001a) and the fact that frequency of words in a language follow a power law (Zipf 1935/65; see chapter four, five and six).

The next part of the chapter focuses on verb generalisations and presents the findings and evidence given in Sethuraman and Goodman (2004a), in that five factors are implicated in early verb generalisation. These factors will be related to the formation and characteristics of a hubterranean structure in chapter nine. The next section will examine how frequency facilitates the learning and generalisation of constructions.

3.6 Verb Generalisation

The degree to which a construction is generalised beyond input may reflect the number of stored exemplars. As Goldberg (2006: 99 caps authors own) points out, "A PATTERN IS CONSIDERED EXTENDABLE BY LEARNERS ONLY IF THEY HAVE WITNESSED THE PATTERN BEING EXTENDED". Consequently, type frequency has a bearing on how extendable a generalisation is (and is related to productivity as discussed in chapter two). Furthermore, "the greater the degree to which previously attested instances fill a semantic space that includes the potential target instance, the more confident speakers will be in using the target instance"(Goldberg 2006:100).

In constraining generalisations, children pay attention to how entrenched a linguistic element is (Braine and Brooks 1995: 368; Goldberg 2006). In an experimental environment, it has been shown that less frequent verbs, such as *strike*, *arrive*, *remove* and *vanish*, tended to be overgeneralised more than highly frequent verbs i.e. *hit*, *come*, *take* and *disappear* by children as young as three years old (Brooks, Tomasello, Lewis and Dodson 1999). Overgeneralisation meant in this case that a fixed transitive verb was used in an intransitive construction and a fixed intransitive verb, in a transitive construction. English and French children,[10] when correcting ungrammatical word order in sentences, were more likely to correct uncanonical word order sentences with a high frequency verb (Matthews, Lieven, Theakston and Tomasello 2004; see Matthews, Lieven, Theakston and Tomasello 2005). Thus, the frequency of the verb has a bearing on productivity and word order knowledge at this age.

Entrenchment is not the whole story concerning productivity, as Goldberg (2006) points out, since verbs can appear in very novel syntactic environments, as Goldberg's examples demonstrate (example 3.2). The verb *sneeze* appears mostly as an intransitive verb and yet it extends easily to the construction in example 3.2a.

---

[10] The mean age of the English and French children in the two training sets were 2;9, 3;9 and 2;10, 3;9 respectively.

Example 3.2 Verbs used in novel syntactic environments. From Goldberg (2006: 94 examples 1-3).

a) She sneezed the foam off the cappuccino
b) She danced her way to fame and fortune
c) The truck screeched down the street

How such creative examples as above can be used, while restricting erroneous generalisations, may be answered by the "idea that specific knowledge always pre-empts general knowledge in production, as long as either would satisfy the functional demands of the context equally well" (Goldberg 2006: 94). Goldberg (2006) uses the comparative forms [adj-er] and [more adj] as an example. Since *better* is a more frequent form of *more good* i.e. it is more entrenched, then the specific knowledge of the lexical comparative *better* pre-empts the pattern [more adj]. If neither form is particularly entrenched neither will pre-empt the other and so both comparatives would be used, as in *more moist/fresh/dull* and *moister/fresher/duller* (Goldberg 2006).

Returning to the creative examples given above in (3.2), naturally each construction has its own particular semantics and/or pragmatics (Goldberg 2006), so how could one construction ever pre- empt another? In answer to this question Goldberg (2006: 96) appeals to the notion of *statistical pre-emption*. That is, "In a situation in which construction A might have been expected to be uttered, the learner can infer that construction A is not after all appropriate if, consistently, construction B is heard instead" (Goldberg 2006: 96). Thus, in example 3.2a, *sneeze* is used in the caused-motion construction because the attributed interpretation of 'causal force' is not pre- empted by another construction. This way of thinking is not unfeasible since a study on 14 month old infants suggests that infants do reason with pre-emption - in experimental settings at least (see Gergely, Bekkering and Király 2002 for details, as cited in Goldberg 2006: 96).

3.6.1 Generalisation to new syntactic patterns (see Sethuraman and Goodman 2004a)

Whilst the entrenchment study (Brooks, Tomasello, Lewis and Dodson 1999) showed that high frequency verbs are less likely to be generalised e.g. the more familiar a child is with a transitive verb, the less likely that verb will be extended to constructions that contravene transitivity, Bowerman (1988, 1996) found that highly frequent verbs such as *go* and *come*, were the source of many overgeneralizations when analysing the speech of her two daughters. Sethuraman and Goodman (2004a) also evidenced that verbs used in "multiple syntactic environments" were the same verbs that children used frequently.

It also appears that the (relative) frequency of a verb occurring in a particular construction correlates with how well young children recall and repeat sentences exemplifying the construction (Kidd, Lieven and Tomasello 2006). The children reproduced complement taking verb (CTV) sentences much more accurately when the CTV was highly frequent e.g. *think* than when a low frequency CTV was used e.g. *hear*. In addition, more corrections to grammatical errors were made in sentences with a high frequency CTV. Kidd, Lieven and Tomasello (2004:58) conclude that "the availability of frequent exemplars provide the basis upon which children can construct a grammar".

When discussing the frequency of a verb, one can refer to its overall frequency within the speech (both the overall token and type frequency with respect to word co-occurrences), its frequency within one construction (the verb-construction token frequency) or how many constructions it appears in (the verb-construction type frequency). The next section looks at how these different kinds of frequency support verb generalisation.

3.6.2 The consistent use of a verb in a specific syntactic pattern (see Sethuraman and Goodman 2004a)

It has been shown that the earliest usage of a word by a child tends to correlate with the most frequent usage of the word in the input (Harris, Barrett, Jones and Brookes 1988). In the caregivers' speech, verbs used frequently in one specific syntactic environment also occur highly frequently in the child's speech in the same syntactic environment (see below Table 3.2 - Goldberg, Casenhiser and Sethuraman 2004).

Since certain verbs appear so often in a particular construction, "the specific formal patterns associated with particularly frequent verbs come to be associated with the meanings of those verbs" (Goldberg, Casenhiser and Sethuraman 2004:25). Highly frequent verbs whose meanings become associated with the patterns they most often appear in, are also more likely to get generalised the earliest (Sethuraman and Goodman 2004a).

Table 3.2 Adapted from Goldberg, Casenhiser and Sethuraman (2004): "Table 4. 15 Mothers' most frequent verbs and number of verb types for 3 constructions" in Bates, Bretherton and Snyder (1988) and Carlson-Luden (1979) corpus (Goldberg, Casenhiser and Sethuraman 2004). For the double object construction, see Goldberg, Casenhiser and Sethuraman (2004: 12). Legend: subject (Subj), verb (V), object (Obj) and oblique (Obl).

construction	mothers	total number of verb types	children
1. Subj V Obj	39% *go* (136/353)	39 verbs	54% *go* (121/224)
2. Subj V Obj Obl	38% *put* (99/259)	43 verbs	31% *put* (16/51)

The distributional pattern of verbs "roughly matched for frequency" i.e. *give, put, like* and *eat* were examined from mother's speech in the Bates corpus and from the child's speech in the Goodman corpus (Sethuraman and Goodman 2004a). Two different corpora were used to show that the results pertained to the English language in general. Verbs used often and consistently in a specific syntactic construction i.e. *give* and *put*, were the verbs that the children extended to other syntactic constructions.

Furthermore, a study has shown that in an experimental setting, children (five to seven year olds) were not only able to associate a newly learnt abstract meaning with a newly learnt syntactic pattern but also generalise the association to novel utterances in which novel verbs appear (Casenhiser and Goldberg 2005; Goldberg 2006). Performance was significantly better if the training was skewed by frequency,[11] "confirming the hypostudy that learning is facilitated when one verbal token accounts for the majority of utterances" (Casenhiser and Goldberg 2005: 503).

A computer simulation i.e. a simple recurrent network, looking at category formation has also confirmed that a high token frequency per category in the input initially helps in forming the respective category (Borovsky and

[11] The skewed frequency condition consisted of five novel verbs, one was heard eight times whilst the other four novel verbs, only twice (**8**-2-2-2-2). In the balanced frequency condition, three of the five novel verbs were presented four times whilst the other two novel verbs were only heard twice (4-4-4-2-2).

Elman 2006). Goldberg (2006: 89, bold is author's own) suggests that this high token frequency effect "may involve a type of **cognitive anchoring**, where a high-frequency type of example acts as an anchor i.e. a salient point of comparison" and also notes how such standards of comparison have been observed in numerical estimations (Tversky and Kahneman 1974).

It is not always the case that a verb used in a consistent syntactic environment gets generalised early. As Sethuraman and Goodman (2004a) point out, both *look* and *get* were extended verbs in the child's speech but not associated with a specific pattern in the mother's speech (Sethuraman and Goodman 2004a). And vice versa: *eat* and *have* appeared consistently in one syntactic environment of the mother's speech, yet were not extended in the children's speech.

3.6.3 Semantic lightness (see Sethuraman and Goodman 2004a)

It has been noted that highly frequent verbs tend also to be "general purpose" verbs or "light" verbs (Clark 1996). These types of verbs i.e. *put, do, make, go* etc are often the earliest verbs to be learnt (see Clark 1996 for discussion of studies on Finnish, French, Japanese and Korean. For Hebrew see Ninio 1999a). Furthermore, such verbs - also known as "pathbreaking verbs" - are cross-linguistically subject to grammaticalisation (Ninio 1999a). In the Goodman corpus, light verbs constituted "82% of the verbs children first extended" (Sethuraman and Goodman 1994: 85).

Not all verbs that are extended early in a child's speech are semantically light, since Sethuraman and Goodman (1994) found that "heavy" verbs like *cut* and *ride* were used in varied syntactic environments in the Goodman corpus (Bates and Goodman 1997). Moreover Sethuraman and Goodman (1994) note that there is also no definitive set of light verbs (Theakston Lieven, Pine and Rowland 2004). Furthermore, the factor of frequency is tightly entwined with the notion of light verbs (Theakston, Lieven, Pine and Rowland 2004).

3.6.4 Multiple environments (see Sethuraman and Goodman 2004a)

Another factor influencing which verbs extend early is syntactic diversity (Sethuraman and Goodman 2004a). It has been suggested that verb meanings are learnable when the input provides a "range of frames" for the verb (Gleitman 1990: 48) and it has been shown in an experimental setting that two year old children can learn the meaning of a novel verb by attending to the uses of the verb in multiple syntactic frames. Parental input facilitates learning, since a study on verb learning in English and Mandarin showed that "parents use a small a set of verbs repeatedly in many contexts and other verbs infrequently" (Sandhofer, Smith and Luo 2000: 578).

Since a child's knowledge of language relies quite heavily on the input, it is not completely unexpected that verbs heard in multiple environments tend to be the first to be extended by the child (Naigles and Hoff-Ginsberg 1998). Studies have shown that verbs used by mothers in multiple syntax ctic frames were also the same verbs used by their child in diverse syntactic environments (Naigles and Hoff-Ginsberg 1995; Sethuraman and Goodman 2004a). However, not all verbs that children extend are used in a syntactically diverse way by their mothers e.g. *ride* and *cut* (Sethuraman and Goodman 2004a: 84).

3.6.5 Extending verbs

Sethuraman and Goodman (2004a: 86) collated the various factors in the early extension of verbs and schematised these factors (Figure 3.1): the use of pronominal frames (see section 3.5.3), verbs that are semantically light, verbs that appear in one consistent frame in the input speech but also verbs that occur in multiple environments in the input speech. These factors are all confounded by the frequency of the verb and all factors, to some extent, explain the generalisation of verbs, although not adequately when considered alone.

Figure 3.1 Sourced from Sethuraman and Goodman (2004a: 86) depicting the different factors influencing the extension of verbs.

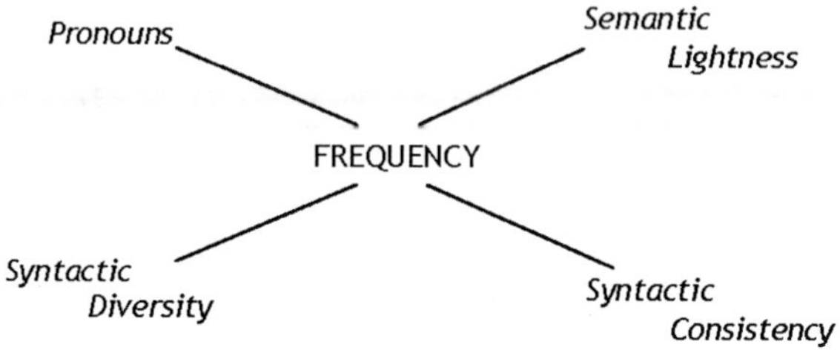

The latter two factors - syntactic diversity and syntactic consistency - are at first glance seemingly paradoxical. This study will show that there is no contradiction and that these factors relate to the hubterranean structure in English adult and child speech (see chapter nine for discussion).

3.7 Optimal exemplars

The developmental path of learning verbs is not linear but an accelerated process. When children learn to use verbs, their initial progress is slow. The second verb occurs some time after the first verb in a particular construction (Ninio 1999a). For example, the below Figure (3.2) shows how many English verbs an English speaking child has learnt between the ages of 1;03.18 - 1;08.08 for the SVO (transitive) and the VO constructions (Ninio 1999a: figure 2).

Note how long the second verb appears after the first in the two constructions and the subsequent increase in the momentum of verb learning (Ninio 1999a). Consequently, the greater the amount of verbs a child uses in a particular construction, the quicker the child will learn to use new verbs in that construction (Ninio 1999a).

It has been suggested that this initial slow period reflects the intensive learning and practise of the construction with its first verbs (Ninio and Keren-Portnoy 2002). This period "establishes the first verbs in the pattern as models of transfer and facilitation for other verbs to be learned in the same combinatory pattern" (Ninio and Keren-Portnoy 2002: 17). Such a study suggests that structures are learnt as a "interconnected system" that facilitates further learning; that is, "the learning of later items leans on the learning of previous items" (Ninio and Keren-Portnoy 2002:4).

Figure 3.2 Cumulative number of different verbs in VO and SVO word-combinations produced by a child learning English, as a function of age (Ninio 1999a: Figure 2).

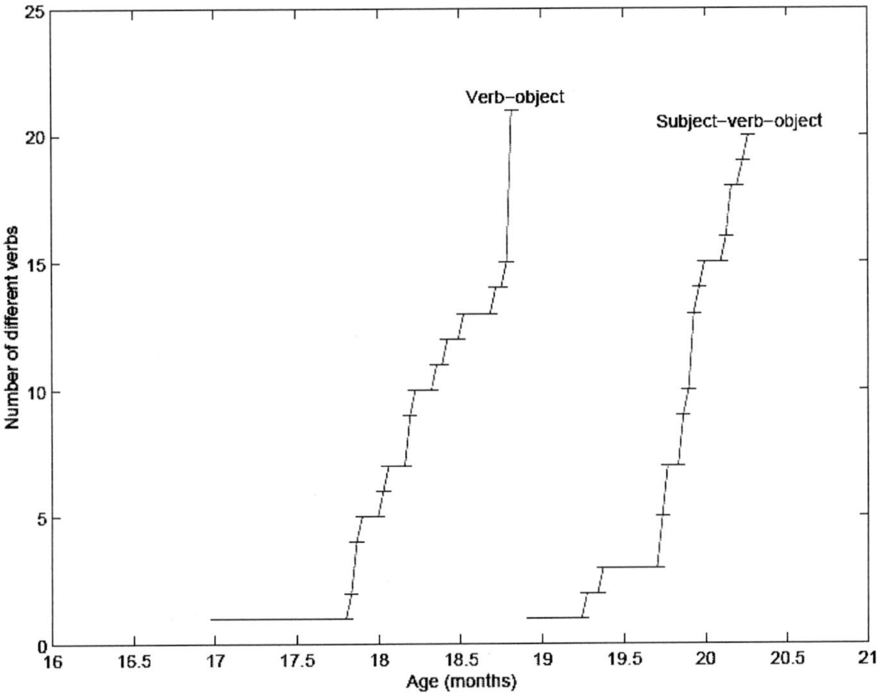

This accelerated learning process is assisted by the parents' usage of a few frequent verbs in representative constructions: either used consistently in one construction as a central example of that construction or used to demonstrate the extendibility of the verb. Furthermore "the highly expressed examples are also extremely central to the system's relationship with the environment" (Hershberg and Ninio in press). These highly frequent exemplars in the input are optimal for learning constructions, forming a part of the general properties of language (Hershberg and Ninio 2004, in press). Not by chance do such correlations occur, since "in natural cognitive systems the inside, the outside, and the process of internalization are tightly interconnected" (Hershberg and Ninio in press).

3.8 Summary

The child's grammar is a set of constructions that consist of exemplars of frozen utterances, partial lexically specific schemas and structures that have graded abstractness. These constructions are linked together by highly frequent elements in the child's speech, that are also highly frequent in the child's input. These elements include familiar names such as *mommy* and the child's name, pronouns and function words/morphemes (see Santelmann and Jusczyk 1998).

The development of these constructions relies on the child's accurate segmentation of the input as well as attending to the linear sequence of those segmented elements. Transitional probabilities assist in this process, the child being able to recognise transitional probabilities at the early age of seven months, a skill that is not specific to language or people but found in monkeys too. Whilst transitional probabilities can assist children in identifying adjacent dependencies, type frequency aids the identification of discontinuous dependencies, albeit within a small word window. It appears that the benefit of discontinuous frames with two anchors in providing categorical information (Mintz 2003) may be language-specific (Erkelens submitted). In fact, it has been shown that analysing the two discontinuous anchors separately i.e. as two separate bigram pairs, provides better categorisation results (Monaghan and Christiansen, 2004).

Children go beyond their experience of the input by utilising general cognitive skills such as intention reading, analogy, schematisation and distributional analysis. Examining verb generalisations in particular, it is found that frequency confounds the factors facilitating the extension of verbs to a new construal. The other factors are eclectic and even seemingly contradictory. However, it will be shown that once interpreted in the theoretical framework of network theory, they gain consistency. Namely the factors all relate to the hub structure of the network theory. Thus, Hershberg and Ninio (2004, in press) comments on optimal exemplar learning are on the mark. There is something to idea that the environmental input and the process of internalisation are "tightly interconnected". Indeed, they are one and the same.

CHAPTER FOUR

NETWORK PERSPECTIVES

4.1 Overview

This chapter will detail the network perspectives adopted in this study. I will start by giving a brief history behind the discovery of small world networks (section 4.2) and will then focus on language networks (section 4.3), particularly those studies with a language learning perspective (section 4.4). The properties characterizing language as a network will be discussed - namely, the hubterranean structure which will later be tied to the functional-lexical (f-l) distinction and basic sequential (syntactic) constituency in language. The explanation as to why hubterranean structure occurs in a network is explicated in section 4.5 (Mathias and Gopal 2001 a&b) and thus will provide an explanation for why (f-l) distinction and basic consistency are perceived in language, a topic that will be picked up in chapter five.

4.2 Small worlds[1]

4.2.1 Introduction

As early as 1736, the mathematical framework to analyse systems as networks was developed by the Swiss mathematician, Euler. Work up until the middle of last century focused on the so-called simple lattice, exemplified by crystalline structure. This type of network is regular in that each node has the same amount of links. Then, around the nineteen fifties, a series of papers (Solomonoff and Rapoport 1951; Erdös and Rényi 1959, 1960, 1961) were produced analysing large random networks – that is, networks with randomly assigned links. Nodes in this type of network have on *average* the same number of links, in that the majority of nodes have the same connectedness but a small amount of nodes may have more

[1] Section 4.2.1 is based on Barabási (2003). Barabási (2003) and Albert and Barabási (2002) discuss small worlds in greater detail.

or less links. Therefore, the connectivity can be expressed in a Poisson distribution; a distribution created when one counts the number of objects in a random area of space i.e. the stars (Panchenko 2005).

Whilst Erdős and Rényi's random network exposition covers ground in expressing complicated networks, their aim was to explore the mathematical world of networks in an abstract sense with less emphasis on its applications (Barabási 2003). Many networks in the world such as transportation systems or social ties involve planning, thought and free will: elements that aren't seen as random. With the advent of computers, further studies of networks have revealed a type of network that shares properties with both random networks and regular networks, the small world network (Granovetter 1973, 1983; Watts and Strogatz 1998). Watts and Strogatz (1998) generated a small world network by randomly rewiring the links in a simple lattice. These random re-wirings criss-crossed the network producing shortcuts from one part of the network to the other. Thus, the network was transformed from a regular network to a small world network. Once all the links had been re-wired, the network became randomly connected (Figure 4.1).

Figure 4.1: Different types of networks are formed at various values of parameter p (Watts and Strogatz 1998: Figure 1: 441)

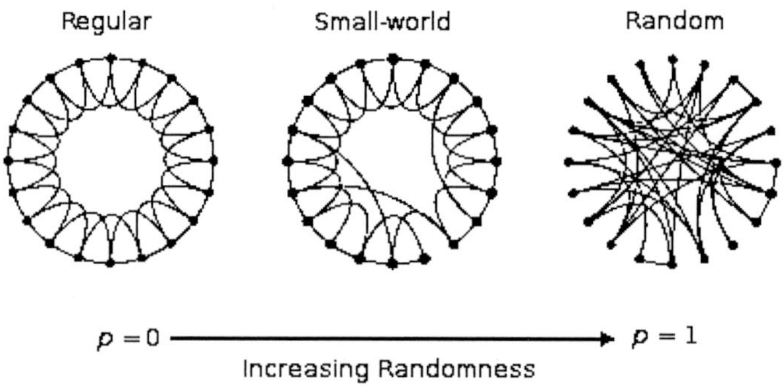

Small worlds have a high number of cliques like a regular network. A clique is formed when two connected nodes are also connected to a 'mutual friend'. The proportion of cliques i.e. clustering co-efficient C is calculated by dividing the total number of links in the network by the number of links (k) possible. Cv is a fraction of the links possible in V's neighbourhood over the actual realised links; whilst C is the average of this fraction for all nodes in a network.

An ordered network obtains a clustering co-efficient (C) of 1 where every node is linked to every other node i.e. a complete graph. A co-efficient of zero indicates an empty graph; one in which the nodes are not connected at all and are independent of each other. Consequently, C may be viewed as a marker of order in the network; random networks having very few cliques and thus a very low C (Watts 1999; Watts and Strogatz 1998). In a simple lattice, and a small world network generated by Watts and Strogatz (1998), the co-efficient C is high and is independent of the network size, whereas in random networks, C is low and decreases with the number of nodes.

Small world networks also share a property with random networks. They both have a small average shortest path whilst simple lattices display large average shortest paths. Indicating the traversal size of the network and also known as the network diameter, the average shortest path is obtained by finding the shortest path between a pair of nodes and then adding up the interconnecting links, and then repeating this procedure for all pairs of nodes. The shortest path of every node pair is then summed and divided by the number of paths to give the average shortest path of the network. In the Watts and Strogatz (1998) model, the path length L increases linearly with the size of the network when a small fraction (*p*) of links are rewired. For a large *p*, the path increases logarithmically (Watts and Strogatz 1998).

Whilst the clustering co-efficient is “a measure of local graph structure” (Watts 1999), the average shortest path distance (L) is a macroscopic measurement of network structure. Using these two metrics, small world networks fall between random and simple lattices. On the one hand, small world networks share a small shortest path with random networks but on the other hand, they also possess a high clustering co-efficient, characteristic of simple lattices.

4.2.2 Hubs

Whilst the Watts and Strogatz (1998) model provided a closer representation of real-world networks, further research into networks with high clustering and low diameter revealed a characteristic power law degree distribution that was not generated by Watts and Stogatz’s small world model nor Erdős and Rényi’s model (Barabási and Albert 1999; Barabási, 2003: 54). The term ‘degree’ refers to the connectivity of a node i.e. the number of links connecting to the node. A power law degree distribution implies that a few nodes will be highly connected i.e. possess a high degree and are known as hubs, whereas many nodes will have few links i.e. a low degree (see Adamic 2000 for a power law tutorial; Adamic and Huberman 2002).

Equation 4.1 The Power Law (Barabási and Albert 1999: 510), where k refers to degree.

$$\mathbf{P(K) \sim k^{-\gamma}}$$

This power law is observed when analysing hyperlinks on the Web. The probability P(k) that "K documents point to a certain Web page" is given in Equation 4.1, where $\gamma_{WWW} = 2.1 \pm 0.1$ (Barabási and Albert 1999: 510). Networks with this type of degree distribution are known as 'scale free'. Scale free networks differ from the small world modelled by Watts and Strogatz (1998). Recall that the clustering co-efficient C of small worlds and simple lattices is independent of the network size. In scale free networks, like random networks, C does depend on the network size *N*. As the nodes increase, C drops; initially for a small amount of nodes, C is five times larger for scale-free networks than for random networks. This divergence in the C value between random and scale-free networks increases as the number of nodes in a network increases (Albert and Barabási 2002).

4.2.3 Path length

With respect to the average path length, it is recognised that "there is no theoretical expression that would give a good approximation for the path length in the scale-free model" (Albert and Barabási 2002: 30) even though the expression below (Equation 4.2) gives a good empirical fit (Albert and Barabási 2002). In real world networks with a power-law distribution, the average path length l_{real} is greater than in random graphs l_{rand} and random graphs with a power-law distribution l_{pow}: "an indication of the non-random aspects of the typology of real networks" (Albert and Barabási 2002: 22).

Equation 4.2: A generalised log form (Albert and Barabási 2002: 30) where *l* is the average path length and *N* is the network size. Graph variables are given as *A*, *B* and *C*.

$$\boldsymbol{l = A \log (N - B) + C}$$

4.2.4 Error tolerance

Scale free networks are generally robust against *random* node attacks due to the formation of hubs (Albert, Jeong and Barabási 2000). For example, even when up to 5% of the entire nodes (1000; 3,000; 20,000) are removed, the network's diameter is still maintained at the same length (Albert, Jeong and Barabási 2000). As many more nodes will have very few links, the chances of hitting a node with few links will be more probable than hitting a hub, as hubs are very few compared to the total nodes. However, if hubs are targeted the network diameter increases dramatically "doubling its original value if 5% of the nodes are removed" (Albert, Jeong and Barabási 2000: 380). Thus, scale free networks are robust to random failures (unlike random networks studied by Erdös and Rényi 1959, 1960, 1961), but are very vulnerable to specific attacks on the hubs of the network.

4.3 Language networks

We now move on to linguistic networks. Many different aspects of language have been represented by networks. As noted in the introduction to this study, a language network has denoted word co-occurrences (Ferrer i Cancho and Solé 2001a; Dorogovtsev and Mendes 2001, 2002; Milo, Shen-Orr, Itzkovitz, Kashtan, Chklovskii and Alon 2002), syntactic dependency relations (Ferrer i Cancho, Solé and Köhler 2004; Ferrer i Cancho 2004, 2006; Ferrer i Cancho, Capocci and Caldarelli 2007; Liu 2008), word usage (Ninio 2006a); association relations (Steyvers and Tenenbaum 2005), the relation of phonological neighbours (Vitevitch 2008), consonant co-occurrences (Mukherjee 2009), "conceptual-semantic and lexical relations" (Fellbaum 1998), as well as hypernymic and meronymic relations (Miller 1990; Steyvers and Tenenbaum 2005).

This study explores the idea that basic syntax - found cross-linguistically - is a property of the general characteristics of a small world with a hubterranean structure. Thus, we are interested in networks that model syntax. There are studies that encode syntactic theory into a network and correlate this theory with graph theoretic measurements. Notably, many studies focus on dependency relations (see above references). This endeavour is interesting and important in understanding how such syntactic relations relate to general system properties.

In contrast, the objective of this study is quite different. Instead of building the syntactic theory into the network and then correlating the syntactic theory with the network properties, this study seeks to explore how syntax emerges from the network in which the very minimum syntactic relations are built in. The only relation encoded in this model is one of precedence which is not specific to language but found in any temporal system.[2] Thus, the study tests the possibility that a type of network with its distinctive characteristics derives syntax i.e. syntax is "for free" (Solé 2005) rather than building in the syntactic scaffolding and then examining the resulting network.

Furthermore, if one understands how this type of network develops these distinctive characteristics, then we have a possible explanation of why the corresponding syntactic properties are found in language. Indeed, if certain abstract syntactic relations such as the f-l distinction and syntactic constituency emerge from the network dynamics of hubterranean networks, one does not need to posit these syntactic relations as stipulated formalisms. Consequently, when building the network from linguistic form, the less syntax explicitly encoded the better.

One argument against utilizing word co-occurrence networks to represent syntactic knowledge is that "The amount of co-occurrences captured (per sentence) that are not syntactic dependencies is about…30%" (Ferrer i Cancho 2005: 63) when using adjacent co-occurrences. On this view, a large portion of knowledge represented by the word co-occurrence network is 'non-syntactic' knowledge in the traditional sense. Taking a word co-occurrence network at face value, one might agree with such criticism. However, it is important to recall that the network links signify frequency of co-occurrence information.

As noted in the section on constituency in chapter two, low frequency co-occurrences equate to a spurious syntactic relation and indicate the end of a constituent. Such information rather than being detrimental is actually useful to the speaker in segmenting the linguistic form. Naturally such information is redundant if one proposes a theory that uses only Boolean syntactic relations, since only 'valid' syntactic relations would be required. However, if a theory is concerned with non-Boolean relations then such graded relations may form an essential part of the syntax of a language.

---

[2] It is also argued that long distance dependency relations are not specific to language either (Ferrer i Cancho, Lorenzo and Longa 2008).

For example, it will be argued that these seemingly redundant and spurious relations representing low frequency combinations contribute to the power law distribution of storage cost values that characterises language (since storage cost is inversely correlated with transitional probability). As such, word co-occurrence networks are not to be seen "as rough approximations" (Ferrer I Cancho 2005: 63) but rather the means to encode the building blocks that derive a more intricate syntax.

Another criticism of word co-occurrence networks is that they fail at representing long distance (LD) dependencies (Ferrer i Cancho 2005: 63). Although LD relations are not *explicitly* encoded in the network, there is the potential for LD relations to be represented by the network. Indeed, word co-occurrence patterns when taken *collectively* can capture non-adjacent relations (see motif 904 chapter 8).[3] Thus, very long distance syntactic relations have the potential to be represented by a word co-occurrence network when more complicated word co-occurrence patterns are able to be computed and extracted.

4.3.1 Frequency distribution

In respect to syntactic networks, small world properties, i.e. high clustering co-efficient and a small average shortest path, have long been observed in word co-occurrence networks. An early account of such properties using collocation relations (between adjacent words and words with one intervening word) was evidenced by Ferrer i Cancho and Solé (2001a). In their (2001a, 2001b) studies, Ferrer i Cancho and Solé examined the frequency distribution of words, building on the work of Zipf (1949) who identified an empirical relation between the distribution of word frequencies in samples of text (Equation 4.3). Zipf (1949) ranked words by decreasing frequency, plotted the rank against the word's frequency and observed a power-law distribution.

Equation 4.3: ranked word frequency where P(r) is the frequency of the rank (r).

$$P(r) \alpha r^{-\alpha}$$

Identifying a power-law is very simple. The graph equation for a power law is given in Equation 4.3. When this equation is normalised i.e. logged,

[3] Naturally to identify a long distance relation, a semantic component in the model is required. Thus the word co- occurrence needs to be grounded (see Steels 2005, on grounding semantics using Fluid Construction Grammar).

Equation 4.3 gives: log (P(r)) = log (C) – α log (r), where log (C) is the intersection on the y-axis of the graph and the exponent -α is the slope of the plot. After normalising the graph, a straight line is observed, suggesting a linear relation between the rank frequency and the rank with an exponent of -1: Zipf's Law. Although written as a law, Zipf's distribution function fits the smaller ranks better (with a deviation in the smallest of the ranks; Tsonis, Schultz and Tsonis 1997) whilst there is divergence in the larger rankings (Ferrer i Cancho and Solé 2001a, 2001b).[4]

4.4 Networks and language learning

The next subsections will briefly discuss the studies that utilize network theory in understanding how children learn their language.

4.4.1 Ninio (2006a)

In Ninio's (2006a) account, the adopted network is slightly different from those discussed above as it is composed of two types of nodes: one representing speakers and another, representing words spoken. The network formed is a bipartite network i.e. one that has two distinct and separate levels of representation. Links between a speaker and a word denote the usage of that word by the speaker. Thus, in this model, the community aspect of speech is highlighted and as such "Children do not invent language, nor do they internalize it; they link to it" (Ninio 2006a:126).

On this view, children are like adults in their drive to "produce a meaningful utterance in a communicative context" (Ninio 2006a:128), and they search adult language for expressions that capture their communicative intent. Ninio (2006a) emphasizes that frequency alone does not dictate which words are learnt first. Although such words are picked up quickly, the words must be useful to the child to express the intended meaning. On

[4] The divergence can be eliminated by positing two separate power laws that capture the distribution of word frequencies: the first power law relating to a kernel of words (consisting of 5000-6000 words), the second to specialized words (Ferrer i Cancho and Solé (2001a, 2001b; Dorogovtsev and Mendes 2001; for a model of a network with weighted links, see Masucci and Rodgers 2007; for Russian see Kapustin and Jamsen 2007; for three regimes see Tuldava 1996). The kernel is shared by every speaker; its size being constrained by the capacity to "store and use" words (Ferrer i Cancho and Solé 2001b:3).

Ninio's (2006a) approach, the principle of preferential attachment captures these two factors (see Barabási and Albert 1999).

Preferential attachment was first used to explain the power law distribution of hyperlinks on the web. People generally link to the most popular site (a node with a high degree) that suits their interest(s). Consequently, a new link preserves the global pattern of the link distribution, in that popular sites get more popular. This type of process is also commonly known as 'the rich get richer'. For language, the most frequently heard words that serve the communicative intentions of the speaker are used more frequently i.e. acquire more and more links. Furthermore, the children's network of links to words will recreate the link statistics of the adult network i.e. the degree distribution being scale free captured by the function of a power law (Ninio 2006a).

Ninio (2006 a&b) also argues that for maternal speech, a power law distribution is also found when attending to frequencies of verb use in an individual construction i.e. the indirect object construction in Hebrew. Consequently, there are a few verbs that are used often in a construction (see Goldberg 2006), with many verbs seldom being employed by the construction. The utterances of children aged 1;04 to 2;08 also displayed the same pattern: the token frequency distribution of verb usage in an indirect construction was a function of a power law (Ninio 2005b, 2006a). Consequently, the frequency distribution of child speech reflects the frequency distribution of adult speech. This finding was also congruent in child speech with the analysis of another construction, the direct object construction (Ninio 2005c, 2006a).

These results point to an account of construction learning where verbs of varying frequencies are learnt initially. Thus, words that suit the child's interest provides a factor that encourages low frequency words to be learnt whilst still attending to high frequency words. It will be shown below and in chapter five that factors such as the 'intention to be as expressive as possible'[5] i.e. a speaker wanting a large vocabulary, and the desire to use energy efficient speech (i.e. choosing frequently used words) optimise to

[5] Children want to be as expressive as possible. Indeed, speakers do not only want to be expressive but sometimes they want to be extravagant with words (Haspelmath 1999). Indeed, when modelling symbolic mapping, the two factors of "maximizing communicative success and expressive power while minimizing effort required" derived Zipf's law (see Ferrer i Cancho and Solé 2003a and Solé, Corominas-Murta, Valverde and Steels 2005: 7).

create power law distributions of frequency, connectivity and the cost of storing two elements together.[6] Thus preferential attachment may be due to an optimisation process. Indeed, it has already been noted that "optimization can explain the selection of preferential attachment strategies in real complex networks" (Ferrer i Cancho and Solé 2003b: 8).

Ninio's (2006a) network approach incorporates a rich linguistic perspective. Ninio (2006a) proposes that children's speech is lexically specific and does not consist of abstract schemas - with the exception of predicate-argument relations. Furthermore, Ninio (2005c, 2006a) maintains that children do not use semantic similarity when learning syntactic constructions. Instead, children "use similarity of form for generalization and transfer of learning, disregarding argument semantics and thematic roles" (Ninio 2006a: 4).

The child's lexically specific utterances are constructed by a merge/dependency operation since Minimalism and Dependency Grammar share the same basic principles e.g. asymmetrical syntactic relations, etc (Ninio 2006a; see Robinson 1970: 275 for an early examination of dependency relations and phrase structure).[7] Although children combine words in lexically specific ways (Ninio 2005a), the words do not form construction islands (contra Tomasello 1992) but develop interconnections between constructions (Ninio 2006a).

Whilst Ninio's (2006a) account of language learning in a network context is commendable since it utilizes a community orientated view whilst retaining the uniqueness of the child's lexicon, this study diverges from some of the basic tenets. Firstly, Ninio (2006a) does not connect words within the network. Speakers only connect to words. This study seeks to explore how well basic syntax derives from network properties and so connections between words are a necessity.

Although this study supports the view that the children's syntactic knowledge is initially item-specific, and generalizations utilise "similarity-based analogy" (Ninio 2006a: 85), semantic considerations play a part in

---

[6] See Ferrer i Cancho and Solé's (2003a) account of optimising vocabulary size and cost of symbolic mapping for both the speaker and the hearer's perspective.

[7] However, a dependency network has certain properties that have yet to be established in phrase structure i.e. degree distribution (see Ferrer i Cancho, Solé and Köhler 2004). On the other hand, hierarchical organisation has been observed in dependency networks (Ferrer i Cancho, Solé and Köhler 2004).

the extension of such generalisations. For example, "constructions that have been heard used with a wide *variety* of types are more likely to be extended broadly than constructions that have been heard used with a semantically circumscribed set of types" (Goldberg 2006: 99).

The perspective that construction knowledge forms the basis of linguistic knowledge is quite different from the Minimalist/Dependency Grammar perspective adopted by Ninio (2006a). Attention to constructions is important since "as a predictor of overall sentence meaning…the construction is at least as reliable and more available than the verb…" (Goldberg 2005). Furthermore, linguistic groupings are not driven initially by a merge/dependency structure but by a simple chunking mechanism based on transitional probabilities. Finally, but also importantly, children may 'link to language' as Ninio (2006a) proposes, but they also internalize language. It's the reason why syntax exists at all. For without the internal process of words and constructions being optimally stored whilst retaining the largest repertoire of expressions, language would lack syntax.

On the other hand, the power law frequency distributions are very amenable to the approach adopted in this study as well as the notion that children have a word choice, driven partly by frequency of usage and partially by the demands of their intended meaning (as noted above). As such, the model (Ninio 2006a) is in part compatible with much of the theory and results within this study.

4.4.2 Corominas-Murtra, Valverde and Solé (2007 a&b)

Around the age of two, children start to produce syntactic constructions e.g. item specific constructions (Tomasello 2003). Interestingly, a syntactic dependency network built from child utterances around this age exhibited a transition from a "tree-like structure to a scale-free, small world syntax network" (Corominas-Murtra, Valverde and Solé 2007:1a). This network was built from two CHILDES corpora (Bloom, Hood and Lightbown 1974; Bloom, Lightbown and Hood 1975; Macwhinney 2000a).

The phrasal structure of the transcription was converted to dependency relations (Corominas- Murtra, Valverde and Solé 2007a; see Corominas-Murtra 2007 for "structural criteria"). Dependency relations were represented by links in the network and nodes denoted words. The initial network was fragmented, therefore, only the largest graph component was analysed. Importantly, the snapshots of this growing network (Figure 4.2) show how the network becomes small world and scale-free in as little as two months.

Figure 4.2: The child network at (a) 25 months (b) 26 months and (c) 28 months (Corominas-Murtra, Valverde and Solé 2007a: 4, Figure 2). Scale free degree distribution is observed at 28 months (c).

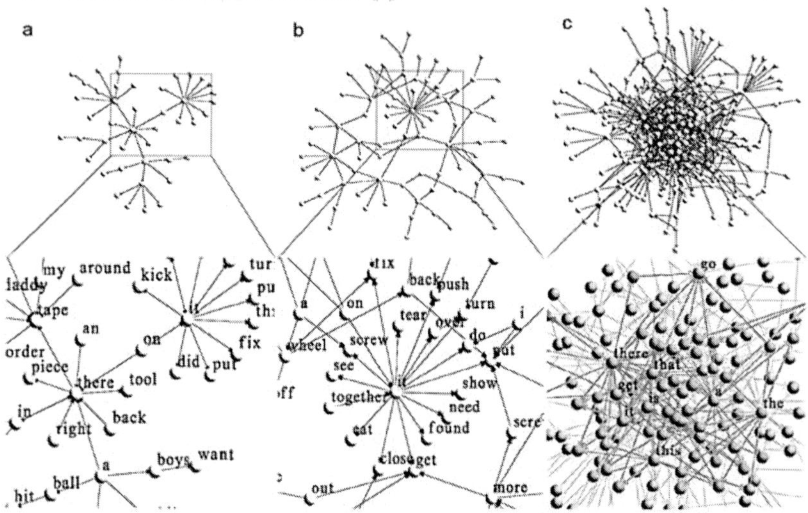

Examining the degrees of the words *it, a, that, the* in this network, it was found that around the transition period of twenty-four months, the degrees of each four words increased rapidly. Furthermore, it is interesting to observe that "*it* is rapidly replaced by *a* as the main hub as soon as purely functional words emerge" (Corominas-Murtra, Valverde and Solé 2007a: 6, Figure 3). In respect to the average smallest path length *L* of the network, the length increases until around twenty-four months, at which point it falls in line with the idea that small world networks have a low *L*. The clustering co-efficient *C* also increases as the small world is approached, and barring a drop at around 25.5 months, regains a high value. As noted earlier, a high C is expected in a small world network.

Around the twenty-four month period, the network displays disassortative mixing (see chapter five), in that there is a tendency for hubs not to connect to one another. Since function words tend to be hub words (see also chapter 5), at this period "strong constraints emerge strongly limiting the syntactic linking between functional words" (Corominas-Murtra,

Valverde and Solé 2007a: 7). Furthermore, the growth of a syntactic network was modelled utilising Zipfs' law i.e. a frequency power law distribution, as the only syntactic prerequisite as well as a "growing working memory" (Corominas-Murtra, Valverde and Solé 2007b: 9).

The resulting networks underwent a sharp typological transition from "tree-like networks to scale-free networks" (Corominas-Murtra, Valverde and Solé 2007b: 7). After the transition, the network exhibited the path length and the clustering co-efficient typical of a small world. Importantly, the degree distribution adhered to a power law. Thus, the power law relation and the small world properties are provided "for free" as a by product of the growth of the network and as such, are "spandrels" of the network (see Corominas-Murtra, Valverde and Solé 2007b: 7; see Solé 2005; Solé and Valverde 2006: 421; for spandrels, see also chapter seven and ten).

4.4.3 Ke and Yao (2008)

A study investigated the growth of networks composed of child utterances from twelve children aged "from 1;8.22 to 2;0.25, with MLUs [Mean Length of Utterances] ranging from 1.06 to 2.27"[8] (Ke and Yao 2008: 73). Both accumulative and staged networks were produced. There were thirty-four accumulative networks representing each visit by the examiner. Each session's utterances were added to last network representing the last sessions' utterances. For the staged networks, the data was divided by MLU into five stages: "[1,1.5] , (1.5, 2] , (2, 2.5], (2.5, 3], and (3,3.5]" with the square bracket "means including the value" and the round bracket "excluding" (Ke and Yao 2008: 79).[9]

For the accumulative networks, the size and the average degree were measured. The network size is the number of nodes in the network and as such is indicative of the productive vocabulary size for each child across the investigation's time frame. The vocabulary size of most children developed in a linear manner - although each child's growth rate was different. The second measurement of average degree is the average

[8] The data was sourced from the Manchester corpus, CHILDES (Theakston et al. 2001).

[9] In respect to network compatibility across the children, the time span was kept constant where possible. "If an MLU stage is particularly long, we divide it further into an early stage and a late stage, and construct one network for each sub-stage" (Ke and Yao 2008: 79-80).

number of links each node has. Network theory distinguishes in-degree and out-degrees. That is, in-degree signifies the links that point towards the measured node and out-degree, the links that point away from the node. Since average in- and out-degree are the same for each network, the out-degree was chosen.

As Ke and Yao (2008) note, this measurement does not have an equivalent linguistic property. Since the average degree relates to the number of words following a given word, "we can interpret the average degree as an integrated index of the productivity of the words in use and of the complexity of the utterances." (Ke and Yao 2008: 76). For each child, there is a linear growth in average degree. However, there is no correlation between the size and the average degree of the networks.

Whilst it is expected that syntactic development i.e. construction learning, correlates with vocabulary size, average degree is not a direct indicator of syntactic development. Even though average degree may be interpreted as an index of productivity, it fails to pick up the breath of constructions used. Since both the amount of constructions used and the productivity of those constructions together provide an overview of syntactic development, it is not surprising that vocabulary size and average degree are not correlated.

Ke and Yao (2008) also investigated staged networks since certain criticisms were raised against accumulative networks. Namely, that children's usage of different words changes and these changes in the child's language are not seemingly reflected well in the accumulative networks. Terms such as "choochoo" and "nana" get replaced by adult forms (Ke and Yao 2008: 79). Furthermore, children overgeneralise until they learn the correct usage e.g. the well-known use of 'goed' instead of 'went'. Thus, Ke and Yao (2008: 79) assume that "These changes are all non- additive in nature and cannot be reflected in the accumulative network model".

However, such changes are able to be recorded in the network, albeit indirectly - not in terms of deleting nodes or links but rather as probabilities representing the link length between nodes i.e. words. Although changes are not registered immediately in the network (as current usage is not indicated by temporal tagging), with time, transitional probabilities and frequency values of words reflect their usage. For example, the frequency value of the word 'choochoo' will become lower in respect to other words in the network over time, as these other words are used more extensively. Furthermore, it is important not to overlook that linguistic knowledge is built from previous language usage. Thus, accumulative networks record this knowledge, even if it turns out to be archaic knowledge.

Returning to the staged networks, Ke and Yao (2008) present the results of four children's staged networks. The size of the networks and the average degree both increase, with the average degree displaying a linear growth. As with the accumulative networks, the rates of growth varied with each child. For the staged networks, Ke and Yao (2008) created networks from the respective mother's speech. Naturally, these networks were larger and have a higher average degree. Indeed, "The mothers' networks have fewer changes over time than the children's" (Ke and Yao 2008: 84).

The vocabulary size of one mother (Carl's) remained fairly stable with a drop in average degree recorded, the other three mother's average degree maintained its value across the staged networks whilst their vocabulary increased along side the growth of their child's network. These differences may indicate a different style of speech, in that Carl's mother spoke in her usual manner to her offspring whereas the other three mothers adjusted their speech "to be child- friendly" (Ke and Yao 2008: 84).

In this study, Ke and Yao (2008) also examined the most highly connected nodes i.e. hubs, in each network. In addition, the notion of authorities was introduced. It has been argued that hubs are nodes that have the largest number of out-degree links, whilst authorities are nodes that have the largest number of in-degree links (Kleinberg 1999). Furthermore, hubs and authorities form a "mutually reinforcing relationship": a better example of a hub is one that points to other good examples of authorities (Kleinberg 1999: 611). The same with authorities - the better the example of an authority, the more good examples of hubs it points too (Kleinberg 1999: 611). From this perspective, if there are many good examples of hubs and authorities, it is predicted that nodes with high degree will link often with other high degree nodes. Such link patterns are found in social networks[10] (Newman 2002) and the phenomena is known as assortative behaviour.

However, as noted earlier syntactic networks are disassortative (Corominas-Murtra, Valverde and Solé 2007 a&b; for word co-occurrence networks,

---

[10] Although, online social networks may display disassortative behaviour, in that highly connected people tend not to interact with each other (see Hu and Wang 2009).

see chapter five and six in this study) and as such, high degree nodes do not extensively interlink in these co-occurrence networks. This point notwithstanding, it is still very interesting to note the high degree nodes and their in-out degrees. To this end, Ke and Yao (2008) observed how the words *the* and *a* were authorities in the mother's network but found that these words were often hubs initially in thc childrcn's networks - becoming authorities later on. The rate and the route of these changes in the child's networks were variable.

4.5 Hubs and small worlds

The above sections have reviewed the acquisition studies that utilize network theory. This chapter will also discuss the theoretical network perspectives on which this study is based. Firstly, a model of how small worlds are formed is presented (Mathias and Gopal 2001 a&b) and is contextualized in respect to a word co-occurrence network. This model is important as it provides an explanation as to why small world networks with hub structure form. This explanation will be presented in this section. Since the study examines the possibility that certain linguistic properties may be shown to derive from small world, hubterranean characteristics, then the explanation as to why hubterranean small worlds exist will also explain why these linguistic properties occur.

Whilst the Mathias and Gopal's (2001 a&b) model is based on global constraints, the local structure of a word co-occurrence network is also very pertinent to examining syntax since syntax is the description of felicitous combinations of linguistic elements whether phrasal or at a word level. Present research utilizes significant link patterns in a network i.e. motifs, to examine the local structure of a network. As such, motifs will provide a potential network metric for syntactic analysis. Chapter 7 will introduce motifs and their identification.

4.5.1 Mathias and Gopal's (2001 a&b) model

Mathias and Gopal (2001 a&b) explain the formation of small worlds through the process of optimising two contrary factors: the physical link length between two nodes and the path length or connectivity between nodes. In real world networks, resources are required to link nodes together; the longer the link length the more costly the link. Thus, the physical length of each link correlates with the cost of 'wiring' the link.

Shorter links would save resources, as would reducing the number of links between nodes. Since links represent relations between nodes, reducing the number of links in the network has a detrimental effect on the networks ability to capture complex relations. Consequently, too much emphasis on cost cutting simplifies the capacity of the network to represent relations. On the other hand, representing complex relations is a costly venture.

Mathias and Gopal (2001 a&b) show that minimizing the wire cost i.e. link length and maximizing connectivity in an optimal way, results in hub formation. Hubs, then, signify a network that optimally captures complex relations in a cost effective way. Although link length is denoted by Euclidean distance in Mathias and Gopal's (2001 a&b) explanation, it is noted that their model is suitable for "networks with different underlying cost metrics" which are not necessary in the physical realm (Mathias and Gopal 2001 a&b: 1). Note that the ensuing discussion does not refer to link length as the number of adjoining links between to nodes (as in the graph distance metric) but refers directly to the (Euclidian) physical distance between two nodes.

4.5.2 Hub formation

Mathias and Gopal (2001 a&b) note that the key to forming small worlds is the development of highly connected nodes i.e. hubs. Utilising a computer simulation, Mathias and Gopal (2001 a&b) investigate the factors affecting hub formation. In their hypothetical network, Mathias and Gopal (2001 a&b) start with a regular graph. Each node is arranged equi-distance on a ring and has an equal number of links. The network size n and the number of links are fixed. Each link is rewired and the distance (W) calculated with Equation 4.4.

Distance W is normalised by calculating the wiring cost when each node is connected to its diametrically opposing node. The path length is given as Equation 4.5 and "normalised by $L(0)$ the path length" in the simple lattice (Mathias and Gopal 2001 a&b). In respect to the terminology used, note that the path length relates to the number of links between two nodes and as such correlates with the connectivity of the network, whereas the link length is the amount of wire required to connect two nodes and correlates with the cost of linking two nodes.

Equation 4.4: Wiring cost (W)) for link e_{ij} is the distance between $node_i$ and $node_j$, where (x_i,y_i) are $node_i$ co- ordinates on the ring (Mathias and Gopal 2001b: 3, Equation 3)

$$W = \sum_{e_{ij}} \sqrt{(x_i - x_j)^2 + (y_i - y_j)^2},$$

Equation 4.5: average shortest path (L) where d_{ij} is the number of links along the shortest path between nodes *i* and *j* (Mathias and Gopal 2001b: 3, Equation 2)

$$L = \frac{1}{n(n-1)} \sum_{i \neq j} d_{ij},$$

Using the below Equation 4.6, the energy of a graph is calculated and the minimal energy graph was sought using a standard Monte Carlo scheme (Press, Teukolsky, Vetterling and Flannery 1992). The parameter λ, was varied depending on whether the average shortest path (L) is minimised or the cost of wiring (W) is minimised.

Equation 4.6: Energy (W) of the network (Mathias and Gopal 2001 a&b: 3, Equation 1)

$$E = \lambda L + (1 - \lambda) W,$$

Figure 4.3: (i) Link length distribution for an n=100, k=4 network optimised network for 0 < λ <1 range (Mathias and Gopal 2001b: 6, Figure 3). (ii) The evolution hubs for an n = 100, k=4 optimised network for 0 < λ <1 range (Mathias and Gopal 2001b: 9, Figure 5).

i.

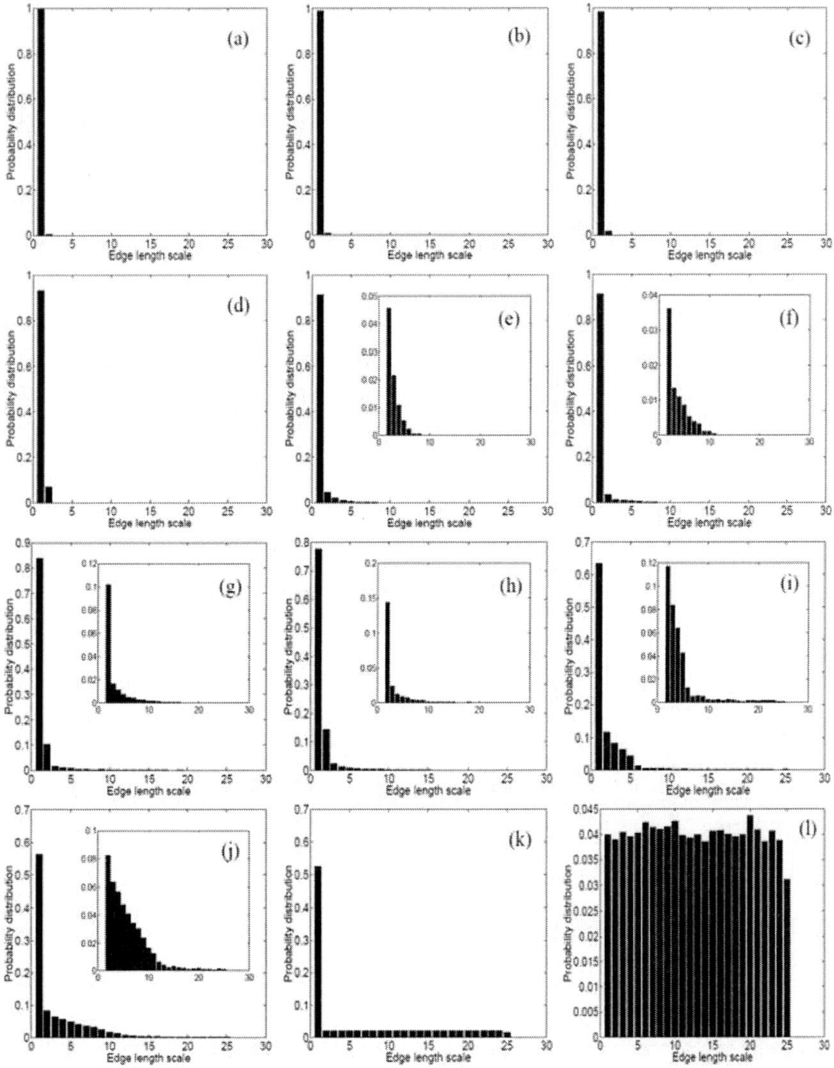

ii.

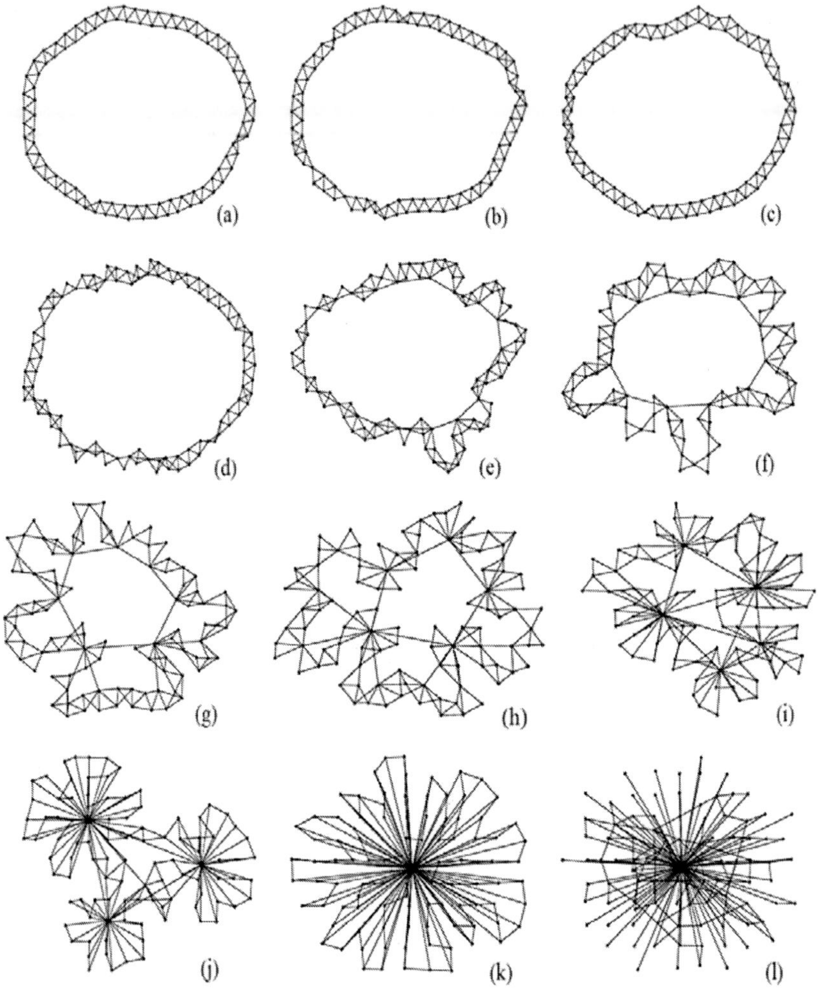

When λ = 0, the length of the wire is as short as possible due to the high cost of wiring. On this parameter setting "a regular network emerges with uniform connectivity and high characteristic path length" (see Mathias and Gopal 2001b: 5, Figure 4.2ii (a)). In contrast, when λ = 1, connectivity of the nodes is maximised since the cost of wire is low. In this case, the average shortest path length (P) is minimised, resulting in a "near random network" (Mathias and Gopal 2001b: 5, Figure 4.2ii (l)). This network is characterised by a 'universal hub'- one single node that is connected to all the other nodes.

When both P and W are minimised, hubs develop in the graphs which assist in lowering P. Notice how the small hubs in Figure 4.2ii (d) are close together to minimize the cost of wiring nodes to the hub node. However, as the cost function is relaxed (traversing Figure 4.2ii from d-j), the hubs start to scatter across the network and reduce in quantity. Consequently, the cost of wiring affects the size and number of the hubs. The size and number of the hubs increase as wire cost decreases as does "the range and number of inter-hub links" until at a low cost of wiring, the number of hubs diminishes (Mathias and Gopal 2001 a&b:10).

Whilst Watts and Strogatz's (1998) model employs shortcuts between nodes to create the small world network, Mathias and Gopal's model emphasises that the 'smallness' of the network results from the formation of highly connected nodes. Furthermore, the link length's pattern of distribution correlates with the decreasing wire cost. Figure 4.2i shows the corresponding link length distribution for each graph in Figure 4.2ii. A single scale length of link is observed in the regular graphs (Figure 4.2i (a)) whilst a universal hub has link "lengths distributed uniformly over the entire length scale range" (Figure 4.2i (l); Mathias and Gopal 2001b: 5). Importantly, the graphs with hubs (Figure 4.2ii (e-j)) have a distribution of link lengths that conforms to a multiple length pattern (Figure 4.2i (e-j)), in line with Kasturirangan (1999).

4.5.3 Language and hubs

This section will now apply Mathias and Gopal's (2001 a&b) model to the word co-occurrence network. As noted earlier in the study, nodes in a word co-occurrence network represent words in the corpus and the linking arrow or arc denotes linear precedency, i.e. the first word points to the adjacent second word. Link length, or the cost of wiring two nodes together, may be represented by the inverse frequency of co-occurrence or transitional probability (TP) of a pair of words (known as a diad or a bigram). The transitional probability that $word_a$ precedes $word_b$ is calculated by counting how many times $word_a$ precedes $word_b$. This figure is then divided by the total number of co-occurrences of $word_a$ in the data set i.e. $word_a$'s token frequency.

It is well-known that high co-occurrence frequencies are linked to changes in linguistic form, correlating with phonetic and phonological reduction and palatalization in words and phrases (Bybee and Schreibman 1999; Bell, Jurafsky, Fosler-Lussier, Girand, Gregory and Gildea 2003; Jurafsky, Bell, Gregory and Raymond 2001;[11] Bush 2001). Such changes are indicative of chunking. As discussed in chapter two, when the TP of a bigram is high, the words in that bigram become chunked together lowering the storage cost of the bigram since they may be represented as one single unit. The link length in the word co-occurrence network, then, represents the storage cost of the words in the bigram. The longer the link length, the more costly the bigram is to mnemonically store.

It has been shown that word co-occurrence networks are small worlds with a hubterranean structure (Ferrer i Cancho and Solé 2001a, 2001b; Dorogovtsev and Mendes 2001; see results chapter five and six for additional evidence). As predicted by Mathias and Gopal's (2001 a&b) model, the hubterranean structure will be formed in a network by optimising the cost of wiring and connectivity of its nodes i.e. reducing the link length and establishing a small average path length (respectively). Indeed, in actual neural structures, neuronal wiring length is minimised by the economical location of neuronal components (Cherniak 1995).

In our case, the cost of wiring equates to the cost of storing a bigram, whilst connectivity of the nodes represents word combinations in the corpora. Thus, the small world of a word co- occurrence network is formed by minimising the storage cost of bigram whilst maximising the variety of bigram i.e. maximising the combinations of adjacent words. The optimisation of these two factors drives the formation of hubs which generates a small world network.

Furthermore, the optimisation process is characterized by the link lengths in the network having multiple scales that follow a power law (Kasturirangan 1999). Thus, it can be predicted from this model that language has multiple values of transitional probabilities that also follow a power law pattern. This prediction will be proven in chapters five and six. Mathias and Gopal's model (2001 a&b) also predicts that as connectivity increases, hubs will less likely co-occur together. That is hub words (in

[11] Jurafsky et al (2001) used both relative frequency ("the frequency of the word divided by the total number of word tokens in the corpus") and transitional probability of the target word defined as the probability of transition from x to y = (the token frequency of x + token frequency of y)/ token frequency of x.

graph 4.2ii i-j) will form low transitional probability bigrams with other hub words. This prediction is borne out as the word co-occurrence network displays disassortative behaviour (see section 5.6).

It is interesting to hypostudy how linguistic form would appear if either of the factors were disregarded in a word co-occurrence network. If storage cost was completely minimised, all bigrams would have the same transitional probability value. Thus, the word order would be extremely rigid. Only fixed expressions would be produced with no combinatorial creativity as network connectivity is very low. If word connectivity was maximised with total disregard to storage cost, one word (a universal hub) would connect with every other word; the bigrams having a very low transitional probability (due to the length of the links being extremely long - see Figure 4.2ii l). As such, fixed expressions would be practically nonexistent and utterances would contain the only obligatory function word with other elements in the utterance being essentially unrestricted.

4.6 Summary

This chapter has briefly reviewed the network theory literature on language learning. Importantly, it has been observed that the development of hubs occurs at the early multi-word stage, correlating with the onset of syntactic structure (Corominas-Murtra,Valverde and Solé 2007 a&b). The degree direction also changes with syntactic development: the highly connected nodes *a* and *the* were classified as authorities in adult speech but hubs initially in child speech, developing into authorities later on (Ke and Yao 2008).

The perspectives of this study share some of Ninio's (2006a) basic principles. Children are motivated by their own interests and the frequency of a word's usage when choosing words to express their communicative intentions. These factors correlate well with the idea that a word co-occurrence network occurs through maximising word combinations (chosen by the speaker) whilst curtailing storage costs (choosing word pairs that frequently co-occur).

This optimisation process was detailed (Mathias and Gopal 2001 a&b) and contextualised in respect to the word co-occurrence network. Furthermore, link lengths in a hub network are distributed by a power law (Kasturirangan 1999) and were correlated with inverse transitional probabilities i.e. storage costs. Thus, it is predicted that this distribution will be found in the word co-occurrence networks under study. This prediction is evidenced in the next two chapters.

CHAPTER FIVE

THE ADULT WORD CO-OCCURRENCE NETWORK

5.1 Overview

If you recall, adjacent word co-occurrences in speech overlap and intertwine to produce a network where nodes are words and whose (precedence) adjacency relations are represented as links. It has been established that a word co-occurrence network has small world properties (Ferrer i Cancho and Solé 2001 a&b). On a local scale, there is a high level of cliques throughout the network i.e. there are many examples of where two linked nodes are linked to a mutual friend. And on a more global scale, the language network exhibits a small diameter i.e. a small average shortest path. These results are replicated for a word co-occurrence network of approximately 5000 words composed of utterances from the BNC World (2001) corpus (section 5.2-3 for details). In addition, the network exhibits a frequency and a degree power law distribution (section 5.4.1 and 5.4.2 respectively).

Evidence will be presented to support the theory that hubs within a word co-occurrence network are formed by optimising the path length i.e. connectivity and the link length of the nodes within the network (see Mathias and Gopal 2001 a&b). As discussed in chapter four, the link length in a word co-occurrence network equates to the storage cost of adjacently occurring words, calculated by establishing the transitional probability that one word adjacently co-occurs with another word and then inversing this value. A distinguishing characteristic of small world networks with hubs is the multiple lengths of the links (Kasturirangan 1999; Mathias and Gopal 2001 a&b). Thus, it is predicted that there will exist multiple values of storage cost in a word co-occurrence network. This power law distribution of storage cost will be empirically verified in the word co-occurrence network composed of adult speech (section 5.4.3).

Networks which exhibit a cost constraint tend not be hierarchical (Ravasz, Somera, Mongru, Oltvai, and Barabási 2002; Ravasz and Barabási 2003). Since a word co-occurrence network is constrained by storage cost, it is predicted that the network will not be hierarchical. This is indeed the case (section 5.5). In addition, this chapter will also establish that the hubs within the network tend not to connect to one another (section 5.6). That is, the network demonstrates disassortative behaviour.

Furthermore, there is a close correlation between degree i.e. node connectivity and the amount of 'shortcuts' i.e. betweenness centrality, for disassortative networks (Goh, Kahng and Kim 2003). This correlation is predicted and confirmed for the word co-occurrence network (section 5.6.1). Certain studies further differentiate highly connected nodes by their in- and out- degree i.e. authorities and hubs respectively (Kleinberg 1999; see also section 4.4.3). The final section in this chapter (section 5.7) identifies these types of nodes within the word co-occurrence network and discusses their linguistic relevance.

5.2 Data

The adult network is formed from conversation transcriptions found in the British National Corpus (World Edition 2001). The utterances used were collected from a variety of public forums e.g. meetings, presentations, debates on television shows, a church sermon, etc (see List 5.1).

List 5.1 The list of conversations from the British National Corpus (2001) used to create the adult word co- occurrence network

D8Y, D90, D91, D92, D94, D95, D96, D97, DCH, DCJ, DCK, F7A, F7C, F7E, F7F, F7G, F7J, F7K, F7L, F7M, F7N, F7R, F7S, F7T, F7U, F7V, F7W, F7Y, F8A, F8B, F8C, F8D, F8E, F8F, F8G, F8H, F8J, F8L F8M, F8N, F8P, F8Q, F8R, F8S, F8U, F74, F76, F77, F78, F81, F82, F84, F86, F87, F88, F89, FL4, FL5, FL6, FL7, FL8, FL9, FLA, FLB, FLC, FLD, FLE, FLF, FLG, FLH, FLK, FLL, FLM, FLP, FLR, FLS, FLU, FLW, FLX, FLY.

The data contains 76069 tokens, of which 5365 are unique types. The type/token ratio was calculated at 7.05% and a total of 31859 word pairs i.e. bigrams were formed (Klarskov Mortensen 2006).

5.3 Data preparation

The SGML tagging code was stripped away by a specially designed computer program (Andrew Smith p.c.), to leave the bare (punctuated) text. Cliticised forms were fused with their dependent word since elements do not intervene between the clitic and the form on which it is dependent. Forms such as *eh* were retained, as phonetic material may arguably be useful in segmenting speech and thus may facilitate a syntactic analysis of speech.

Next, the text was compiled into one file and the token/type values and the transitional probabilities of each bigram were calculated using a specifically designed program (Klarskov Mortensen 2006). The transitional probability (or joint probability) relates to the probability that a word will co-occur with another word given its overall occurrence. The equation is given in 5.1. As noted above, the storage cost of each bigram corresponds to the inverse value of its transitional probability as given by Equation 5.1. Bigrams formed across commas i.e. pauses in speech, were rejected.

Equation 5.1: Transitional probability of adjacent word$_X$ to word$_Y$, P(Word$_{X,Y}$), where T(f$_X$) and T(f$_Y$) is the token frequency of word$_X$ and word$_X$ respectively (Jurafsky, Bell, Gregory and Raymond 2001)

$$P(Word_{x,y}) =$$
$$(fy) \mid T(fx)$$
$$T(fx)$$

5.4 Small world properties

The aim of this section is establish the small world characteristics of the BNC speech data:

a small average shortest path (L)
a relatively high clustering co-efficient (C)
a power law distribution for frequency
a power law distribution for degree
a power law distribution for storage cost values

Using PAJEK, distance *L* between nodes in the adult language network (N_{adult}) was calculated and found to be small i.e. 3, even when compared to the *L* in a random network (N_{random}) with an equivalent number of nodes and arcs (see Table 5.1). The clustering co-efficient in N_{adult} was also higher than the equivalent N_{random} reading, as expected for a small world network.

Table 5.1. Small world properties of the adult language network N_{adult} compared to the equivalent random network N_{random} with 5365 nodes and 31602 arcs. Multiple lines were removed for the calculation in PAJEK (de Nooy, Mrvar and Batagelj 2005).

Network	Average Shortest path (L)	Clustering co-efficient (CC1)
Adult language (BNC) N_{adult}	3.3	0.0013071 (1.3 x 10^{3})
Random N_{random}	5.1	0.0005531 (5.5 x 10^{4})

5.4.1 The frequency distribution

Ranking the words in the BNC text by descending frequency, the ten most frequent words in this text are: *the, and, to, I, a, that, of, you, it, we*. A LOG-LOG rank versus frequency plot presents a slope or exponent of -1.4 (R^2= 0.9799), which is slightly steeper than the Zipfian -1 exponent. Recall that Ferrer i Cancho and Solé (2001 a&b) found two regimes diverging at approximately 5,000 words (chapter four, footnote 5). Since the word total is just over 5000 words, there is no opportunity to establish two regimes in this data set.

Figure 5.1 A log-log plot of the rank v frequency of the words in N_{adult}

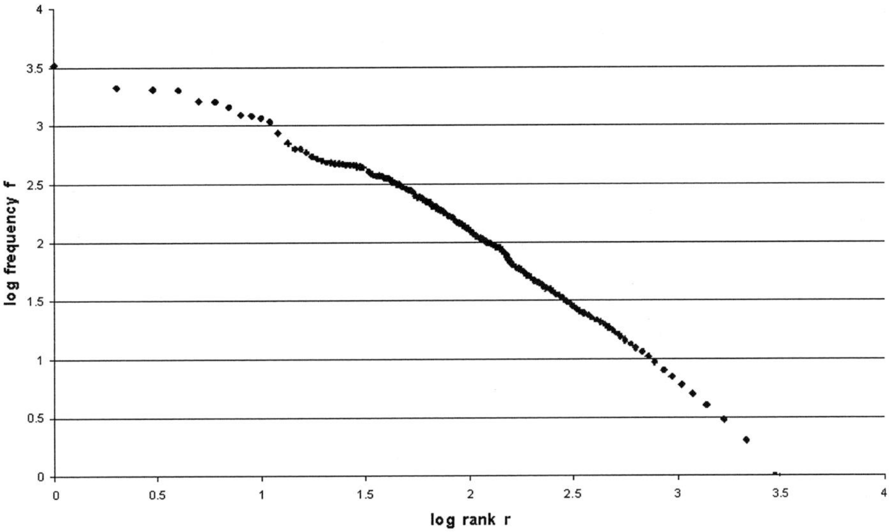

5.4.2. The degree distribution

The connectivity (or degree) of the words in the BNC text was calculated by PAJEK (de Nooy, Mrvar and Batagelj 2005) and plotted as a cumulative degree distribution[1] (Newman 2005). A linear line represents a power law distribution (Newman 2005). The degree distribution is captured by two power law regimes with exponents -1.0 and -0.7 (regime 3 and 2 respectively) and a truncated tail (regime 1). The truncated tail indicates that there is movement away from the scale-free behaviour for nodes that have a high degree i.e. function words. Why this may be so is discussed in the next section 5.4.3. Note that the ten most connected words (*the, and, to, of, a, that,* ***in, for****, it, i*) correlate very closely with the ten most frequent words (*the, and, to, I, a, that, of,* ***you****, it,* ***we***). The bolded exceptions are still in the twenty most connected and frequent words e.g. *you* has a connectivity ranking of 11.

[1] The cumulative distribution is "the probability that the random variable *X* takes on a value less than or equal to *x*". see Wikipedia, http://en.wikipedia.org/wiki/Cumulative_distribution_function.

Figure 5.2 The first regime of the cumulative degree (K) distribution - the truncated tail. Where P(K) is the cumulative distribution.

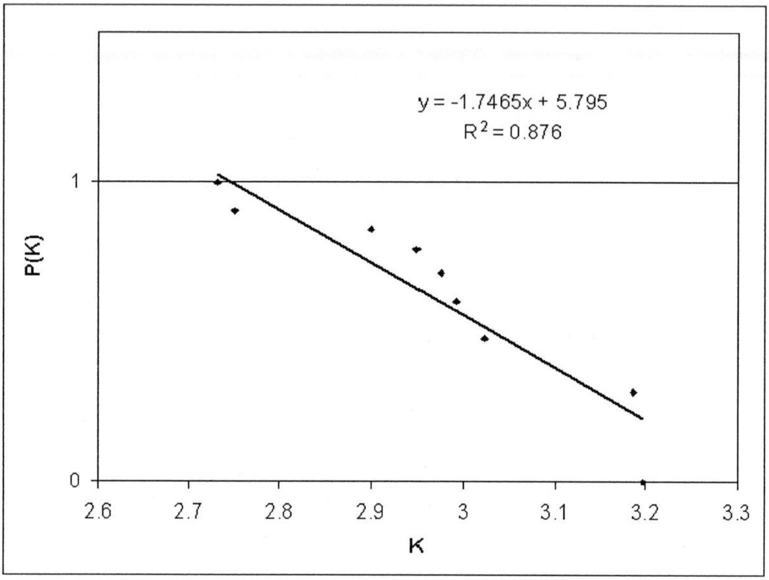

Figure 5.3 The cumulative degree (K) distribution P(K) is captured by two power law regimes (a) the second regime with exponents -1 and (b) the third regime with exponent -0.7.

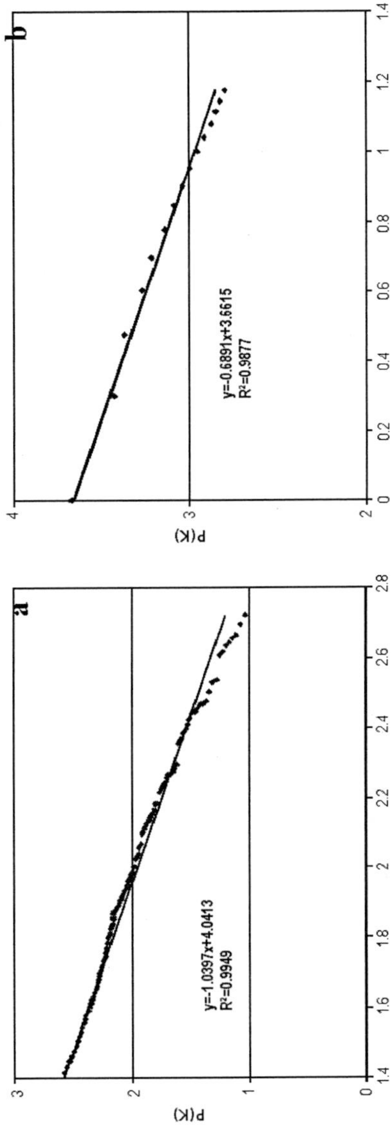

5.4.3 The distribution of storage cost values

In respect to the Mathias and Gopal (2001 a&b) model, the link length of the word co-occurrence network is represented by the storage cost of a bigram i.e. the inverse transitional probability value (TP) of the bigram. The TP values for all the bigrams in the network were inverted and then plotted as a cumulative distribution. In the word co-occurrence network, five different regimes were observed, four of which are power law distributions. The further details of the regimes are given in the following sections (5.4.3.1-3). See also appendix A.1 for examples of word pairs in each regime.

Equation 5.2 The frequency of bigrams or word pairs (N(S)) which have a storage cost (S) may be expressed in a cumulative distribution plot expressed by the below equation (see Solé 2006).

$$N_{>}(S) = \int_{S}^{\infty} N(S)dS$$

The power law distribution of storage cost supports the hypostudy that the hubterranean small world of linguistic form arises from the optimisation process that minimises the storage cost of a bigram and maximises the word connectivity (see Kasturirangan 1999; Mathias and Gopal 2001 a&b). These two factors are very general principles: maximal storage efficiency and maximal connectivity, equating to the desire to be as expressive or "extravagant" (Haspelmath 1999) with language as possible (see also Ferrer i Cancho and Solé 2003a for a symbolic mapping approach that utilises least effort whilst maximising expressiveness).

As you recall, storage cost is inversely proportional to transitional probability, and transitional probability is in turn representative of the cohesion between two linguistic elements. Thus, words in a bigram with a low storage cost will be highly cohesive, in that they will form a type of syntactic constituent. Indeed, frequency of co-occurrence is implicated in the formation of a traditional syntactic constituent as noted in chapter two (see Bybee 2002a). Furthermore, transitional probabilities are also attended to by adults and infants as young as seven months when segmenting an artificial language (Thiessen and Saffran 2003). Thus, the power law distribution of storage cost values is a syntagmatic relation represents constituents with multiple values of cohesiveness. As such, this type of syntactic constituency, driven by transitional probabilities, displays gradience (see Bybee 2002a:111). The five regimes in the storage cost distribution represent bands of cohesion i.e. bands of gradient constituents. The significance of these bands and why they occur is beyond the scope of this study and a matter for future research.

This power law distribution of storage cost represents a non-atomic and substantive[2] syntactic relation. It is argued that since the power law is formed by an optimisation process, then the relation "cannot be defined in terms of other units in the theory" (Croft 2001: 47); that is, it can be construed as a primitive. This type of syntactic relation contrasts with the atomic and abstract relations traditionally used in linguistic analyses e.g. head-dependency relations. The degree power law distribution is another non-atomic and substantive syntactic relation. In this case, the syntactic relation pertains to the quantitative nature of word combinations or more specifically, the distribution of type frequency. In fact, type frequency, as noted in chapter two, has been implicated in grammaticalisation and promotes "the storage of an element as a productive template" (Rostila 2004:14).

Furthermore, the power law distribution of (token) frequency may arguably be classified as a non- atomic syntactic relation since high frequency usage entrenches a linguistic element which then becomes resistant to regularisation e.g. recall the highly frequent but irregular verbs such as *went*.. Thus, high token frequency can reinforce irregularity or asymmetry within syntax whereas elements that are used less frequently tend not to resist regularisation. In addition, frequency is implicated in the f-l distinction (see section 2.8) and grammaticalisation e.g. high frequency "conventionalize[s] the pragmatic inferences involved in grammaticalization processes by offering them a form to be anchored on" (Rostila 2004: 3).

Thus, the three power laws are non-atomic syntactic primitives. Each power law (frequency, degree and cohesion) captures a very basic syntactic property of linguistic form (respectively): the strength of a word's representation i.e. entrenchment; paradigmaticity i.e. productivity and syntagmaticity i.e. syntactic constituency. Furthermore, these non-atomic syntactic relations are not specific to language but characterise a

---

[2] Substantive is used in this context to denote specificity in contrast to schematic or abstract entities. In this case, the specificity relates to a quantitative property rather than the semantics or phonetics of the elements.

hubterranean small world. As such, the child word co-occurrence network is expected to exhibit these syntactic relations and indeed in does (see chapter six). Thus, certain relations in syntax show continuity within the language learning process opposing the general claim that usage-based language learning approaches are discontinuous.

5.4.3.1 First regime - the truncated tail

In respect to cohesion, the bigrams in the first regime represent the least cohesive syntactic constituents. Notice how the first word of the bigrams in this regime are highly frequent and have a high degree: *the, and, to, i, a, that, of.* Recall that the most frequent words in the text are: *the, and, to, i, a, that, of, you, it, we*. The distribution of storage cost S decays faster than that of a power law, as shown by the non-linear relation observed in the log-log graph (Figure 5.4).

Figure 5.4 The cumulative distribution P(S) of bigrams with storage cost (S) in the first regime of the adult network.

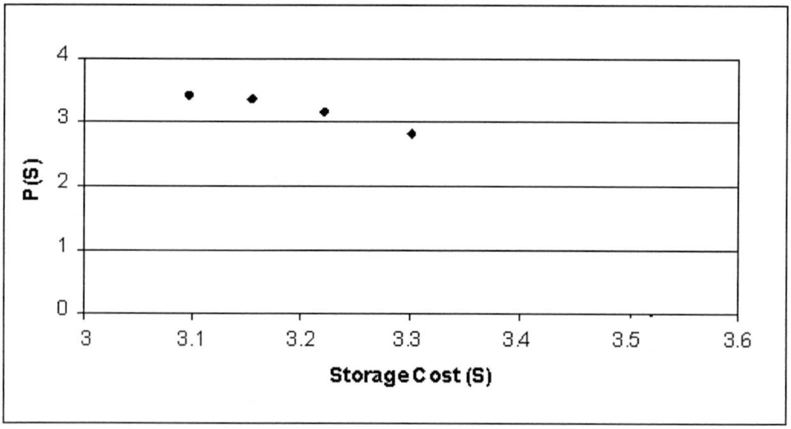

This type of distribution is found in the power law distribution of connectivity in broad scale or truncated scale-free networks (Amaral, Barthélémy and Stanley 2000): there is a power law regime with a tail that decays (either exponentially or Gaussian-like). For example, certain social networks such as the movie-actor network displayed a truncated scale-free distribution (Amaral, Barthélémy and Stanley 2000: 11150).

Constraints causing the truncation in the distribution of connectivity may be manifold. It has been suggested that the aging of nodes, the cost imposed on link formation and "limits on information and access" concerning the nodes, all limit the scale-free nature of a broad scale system (Amaral, Barthélémy and Stanley 2000; Amaral and Ottino 2004). In respect to the word co-occurrence network and the power law distribution of storage cost, the truncation is likely to be due to a limit on the cost of storing a bigram. After all, if there were no constraints on storage cost then the optimisation process that produces the power law would be unnecessary.

Amaral, Barthélémy and Stanley (2000) note, citing Stanley (1971), that "possible distributions of connectivity of the small-world networks have an analogy in the theory of critical phenomena". Truncation signals the movement away from the critical point of a system i.e. a point at which a change of state occurs. In respect to the storage cost distribution, the truncation suggests that there is a slight departure from the most optimal storage cost, thus allowing bigrams to be slightly more costly.

Notice how the first word in the bigrams in regime one are all traditionally classed as function words i.e. *the, and, to, i, a, that, of* and occur in the ten most highly connected and frequent words (see section 5.4.2). These bigrams, which have the highest storage cost, represent less frequent word combinations. Such bigrams may potentially correspond to the co-occurrence of novel words with words that are highly connected; that is, words that have a high type frequency, such as function words that facilitate the identification of constructions. Consequently, the truncation of the storage cost power law relates to infrequent function word combinations which arise through the novel extension of constructions. As such, this type of linguistic creativity limits the optimisation of the word co-occurrence network.

5.4.3.2 Second, third, fourth and fifth regime

In respect to the power law distribution of storage cost, the second regime has the largest exponent i.e. - 0.74. It is interesting to observe that the first word in each bigram formed in the first regime, are nested in the second regime. That is, they appear as a first word of a bigram in the second regime (see the bolded words in example 5.2 below. The italicized words are found in both the second regime heading bigrams and are words that have a frequency ranking of 24 or above.

Example 5.1 Words in the BNC text with a frequency ranking of 1-24. Bolded words indicate first words of bigrams in regime 1.

> **the**, **and**, **to**, **i**, **a**, **that**, **of**, you, it, we, in, is, they, was, on, erm, have, for, it's, be, there, this, think, but.

Example 5.2 The first word in bigrams from the second regime. Bolded words indicate first words of bigrams in regime 1. Italics denote first words with frequency ranking of 24 and above.

> ***the***, ***and***, ***to***, ***i***, ***a***, ***that***, ***of***, *you*, *it*, *we*, *in*, *is*, *they*, *was*, *on*, *erm*, *have*, *for*, *its*, *be*, *there*, *this*, er, yeah.

Figure 5.5 The second regime of the power law in the adult network. The cumulative distribution P(S) of the storage cost (S).

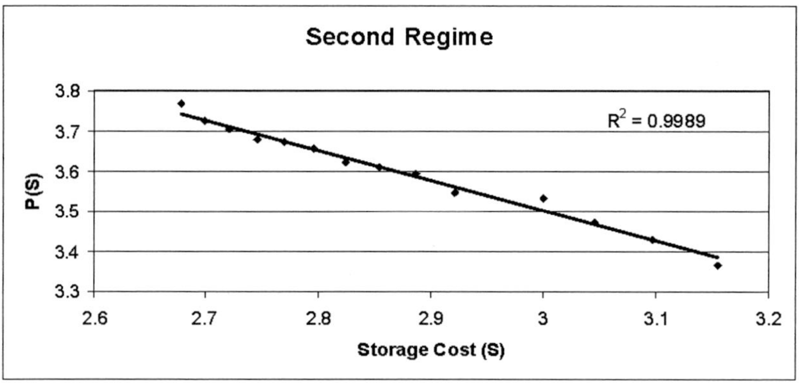

The words in the BNC text which are ranked higher than 24 correspond to the first words of bigrams in the second regime. The exceptions are *er* and *yeah* which are ranked 29 and 34 respectively, and appear in the regime, whereas the words *think* and *but* ranked 23 and 24 are not found in this regime.

The power law for the third regime has an exponent of -0.48, with an R-squared value of 0.9989 (Figure 5.6). There are 142 different words that begin a bigram in this regime. Interestingly, *all* the first words of the bigrams in the second regime are found in the third regime. See Table 5.2 below, for details on the actual first words of a bigram in each regime.

Figure 5.6 The third regime in the adult network. The cumulative distribution P(S) of the storage cost (S).

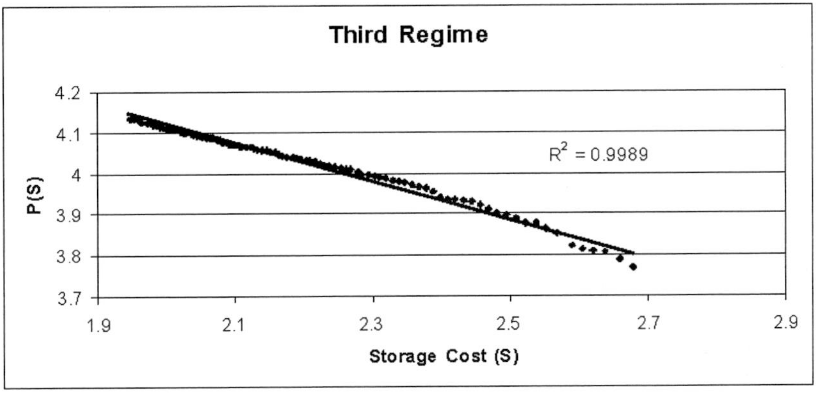

A power law with an exponent of – 0.25 captures the data in the fourth regime with an R-squared value of 0.9979 (Figure 5.7). Of the 596 words in this regime, all the words starting a bigram in the third regime (142) are present in the fourth regime.

Figure 5.7 The fourth regime of the power law in the adult network. The cumulative distribution P(S) of the storage cost (S).

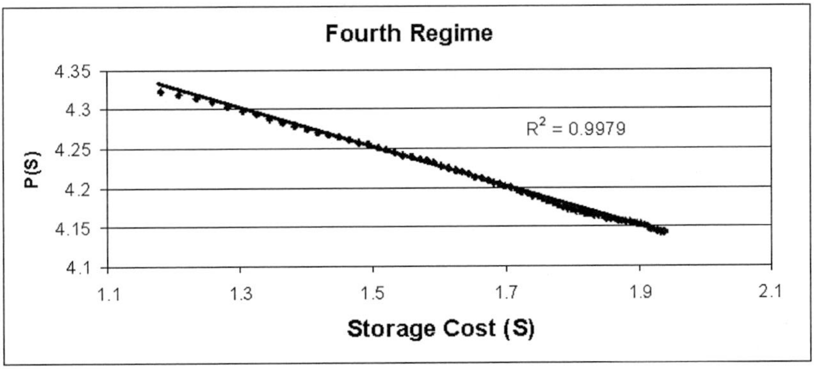

In the fifth and last regime, the bigram distribution does not fit so tightly to the linear function (Figure 5.8). Compare the R-squared value of this regime (0.9704) with the other regimes i.e. 0.9989 and 0.9979. The slope is also very shallow i.e. - 0.11. Even though this regime has approximately 4,500 different words which begin a bigram, not all the words in the fourth regime are found in this last regime. Thus, the nesting ends at the fourth regime. This may be due to the nature of the first words in the bigrams in regime one, in that their type frequency is very high i.e. they have a high degree, which inhibits the formation of very low cost bigrams. However, some high (token and type) frequency words such as *i* and *of* do form low cost bigrams representing strong collocations. For example, *i-* bigrams in the 5th regime are: *i-think*, *i-mean*, and *i-don't*.

Figure 5.8 The fifth regime of the power law in the adult network. The cumulative distribution P(S) of the storage cost (S).

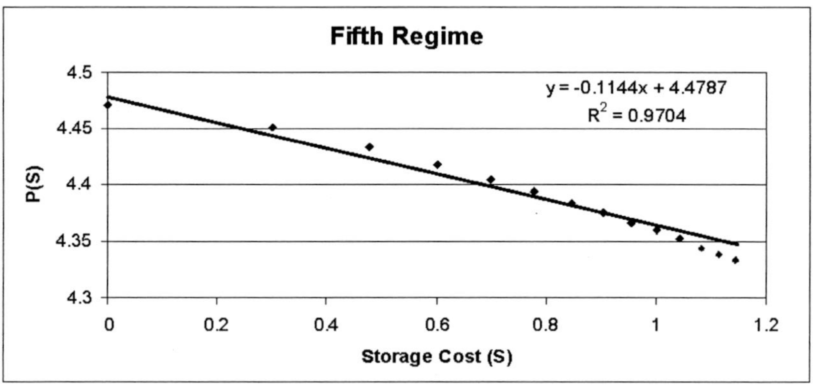

Table 5.2 The first words of the bigrams are shown with their respective regime. The words which occur in the previous regime are bolded. The fifth regime is not present as the number of words in this regime numbered approximately 4,500 – too many to include. Power law regimes are indicated by (PL).

REGIME	COST	NO. OF WORDS	FIRST WORD IN BIGRAM
1st	3333 - 1429	7	the, and, to, i, a, that, of
2nd (PL)	1428 - 477	24	it , you, we, **the,** in, **and, i,** erm, **a, that,** is, **of,** they, was, on, have, for, its, **to,** er, there, this, be, yeah,
3rd (PL)	476 – 88	142	**a,** about, actually, all, also, an, **and,** any, are, as, at, back, **be,** because, been, but, by, can, come, cos, could, council, did, didn't, do, doing, don't, **er, erm, for,** from, get, go, going, good, got, group, had, harlow, has, **have,** haven't, he, here, how, **i,** if, i'll, i'm, **in,** into, **is, it, it's,** i've, just, know, last, like, lot, make, me, mean, meeting, mm, more, much, my, no, not, now, **of,** oh, **on,** one, only, or, other, our, out, people, perhaps, put, quite, really, right, said, say, see, she, should, so, some, something, sort, take, **that,** that's, **the,** theatre, their, them, then, **there,** there's, these, **they,** thing, things, think, **this,** time, **to,** two, up, us, used, very, want, **was,** way, **we,** well, were, we're, we've, what, when, where, which, who, why, will, with, would, **yeah,** year, yes, **you,** your,
4th (PL)	87	596	**a,** able, **about,** action, actual, **actually,** address, after, again, ago, agree, ah, **all,** allowed, along, already, alright, **also,** although, always, am, amnesty, **an, and,** another, answer, **any,** anybody, anyone, anything, anyway, **are,** area, around, arts, **as,** ask, asked, asking, **at,** available, away, **back,** bad, bar, basically, **be, because, been,** before, being, best, better, between, big, bishop's, bit, board, body, book, bring, british, brought, building, bus, business, **but,** buy, **by,** c, call, called, came, campaign, **can,** can't, car, case, central, centre, certainly, chair, chairman, change, check, children, chris, **come,** comes,

			coming, committee, community, companies, company, concerned, contact, **cos,** cost, **could,** couldn't, **council,** council's, country, couple, course, cut, d, date, day, days, deal, decided, department, details, development, **did, didn't,** different, difficult, discuss, **do,** does, doesn't, **doing,** done, **don't,** door, down, during, e, each, east, eh, eight, either, else, end, enough, **er, erm,** even, evening, every, everybody, everything, example, facilities, fact, family, far, feel, few, fifty, fill, financial, find, first, five, food, **for, form,** forty, found, four, foyer, friends, from, front, full, fund, funding, furniture, future, gave, general, **get,** gets, getting, give, given, **go,** goes, **going,** gonna, **good, got,** government, great, **group,** groups, **had,** hall, hand, happen, happened, happening, happy, **harlow, has,** hasn't, **have, haven't,** having, **he,** health, hear, heard, help, her, **here,** he's, him, his, home, hope, hopefully, house, housing, **how,** hundred, **i,** i'd, idea, **if, i'll, i'm,** important, in, information, interest, interested, interesting, **into,** involved, **is,** isn't, **it,** it'll, **it's, i've,** john, **just,** keep, kept, kind, **know,** lady, large, **last,** later, leaflet, leaflets, leave, left, let, letter, letters, **like,** list, little, live, local, london, long, look, looked, looking, **lot,** m, made, major, **make,** making, many, march, market, marketing, mary, material, may, maybe, **me, mean,** means, meet, **meeting,** meetings, members, mend, mention, mentioned, might, mind, minutes, moment, money, month, months, **more,** morning, most, mr, **much,** must, **my,** myself, name, national, need, needed, needs, never, new, news, next, night, nine, nineteen, ninety, **no, not,** nothing, **now,** number, obviously, o'clock, **of,** off, office, officer's, often, **oh,** oil, okay, old, **on,** once, **one,** ones, **only,** open, **or,** organisation, organisations, **other,** ought, **our, out,** outside, over, own, p, paid, paper, part, particular, particularly pass, past, pay, pensioner's, **people, perhaps,** person, phone, pick, place, places, plastic, playhouse, please, point, points, policy, poll, possible, post, pounds, prepared, press, prisoner, prisoners, probably,

			problem, problems, produce, produced, programme, provide, public, **put,** question, **quite,** r, rather, read, **really,** reason, recently, recycling, released, remember, report, restaurant, **right,** ring, road, room, round, run, running, s, **said,** same, **say,** saying, says, scheme, school, schools, seats, second, secretary, section, **see,** seems, seen, sell, send, sent, september, service, services, seven, shall, **she,** she's, shops, short, **should,** shouldn't, show, since, situation, six, small, **so, some,** somebody, someone, **something,** somewhere, sorry, **sort,** sounds, speak, speaker, spoke, st, stage, stall, start, started, still, stortford, studio, stuff, such, support, suppose, sure, system, **take,** taken, talk, talking, tax, tea, team, tell, ten, than, thank, **that, that's, the, theatre, their, them, then, there, there's, these, they,** they'd, they're, they've, **thing, things, think,** thinking, thirty, **this,** those, though, thought, thousand, three, through, thursday, **time, to,** today, together, told, tonight, too, took, top, touch, town, trust, try, twenty, **two,** under, understand, unit, **up, us,** use, **used,** useful, using, usually, **very,** views, **want,** wanted, wants, war, **was,** wasn't, water, **way, we,** wednesday, week, weeks, **well,** we'll, went, **were, we're,** weren't, **we've, what,** whatever, what's, **when, where,** whether, **which,** while, **who,** whole, who's, **why, will,** wish, **with,** within, without, women, women's, won't, work, working, world, **would,** wouldn't, write, writing, written, wrote, **yeah, year,** years, **yes,** yet, **you,** you'd, you'll, young, **your,** you're, youth, you've

5.5 Hierarchical modelling

In the last section, evidence was presented to support the hypostudy that the small world structure of the word co-occurrence network is formed by the optimisation of storage cost and connectivity. Networks that are constrained by cost tend not to exhibit hierarchical organisation (Ravasz, Somera, Mongru, Oltvai and Barabási 2002; Ravasz and Barabási 2003). Hierarchical organization in a network was introduced to combine the notions of modularity and scale-free behaviour.

Modularity in a network implies there is an average number of links connecting to any given node in the system. However, real modular networks do exhibit both a high clustering co-efficient *and* scale- free properties i.e. there is no average node connectivity (Ravasz, Somera, Mongru, Oltvai and Barabási 2002; Ravasz and Barabási 2003). To address this issue of how the multiple node connectivity can be combined with modular structure, the notion of hierarchical organization was established (Ravasz, Somera, Mongru, Oltvai and Barabási 2002; Ravasz and Barabási 2003).

Figure 5.9 The networks on the left show the schematic version of (a) a scale-free network (b) a modular network and (c) a hierarchical network. The networks on the right correspond to a "typical configuration" respectively (each with 256 nodes). (Ravasz, Somera, Mongru, Oltvai and Barabási 2002; Ravasz and Barabási 2003: Figure 1)

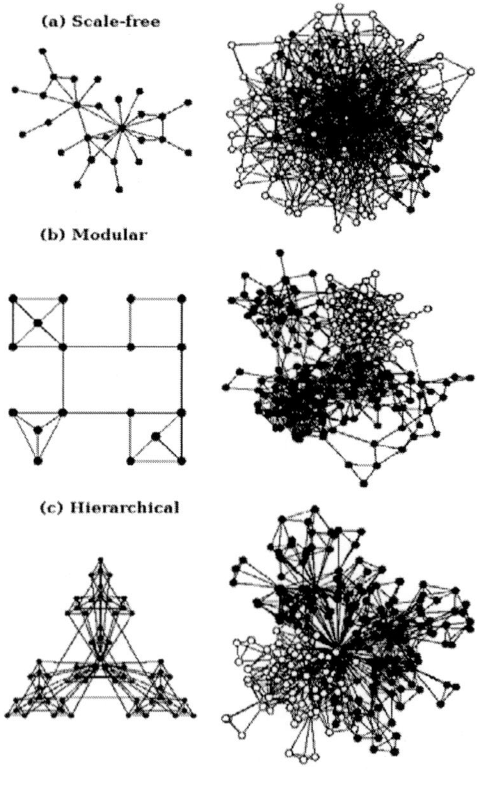

For example (Figure 5.9c), to create a scale-free but highly clustered network, four fully connected nodes are replicated three times. The central node of the initial four nodes is then connected to the three external nodes of the replicas, generating a "16 node module". Next, the sixteen node module is reproduced three times - again the external nodes of the new module connecting to the middle node of the old sixteen node module. Compare this hierarchical organization (Figure 5.9c) with a modular network that is not scale free (5.9b) and a scale free network that is not modular (5.9a) (Ravasz, et al 2002: Figure 1).

To ascertain whether a network displays hierarchical structure, the connectivity of a node is plotted against the clustering co-efficient of the same node, for the whole network. When the data is normalized i.e. logged, a linear slope of -1 is indicative of hierarchical clustering (Ravasz and Barabási 2003: Figure 3b, see Figure 5.11). It is important to note that "geographically organised webs", that is, networks that have a cost constraint on their link length such as power grids and internet router-level connections, tend not to display hierarchical organization (Ravasz and Barabási 2003:1).[3] In these cases, the clustering co-efficient is independent of degree (Ravasz and Barabási 2003: 4).

In regard to the adult word co-occurrence network, the connectivity value and clustering co-efficients for all the nodes in the network were calculated by PAJEK software (de Nooy, Mrvar and Batagelj 2005) and plotted (Figure 5.10). As expected for a network constrained by (storage) cost, an exponent of -1 is not observed; compare Figure 5.10 with Figure 5.11 (Ravasz and Barabási 2003). However, contrary to power grids and internet router level connections, the clustering co-efficient C(K) is not entirely independent of connectivity size (degree K). As C(K) increases, the K also increases, plateauing just below the threshold of 100 connections.

[3] However, the property of nesting associated with hierarchical modelling did occur in the storage cost regimes. Since hierarchical organisation is linked to the property of degree in a non-cost orientated network, perhaps cost sensitive networks also exhibit hierarchical structure, albeit in the link lengths of particular nodes. This issue is a matter for future research.

Figure 5.10 Log-log plot of connectivity (degree (k)) against the clustering co-efficient for the adult network. The line dissecting the squares represents the slope of -1. A slope of +0.57 is indicated.

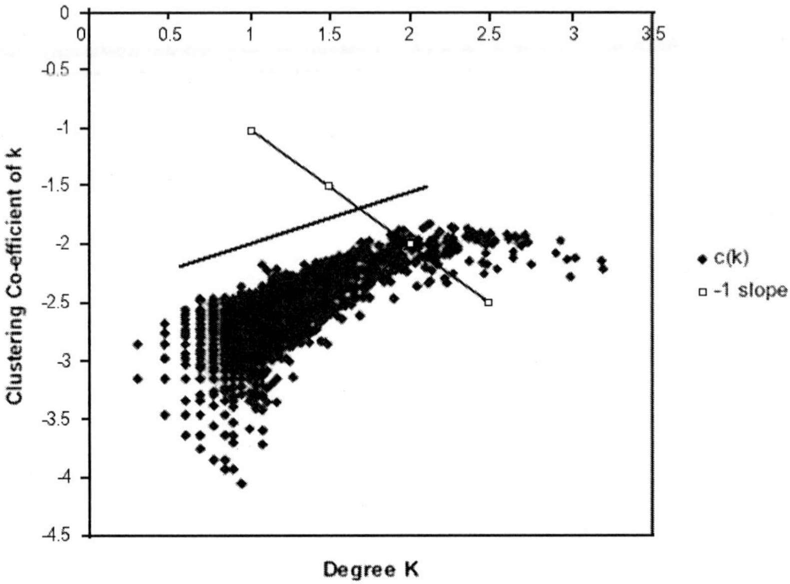

Figure 5.11 Log-log plot of connectivity (degree (k)) against clustering co-efficient for a semantic network "connecting two English words if they are listed as synonyms in the Merriam Webster dictionary" (Ravasz and Barabási 2003: Figure 3b) The line indicates a slope of -1.

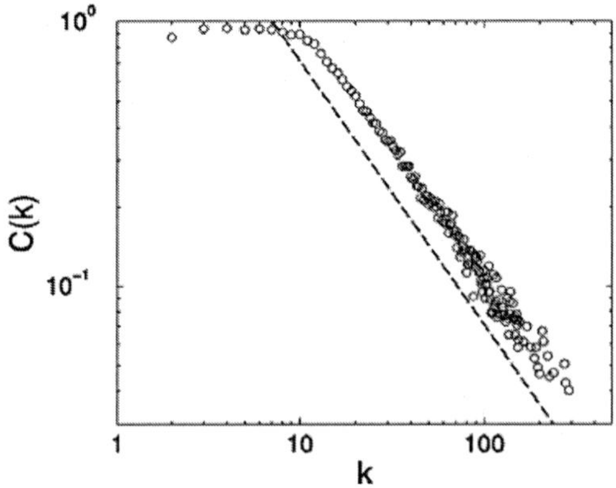

5.6 Assortative versus disassortative mixing

In this section, it will be shown that hubs in the word co-occurrence network tend not to connect to other nodes with a high degree. This tendency is known as "disassortative mixing" as noted in chapter three and is displayed in other types of networks such as the www, information processing systems and biological systems (see Table 1, Newman 2002). In contrast, social systems such as those formed through collaborations e.g. film actors, company directors, physics co-authorship display "assortative mixing" (see table 1, Newman 2002). That is, the hubs tend to connect to other hubs.[4]

To ascertain which type of mixing is found in the word co-occurrence network, the degree of the first word in the bigram is correlated with the degree of the second word i.e. the Pearson co-efficient. A positive value indicates assortative mixing whilst a negative value, disassortative (Newman 2002).[5]

The co-efficient for the word co-occurrence network is -0.26, indicating that there is not a strong positive correlation between the counts of the first and second word in the bigrams. As such, the network displays disassortative mixing.

As a comparison, a Pearson co-efficient of -0.065, -0.156, and -0.276 were recorded for the www, protein interactions and the food web of Little Rock Lake, Wisconsin, all of which have disassortative mixing (Newman 2002). In a language context, disassortative mixing has also been found in syntactic dependency networks derived from adult Czech, German and Romanian utterances (-0.06,[6] -0.18, -0.2, respectively; see Ferrer i Cancho, Solé and Köhler 2004) and child utterances (Corominas-Murtra, Valverde and Solé 2007 a&b).

---

[4] However, online social networks may display disassortative mixing (see Hu and Wang 2009: Table 1).

[5] A zero value indicates that no assortative mixing occurs (Newman 2002).

[6] Note that in the Czech corpus certain function words were not linked i.e. prepositions, reflecting the higher mixing value (Ferrer i Cancho, Solé and Köhler 2004).

5.6.1 Disassortative networks: the correlation between degree and betweenness centrality

It has been argued that degree and betweeness centrality (BC) correlate closely for disassortative networks (Goh, Kahng and Kim 2003). The concept behind BC captures the idea that certain nodes in networks act as intermediaries, creating shortcuts through the web of links. In a word co-occurrence network, the most obvious example is the word *and* that may co-occur between or link two semantically diverse words.

If a word has a high BC, this means that many shortcuts or shortest paths form through the word. In linguistic terms, a high BC may infer a word is used often with different collocations. Since shortcuts aid connectivity, it is plausible that these words take a prominent role in anchoring patterns in the language network i.e. establishing a highly productive construction.

In Equation 5.3 (Everitt and Borgatti 1999), the amount of geodesics (shortest paths) that pass through a given node or group provides a measure of ***betweenness***, where C denotes a subset of a

Equation 5.3 Betweenness (Everitt and Borgatti 1999)

$$\sum \frac{G_{u,v}(C)}{G_{u,v}} \qquad u, v \notin C$$

In practical terms, *betweenness* is determined by a series of computation steps (Everett and Borgetti, 1999). Firstly, the number of geodesics between every pair of non group nodes is calculated and a matrix of node by node counts produced. After removing the link counts that include group nodes, a new matrix is recreated so that the ratios may be summed. Individual and group betweenness may be normalised by the division of *each value by the theoretical maximum*. The theoretical maximum obtains when the identification of the groups vertices results in the collapse of the links into a single vertex that centres itself within a star shaped graph (Everett and Borgetti 1999). Therefore, in the normalisation of a group, betweenness of centrality C’ $_B$ (C) is defined by Equation 5.4.

Equation 5.4 Betweenness Centrality (Everett and Borgetti 1999)

$$C'_B(C) = \frac{2C'_B(C)}{(|V|-|C|)(|V|-|C|-1)}$$

As noted earlier, in disassortative networks, the BC and the degree correlate well in a disassortative network. The degree (both the in- and out- degree) and BC values of the BNC adult text were plotted logarithmically and it can be observed that for words with a high degree, degree and BC are closely correlated. However, as the degree decreases, the correlation breaks down.

Figure 5.12 The degree of a word plotted against the betweenness centrality (BC) value.

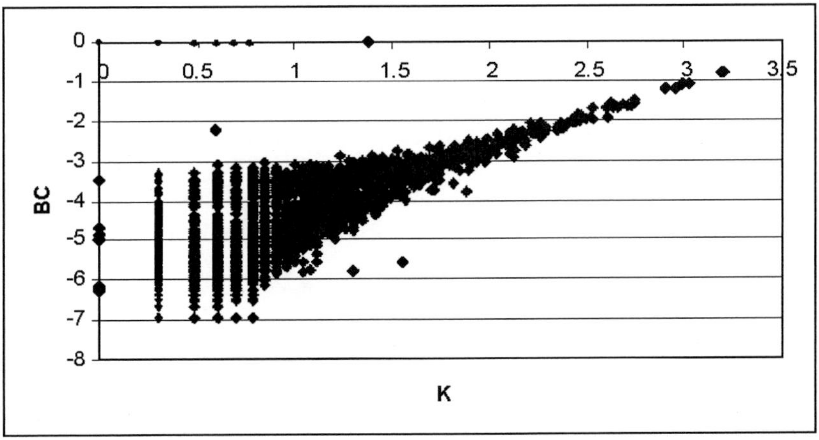

It has also been observed that the BC distribution in scale free networks is captured by a power law function (Goh, Oh, Jeong, Kahng and Kim 2002).[7] The distribution of BC in the adult BNC text follows two distinct power law regimes. The first regime accounts for 75% of words in the network (4024 words) and has an exponent of -0.15. The second regime

[7] When the degree exponent for a given network is between 2 and 3, it is predicted that the BC distribution exponent will be between -2 and -3. '-2' for class I and '-2.2' for class II (Goh, Oh, Jeong, Kahng and Kim 2002). However, the degree exponent for the word co-occurrence network was calculated at -1 and -0.7.

exhibits a slope of -0.88 (including 1341 words).

Figure 5.13 The distribution of betweenness centrality for the adult network.

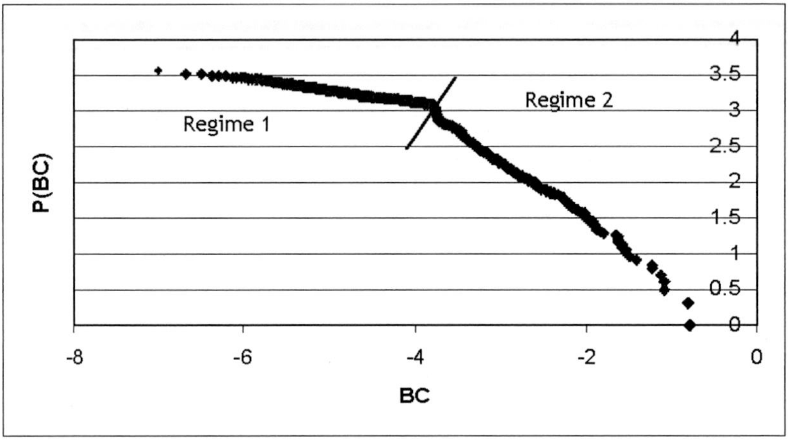

Typically, hubs have both a high degree (connectivity) and a high BC value. Twenty words with the highest BC value are compared with the twenty most highly connected words (Table 5.3). Bolded font identifies the words that are ranked twenty or higher for BC values and degree. Italicised words denote the twenty most frequent words. Sixteen of the most highly connected hubs are the most frequent words and have the highest BC values in the network: *and, the, of, to, a, that, in, for, it, on, you, erm, i, is, was, be*. As will be shown in chapter 8, these words anchor many large link patterns.

Table 5.3 The twenty hub words are compared with words with the highest betweenness centrality and the twenty most frequent words. Bolded font identifies the words with the largest BC values and hubs (ranked twenty or above). Italicised words denote the twenty most frequent words.

Twenty most highly connected hubs	**be**, ***erm***, ***er***, ***on***, ***of***, ***and***, ***you***, *we*, ***was***, ***a***, ***to***, this, ***the***, ***that***, ***in***, ***i***, *have*, ***for***, ***it***, ***is***,
Twenty largest BC	***and***, ***the***, ***of***, ***to***, ***a***, ***that***, ***in***, ***for***, ***it***, or, ***on***, ***you***, ***erm***, ***i***, with, ***is***, ***was***, **er**, were, **be**,
Twenty most frequent words	*the, and, to, i, a, that, of, it, you, we, in, erm, is,* they, *was, on, have, for, it's, er*

5.7 Hubs and authorities

If you recall, highly connected nodes may be distinguished by the direction of the majority of their links. Hubs have a high out-degree i.e. many links point away from the hub, whilst authorities have a high in-degree i.e. many links point towards the authority (see Kleinberg 1999; see section 4.4.3). Since the word co-occurrence network is disassortative, there will not be many good examples of hubs and authorities. Nevertheless, the adult word co-occurrence network is directed i.e. the link direction represents precedence relations between adjacent word pairs, and as such it would be relevant to briefly examine the hubs and authorities within the word co-occurrence network.

The ten most highly connected hubs and authorities in the adult network were calculated by PAJEK – see Table 5.4 (de Nooy, Mrvar and Batagelj 2005). Most of the hubs are also the most frequently used words, whilst only one authority i.e. *it*, is included in the ten most frequent list. Interestingly, of the ten hubs and authorities, none are both a hub and an authority. The first word to be both a hub and an authority is the word *it*, appearing in the eleven most connected hubs and authorities. Ke and Yao (2008) study - using PAJEK (de Nooy, Mrvar and Batagelj 2005) - found that the words *the* and *a* were authorities in the mother's speech (as well as the words *you*, *it*, *that*, *your* and *in*). However, in this adult word co-occurrence network most of the words are hubs, apart from *your*, and the word *it* being both a hub and an authority. The difference between the two studies warrants further investigation but is beyond the scope of this study.

Table 5.4 Bolded text indicates the words that are also in the ten most frequently used words in the adult text.

Ten largest hubs	*erm,* ***of, and, you, a, to, the, that,*** *in,* ***i***
Ten largest authorities	*friends, bin, point, evening, don't, year, yeah, treasurer,* ***it,*** *is*
Ten most frequent words	***the, and, to, i, a, that, of, you, it,*** *we*

Plotting the ratio of words which are both a hub and an authority (H&A) compared to either hubs or authorities, a linear relation is observed with a positive slope of 0.14. Therefore, the number of hubs and authorities increase uniformly with the number of words which are both a hub and an authority. The hub and authorities were limited to 450 hubs and authorities since the corresponding words had a minimum of 22 connections. That is, 1.3% of the maximum connectivity (i.e. 1573).

Figure 5.14 For the adult's network, the amount of words which are both hubs and authorities (H&A) is plotted against the number of hubs chosen to be represented (the amount of hubs chosen equalled the number of authorities also represented: H=A).

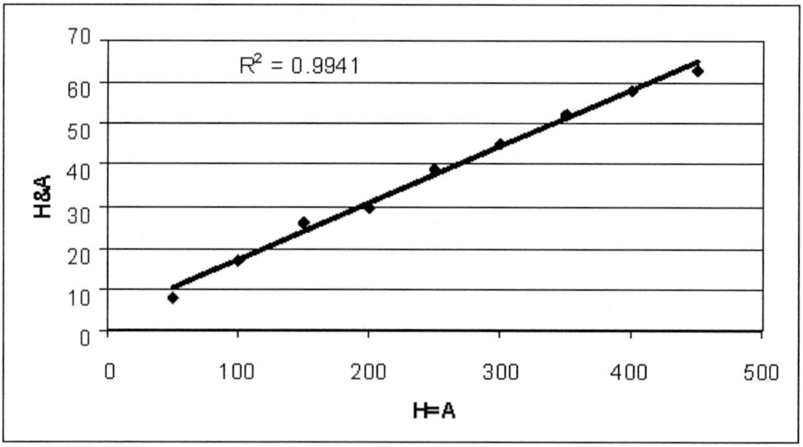

Table 5.5 below gives a snapshot of the types of words that form hubs and authorities or both. The number of permitted hubs and authorities is given in the leftmost column.

Table 5.5 The breakdown of hubs and authorities in the adult co-occurrence network. As the number of hubs chosen to be represented in the network increases (first column), so too does the number of words given as both hubs and authorities (fourth column).

Number of H=A	Hubs (H)	Authorities (A)	H & A
10	erm, of, and, you, a, to, the, that, in, I,	friends, bin, point, evening, don't, year, yeah, treasurer, it, is	
15	erm, er, on, of, and, you, was, a, to, the, that, in, I, it, is	It, is, friends, bin, point, evening, don't, year, yeah, who, treasurer, three, has, seat, seat has, three	It is
30	Be, erm, er, or, one, on, of, about, and, you, with, we, was, a, know, to, but, it's, this, there, the , that, in, I, so, have, For, it, is, they	For, it, is, they, gentleman, friends, bin, rally, point, evening, many, don't, low, s, your, year, yeah, written, who, well, various, c, treasurer, three, support, straight, correct, has, seat, all	For, it, is, they
60	Got, right, get, are, from, been, actually, people, be, erm, er, or, one, on, of, now, about, not, no, more, mean, and, do, you, with, like, what, were, we, was, can, a, know, just, to, but, it's, this, think, them, the, that's, that, in, if, I, some, so, as, have, had, for, yeah, well, it, is, they, there, at, all	For, yeah, well, it, is, they, there, at, all, consultation, resources, gentleman, gas, bishop's, friends, bin, rally, benefits, point, far, pieces, evening, s, bar, many, don't, low, your, year, written, working, limit, will, light, who, leisure, waste, leaflets, details, various, c, by, treasurer, three, thirty, thick, debate, support, straight, st, bread, smaller, shop, correct, sheet, has, hand, book, seat, savings	For, yeah, well, it, is, they, there, at, all

100	said, got, good, right, get, are, from, really, been, actually, people, because, out, our, other, be, erm, er, or, any, one, on, oh, of, not, no, my, much, more, meeting, mean, me, and, do, an, you, yes, would, with, like, which, when, what, were, we, was, can, a, very, know, up, two, just, to, but, time, this, think, things, thing, these	for, yeah, well, it, is, they, there, at, all, consultation, girls, resources, gentleman, gas, fund, function, regional, conclusion, bishop's, friends, bin, rally, benefits, present, feet, potter, point, far, pieces, clothes, evening, open, s, eh, bar, mm, action, many, make, b, low, year, written, working, limit, light, why, whale, difficult, leisure, waste, leaflets, details, designs, various, labour, c, turning, demand, treasurer, trade, three, thirty, thick, debate, into, telephone, t, d, support, india, cut, straight, storage, ice, already, st, household, hope, bread, hill, smaller, small, sixteen, shop, correct, sheet, has, hand, book, seat, savings	For, now, about, don't, your, yeah, will, who, well, by, it's, it, is, they, there, at, all

Apart from a few exceptions, most of the words forming hubs are function words. Included in the list are the notoriously hard to categorise interjections such as *yeah, yes* and *erm*, plus a few discourse markers, i.e. *well*, *also*. There are also words which are leitmotives for certain lengthy dialogues. Due to their high frequency of usage, these words become hubs e.g. *theatre* (from an extensive public meeting on the Harlow Theatre) and *people* (referenced to by a public speaker on the subject of pre- war life). Authorities tend to represent lexical words. There are a few very significant exceptions: *don't, it, is, why*. Most of the authorities that represent function words may also be classed as hubs, e.g. *it* and *is*. Nevertheless, it is not admissible to correlate the function and lexical category with the notion of hub and authority for various reasons briefly discussed below.

Firstly, and perhaps the least significant point, is that a word may be both a hub and an authority. In network theory, this overlap is not paradoxical as the notion relies on the link direction. Indeed, certain linguistic theories allow a word to be a functional and also a lexical word depending on the context of the usage (see chapter 2 example 2.4). However, a more troublesome problem is that the correlation between the notions of hub-authority and functional-lexical only stands in VO languages. In an OV language an authority would be a function word and a hub, a lexical word. Thus, only the contrast in the link direction would distinguish the notion of a function word and a lexical word.

The third critique, which is the most critical, concerns the definition of a good hub and a good authority. A good example of a hub connects with many good authorities. In other words, many lexical words connect to a function word which is admissible in the data. However, a good example of an authority word connects with many good hub words. Thus, a good example of a lexical word would be one that occurs with many function words. Although a few function words do occur with a lexical word, this amount is small when compared with the amount of lexical words that occur with a function word. Thus, there is a type frequency asymmetry inherent in the functional-lexical (f-l) distinction that is not captured by the hub-authority distinction. Consequently, there is not a strong correlation between the f-l distinction and the hub-authority distinction. Chapter 8 will show how the notions of anchor and automorphic relations within a certain link pattern better capture the f-l distinction.

5.8 Summary

As predicted, the adult co-occurrence network is a small world network, possessing a low average shortest path and a higher clustering co-efficient compared to a random network counterpart. The frequency and degree power law distributions were verified for the word co-occurrence network. The distribution of link lengths in a network is captured by a multiple-regime power law with a truncated tail. The truncation of the tail is argued to be resultant of the optimisation process being constrained by high (storage) cost bigrams that represent novel and/or infrequent word co-occurrences whereas within regime five, strong collocations such as *i-think* at found. Thus, the first and last regime include novel and formulaic combinations respectively (see section 9.3.2). It is argued that the power law distribution of storage cost, in addition to the degree and frequency power law distribution represent non-atomic, primitive and substantive syntactic relations. As such, the storage cost, degree and frequency power laws capture basic syntactic properties such as syntagmaticity, paradigmaticity and entrenchment respectively.

The hubs, which are the key to small world formation, result from optimising the cost of each link length i.e. the storage cost represented by an inverse transitional probability value and the word connectivity. Furthermore, the hypostudy that cost limits the distributional pattern of words in language is supported by the lack of hierarchical structure (in respect to the degree of nodes) within the network. The word co-occurrence network displays disassortative behaviour, and as predicted for this type of network the betweenness centrality value of a node and its degree correlate – although the correlation is very weak for poorly connected nodes.

Hubs and authorities were identified in the network. As the number of hubs and authorities examined increases, the amount of nodes that are hubs *and* authorities also increased linearly. Although hubs tend to represent function words and authorities, lexical words, the f-l distinction does not correspond well with the hub-authority distinction (respectively). It will be shown that the f-l distinction is better described by the notions of anchor and automorphic relations within a certain link pattern as discussed in chapter seven and eight. However, the next chapter examines a word co-occurrence network composed of child utterances using the metrics adopted in this chapter.

CHAPTER SIX

THE CHILD WORD CO-OCCURRENCE NETWORK

6.1 Overview

This chapter presents the analysis of a word co-occurrence network composed of child utterances. Firstly, the small world characteristics of this network will be established i.e. the smallest average shortest path and the clustering co-efficient (section 6.4), as well as identifying the power law of frequency (section 6.4.1) and degree (section 6.4.2). Furthermore, the power law of link length (the storage cost of a bigram) will also be evidenced (section 6.4.3).

In line with the adult word co-occurrence network, the network does not display hierarchical organisation (section 6.5) but is disassortative (section 6.6). Typical of disassortative networks, the child word co-occurrence network exhibits a correlation between betweenness centrality (Goh, Kahng and Kim 2003) and degree albeit a weaker one compared to the adult network (section 6.6.1). Lastly, the authorities and the hubs are identified (section 6.7). Comparisons with the adult network are made throughout the analysis.

6.2 Data

The child network is formed from fifty-six CHAT conversations by Naomi, an American infant aged between 1;2.29 and 2;3.29, transcribed by Sachs (Sachs 1983; MacWhinney 2000a) and sourced from the CHILDES internet database (MacWhinney 2000a). The utterances have an MLU of equal to or less than 3.5 and were collected in a range of scenarios throughout various times of the day: breakfast, bath time, free play, reading books, birthday party, etc. The data contains 17469 tokens, of which 1016 are unique types. The type/token ratio was calculated at 6% (very close to the adult ratio of 7%) and a total of 3061 bigrams were formed (Klarskov Mortensen 2006).

6.3 Data preparation

The CHAT tagging code was stripped away by hand, leaving the bare (punctuated) text. The phonetic transcription of nonsense words or imitations was retained. Linguistic elements that are not usually classified as words such as *uh, ah* and *uhs* were not removed since these may constitute proto-morphological fillers (see Peters 1993; Peters and Menn 1993).

The types of repetition shown in example 6.1 are treated as one word/morpheme (6.1a. *it* and b. *milk*) and so the repetitions are deleted. However, example two type repetitions are retained (6.2). Corrections such as 6.3 are maintained. Sentences with uninterpretable words signified by XXX, or unrecognisable words YYY are omitted unless the phonetics of the unrecognisable form YYY is given in the 'pho' tier in the CHAT transcription. In which case the sentence is saved with the phonetic transcription substituted for YYY. The contracted copula and the nominal constituent are kept together, consistent with the adult analysis (example 6.4). Signs [#],[##] and non- linguistic communication, such as whistles ($<text>), are deleted as is background chatter [+bch]. Compounds (+) are realised as one word. Guess [?] and one word utterances are included. The brackets are taken out leaving the text in the sentence (example 6.5). A note is taken in the original data sheet.

Example 6.1 Repetition type 1

a) *CHI: <it, it, it> [/] it
b) *CHI: [milk] [/][milk] [/][milk] [/]

Example 6.2 Repetition type 2

*CHI: it
*CHI: it
*CHI: it

Example 6.3 Corrections

*CHI: I wanted …[//] I thought

Example 6.4 Contraction

'That's' becomes 'thats'

Example 6.5 Bracketing

'mou(se)' becomes 'mouse'

Next, the text was compiled into one file and the token/type values and the transitional probabilities of each bigram were calculated using a specifically designed program (Klarskov Mortensen 2006) in exactly the same procedure as the adult network as described in chapter five.

6.4 Small World Properties

The average shortest path distance (L_{child}) between nodes in the child language network (N_{child}) measured 11 and is considerably larger than the diameter of the adult network L_{adult} i.e. 3.3. However, in a random network ($N_{random\ child}$) with an equivalent number of nodes (1346) and arcs (3061), the path length is 14 (higher than L_{child}). The clustering co-efficient in N_{child} was twice as high as the equivalent $N_{random\ child}$ clustering co-efficient. Thus, the child network exhibits characteristics of a small world network.

Table 6.1 Small world properties of the adult and child language network N_{adult} and N_{child} compared to the equivalent random networks ($N_{random\ adult}$ = 5365 nodes and 31602 arcs whilst $N_{random\ child}$ = 1346 nodes and 3061 arcs). Multiple lines were removed for the calculation in PAJEK (de Nooy, Mrvar and Batagelj 2005).

Network	Average shortest path (L)	Clustering co-efficient (CC1)
Adult language (BNC) N_{adult}	3.3	0.00131 (1.3 x 10^{3})
Random $N_{random\ adult}$	5.1	0.00055 (5.5 x 10^{4})
Child language (BNC) N_{child}	11	0.0024 (2.4 x 10^{3})
Random $N_{random\ child}$	14	0.0012 (1.2 x 10^{3})

6.4.1 The frequency distribution

The ten most frequent words in the child text, in descending order, are: *I, it, what's, this, that, want, uh, down, the, daddy*. Compare with the adult counterparts: *the, and, to, I, a, that, of, you, it, we*. Whilst both the adult and child most frequent words include *i, it, that, the,* the child's list includes words that are highly frequent in the adult network e.g. *what's, this, want.*

The list of word frequencies were ranked and a LOG-LOG rank versus frequency plot was produced. The recorded slope of -0.7 is shallower than the expected Zipfian -1. However, Zipf himself noted a range of rank-frequency slope exponents α for children's speech, between -0.76 < α > -1.31 (Fischer material, Zipf 1965: 142-143: table 4-3). The children were aged between 22 months and 59 months and the lowest slope was observed in the 27 types and 64 tokens spoken by a 27 month old girl. In Figure 6.1, under 380 words, the slope is -1, and in the last part of the plot -2.6.

Figure 6.1 A log-log plot of the rank (r) versus frequency (f) of the words in child network

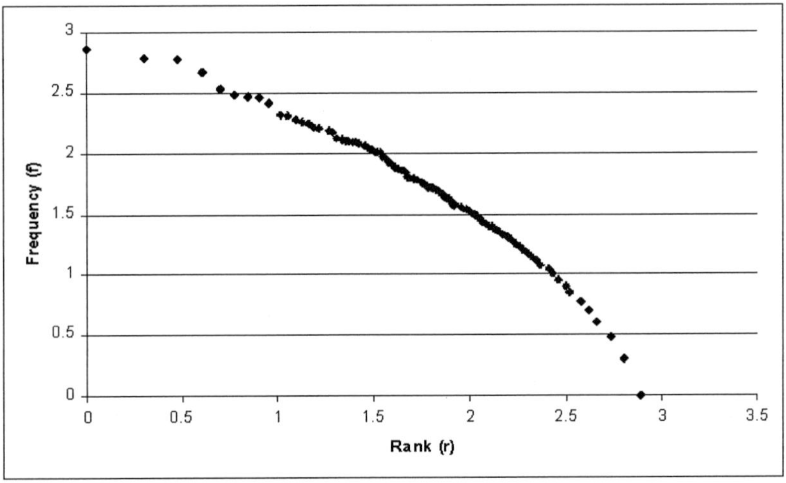

6.4.2 The degree distribution

Using PAJEK, the connectivity of the words in the network was established (de Nooy, Mrvar and Batagelj 2005). Like the adult degree distribution, there are two power law regimes with exponents -1.1 and -0.7 (regime 2 and 3) and a truncated tail (regime 1). Compare the adult regimes exponents of -1.0 and -0.7 (regime 2 and 3 respectively). In the last chapter, it was argued that the truncation is due to parts of the network moving away from the most optimised state through the introduction of new connections to the network. The words in the tail are virtually all function words e.g. {*the, it, I, uh, a, you, my, on, that, is, this, want, in, up*}. Function words facilitate the identification of constructions, and so it is quite likely that a new word will connect to a function word, increasing its degree.

Figure 6.2 The degree distribution regimes. (a) Regime three has an exponent of -0.7 whilst in (b), regime two has an exponent of -1.1. (c) The truncated tail in regime one.

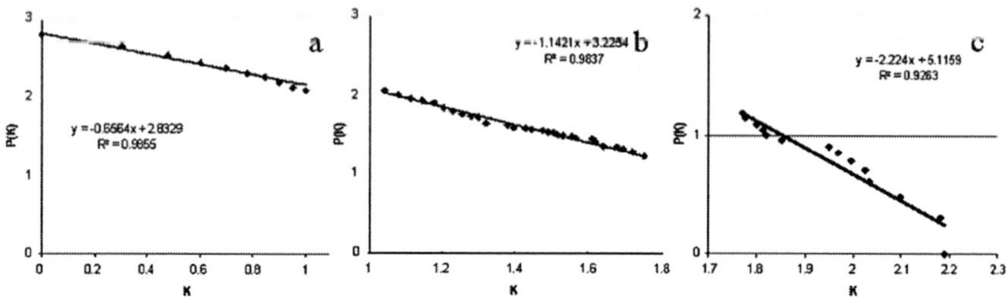

When the degree and frequency ranking are compared, the correlation is weaker in the child network than in the adult network (Figure 9.3). When comparing the degree and frequency of the twenty-five highest ranked words in the child network, a Pearson co-efficient of 0.43 was observed, compared to 0.95 for the adult network (section 9.3.1, Figure 9.3).

Figure 6.3 A comparison of frequency and degree for the twenty-five highest ranked words in the child network.

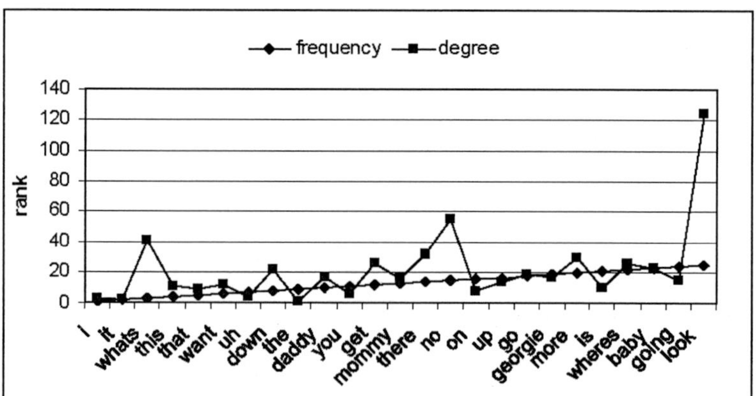

6.4.3 The distribution of storage cost values

The transitional probabilities of each bigram in the child network were calculated. Inversing the values, we obtain the storage cost of each bigram. Plotted against the cumulative distribution, a multi-power law distribution is observed, in line with the adult network. As such the child word co-occurrence network is also a hubterranean small world network, formed through the optimisation of maximal word combinations i.e. connectivity, and minimal storage cost i.e. the cohesion of a given bigram (see Mathias and Gopal 2001 a&b).

Equation 6.1 The frequency of bigrams or word pairs (N(S)), which have a storage cost (S), may be represented by a cumulative distribution plot and is expressed by the below equation (see Solé 2006).

$$N_{>}(S)=\int_{S}^{\infty} N(S)dS$$

Akin to the adult network, the child network displays five regimes. The regimes from the second to the fifth may be captured by a power law function, each with a different exponent. There is a fast decay in the first regime, in which the highest storage cost bigrams reside. In this regime the power law does not hold and is comparable in this respect to the first regime in the adult network. See appendix A.2 for examples of word pairs in each regime.

The storage cost power law corresponds to the multiple scales of bigram cohesion found within the word co-occurrence network which in turn represents the varying strengths of syntactic constituents. As such, the child utterances exhibit the same broad constituency characteristics as the adult network even though at first glance, the child's vocabulary and construction repertoire are smaller and their constructions, less abstract. The five regimes in the logged cumulative plot are given below. Table 6.2 provides the number of first words forming bigrams in the given regime with details on the specific words.

6.4.3.1 The five regimes

The five regimes are given in figures 6.4-6. As noted previously, the first regime truncates the power law distribution since the high storage cost of low frequent word co-occurrences moves the network away from the most optimal point. Thus the optimisation of the child network, like the adult network, is constrained by novel or infrequent word combinations.

The second regime has the steepest slope at -1.2, with an R-squared rating of 0.99. As with the adult network, the first word of the bigrams in regime 2, are found as first words of bigrams in regime 1.

Figure 6.4 The first (a) and second (b) regimes in the Sachs (1983) child language network. Both plots show the cumulative distribution P(S) and the storage cost (S).

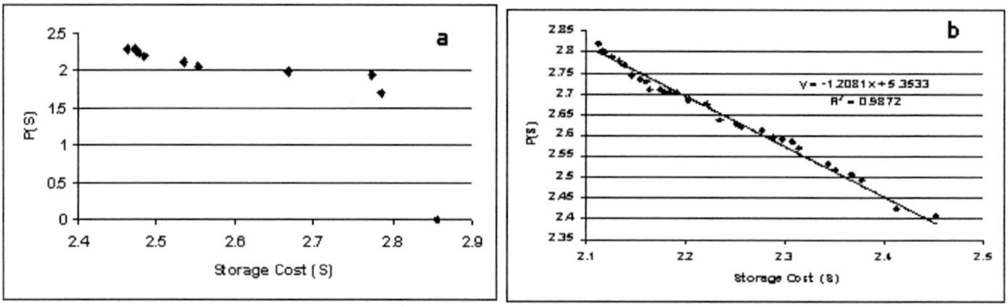

Likewise in the third regime, the first words of the bigrams are also the first words of second regime bigrams. The slope of the graph is shallower in this regime, the exponent being -0.6, with an r-squared value of 0.99.

Figure 6.5 The third (a) and fourth (b) regimes in the Sachs (1983) child language network. Both plots show the cumulative distribution P(S) and the storage cost (S).

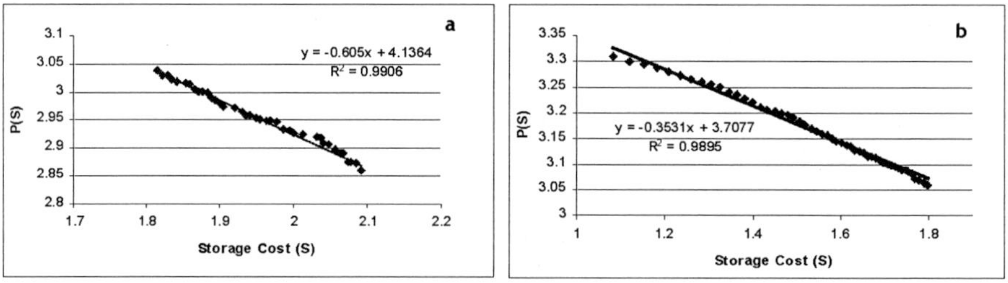

In the fourth regime, most of the first words that form bigrams in regime three (consisting of 208 words) also form a first word in a regime four bigrams. The four exceptions are *book, down, yeah, yup*. The 4th regime's shallow slope has an exponent of -0.35 (r-squared 0.99).

Figure 6.6 The fifth regime in the Sachs (1983) child language network. The graph shows the cumulative distribution P(S) and the storage cost (S).

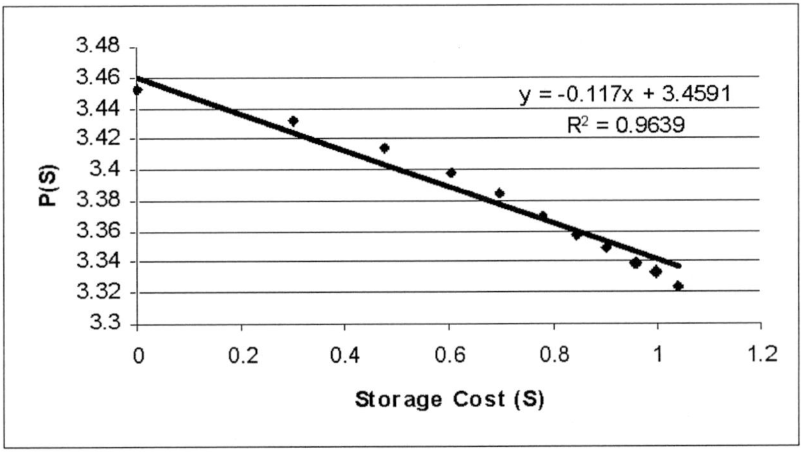

Lastly in the fifth regime, 77 of the 208 words that occur as a bigram's first word in the fourth regime, do not have counterparts in the fifth regime bigrams e.g. *uh, up, on, daddy, go, off* (the fifth regime being composed of 594 bigrams). The bigrams in this regime have the lowest storage cost. That is, the cohesion of these bigrams are the strongest in the network. The power law fit is the poorest of all the regimes with the shallow slope of -0.12 having an r-squared value of 0.96.

The below table (6.2) provides each bigram's first words at a given storage cost found in a given regime. The bolded text indicates the inclusion of these words in the previous regime.

Table 6.2 The first words of the bigrams are shown with their respective regime. The words which occur in the previous regime are bolded. Power law regimes are indicated by the abbreviation (PL).

regime	cost	no. of words	first word in bigram
1st	719-290	7	i, it, whats, this, that, want, uh
2nd (PL)	285-129	27	down, the, **i, this,** daddy, you, get, it, **whats,** mommy, there, no, up, **that,** on, go, georgie, **want, uh,** more, is, wheres, baby, going, look, in, a,
3rd (PL)	124-65	52	im, **it, i,** my, **this,** at, **that,** nomi, **daddy, you,** juice, do, dont, **get, want,** doing, **whats,** see, **mommy, uh, down, there,** like, **no,** here, **the, up, on,** book, **go,** thats, **georgie,** kitty, yeah, to, me, sit, and, **more, is, wheres, baby,** doggie, **going,** yup, put, can, **look,** where, hi, **in, a,**
4th (PL)	63-12	212	what, another, **im,** recorder, **i, my,** oh, **uh,** take, fall, **at, up, on, nomi, daddy, go, you,** off, **juice,** draw, **do,** flower, eat, **dont,** one, **georgie,** coffee, birdie, **this, the, get,** jenko, **want,** its, **doing,** theres, read, **see, there,** broken, **is,** lie, georgies, **wheres,** sleep, **no, going,** ah, **look,** peanutbutter, **in, a, that,** some, open, got, **thats,** diaper, **whats, kitty, to,** moon, **me,** uhs, toast, tired, hot, fish, cant, out, jump, **and,** wanna, sun, cookie, **more,** red, he, green, find, e, did, **it, baby, doggie,** whos, **put,** piggie, **can,** all, **where, mommy,** wash, agra, sleeping, man, girl, drink, **like,** gonna, push, boy, cry, cocoa, foot, come, broke, blanket, okay, play, nightnight, hurt, good, byebye, lying, little, gammy, banana, are, smoke, over, for, eating, bug, too, need, mouth, kangaroo, chicken, **here,** washing, paper, make, have, dirty, clean, chair, **hi,** sandy, how, be, stars, heres, hat, toys, throw, **sit,** close, water, tail, george, dish, change, boo, oig, nomis, name, her, gone, coming, cereal, truck, eh, dropped, camel, bye, um, these, shoes, mouse, mm, kittys, four, flowers, f, spank, not, horsie, dog, dinner, crumb, careful, babys, turn, table, pour, monkey, julia, hair, ear, blue, because, beanice, yogurt, yes, two, toy, six, say, ow, milk, light, lamb, kiss, holding, him, hard, goldie, daddys, coco, brush, baloney,

5th (PL)	11-1	594	**in, a, find, take, want, too, need, like, kangaroo,** with, shoe, round, **put,** puppy, **piggie,** penguin, **make,** lollipop, his, **have,** eight, **draw, dirty,** color, **clean,** back, **where, my, what, some, got, be, another, you,** we, sitting, singing, **recorder, its,** horse, **heres,** has, **gonna,** cheese, bring, boat, **dont, theres, see, toast, throw, fish, close, can,** under, eyes, **in, georgies, and, wanna, sun,** stand, playing, peach, nine, move, moo, mommys, lolly, hands, getting, **foot,** foo, fix, dress, drawing, **change, boo, ah, to, play, good, the, nomis, jenko, her, gone,** cheerios, **little, smoke, wash,** right, poor, hug, home, finished, feel, empty, **eh, eating, dropped,** drinking, donkey, **moon, at, eat,** sugar, **nomi, cant, um, these, mouse, kittys, four, flowers, f, washing, im, get,** youre, years, why, wait, told, toaster, stop, stick, **spank,** skunk, putting, piece, mine, meow, looking, jenkos, **how,** hello, **he,** golfball, gaga, fixing, drop, **cocoa,** carrot, bottle, bear, **babys,** an, **wheres, turn, pour, julia, because, beanice, read,** yellow, **two,** throwing, taking, skunks, **six,** record, pushed, pretty, pink, pants, pajamas, **over,** ma, **light, lamb, kiss, holding,** hey, had, give, girls, finger, **daddys,** could, climb, bumped, **bug,** boom, bam, **is, do, e,** watch, slide, **all, sit, broke, sandy,** your, will, white, trying, todd, stuck, shopping, **shoes,** ribbon, pick, orange, mom, hes, fell, face, **diaper,** birdies, beef, **toys, open, not, going, tail,** hold, **george, come, hurt,** feet, **byebye, coming, are,** was, upstairs, turning, **truck,** todds, time, swish, store, side, shredding, shes, sheep, seeing, reading, of, nice, naomi, mun, matzo, lipstick, icky, happen, haha, found, **for,** fingers, donkeys, dollys, cup, candle, brown, blowing, black, being, ann, **whats, blanket, i,** raining, big, wee, jacket, **hot, chicken,** sleeper, **push,** who, wawa, uhuh, three, tell, so, riding, ready, pull, piggy, peekaboo, other, na, morning, mary, love, lab, knees, happy, ga, funny, fooing, fly, flied, fixed, finish, eyebrows, does, choochoo, changing, bugin, bowl, berries, bad, ba, aw, ain, **more,** talk, **uhs,** windowshade, michael, maybe, **milk,** them, **dog, hi,** hand, **drink,** zip, wrap, whoo, way, wat, tweet, tinkertoys, tickle, they, tet, swing, standing, sneeze, sing, shh, she, salt, s, radio, played, peep, peel, passey, oug, old, moving, mans, made, kristen, hurting, hear, having, harry, glass, geh, footie, food, fen, ears, du, dope, diapers, dey, dee, cups, cool, cook, coloring, climbing, cleaning, cleaned, children, by, bump, bit, behind, bee, agras, 0, **oig,** clothes, closed, window, dat, almost, woof, walking, touch, ten, tape, sliding, munuu, cooled, **did,** vitamin, child, bumpy, bay, wearing, ih, **look,** wake, throwed, **whos, lie,** lay, **lying, fall,** yell, yankeedoodle, writing, wont, windows, which, whered, when, were, went, waving, waters, watching, waiting, used, use, u, turned, tummys, try, tree, toydee, took, toiletseat, toilet, tipping, thumb, those, then, thank, tha, tenne, tear, tastes, taste, taped, swept, sweater, stone, stays, stay, squeezing, spill, soaking, snapped, slow, slamming, sittinging, shuttinging, shaking, seven, scratch, scare,

			sandys, sandals, said, sa, rubbing, rub, rooster, reng, rasp, ran, pushing, pulling, powder, pouring, pop, pins, peggie, peeling, peek, pat, papa, own, oun6, o:h, o, nuts, niaw, new, naughty, nain, mum, mu, moth, monkeys, may, mag, lyinging, loving, locking, licking, lets, leave, laying, laugh, kugu, kugiz, kristens, knock, kno, kimberlys, kem, kaun, kaem, just, julias, ing, idee, hup, hugging, horsies, ho, hats, happened, hanging, hanger, hang, grow, goose, goo, golf, gob, from, flopping, fleece, fine, finding, finded, feets, feeling, feed, feather, fe, falled, er, enjoy, en, eigh, ebae, eated, dropdrop, drawed, doors, doesnt, doed, deres, crush, covers, comforting, comb, cokey, cocos, co, carry, cannot, calves, called, buy, buttons, burn, bringing, boys, boyge, bouncing, boots, birthday, bird, bimbim, bent, been, bathrobe, bang, awl, awful, attacking, ate, as, anug, anek, ands, air, ag, 6wal, 6

Thus, the nesting of the bigram's first words in a subsequent regime is curtailed earlier in the child network (regime four) than the adult network (regime five).

6.5 Hierarchical modelling

To ascertain whether the child language network displays hierarchical structure, the connectivity of words and their respective clustering co-efficient values were plotted (Figure 6.7). Compare the child clustering distribution data (Figure 6.7) with the internet router data, another non-hierarchical network (Figure 6.8). The angular line in each figure depicts a slope of -1. It is clear that the child network, in line with the adult network, does not exhibit hierarchical structure as the data does not display a slope of -1. As previously noted, networks constrained by cost tend to be non-hierarchical (Ravasz and Barabási 2003).

Figure 6.7 Log-log plot of connectivity (degree (K)) against the clustering co-efficient C(K) for the child network. The line indicates a slope of -1

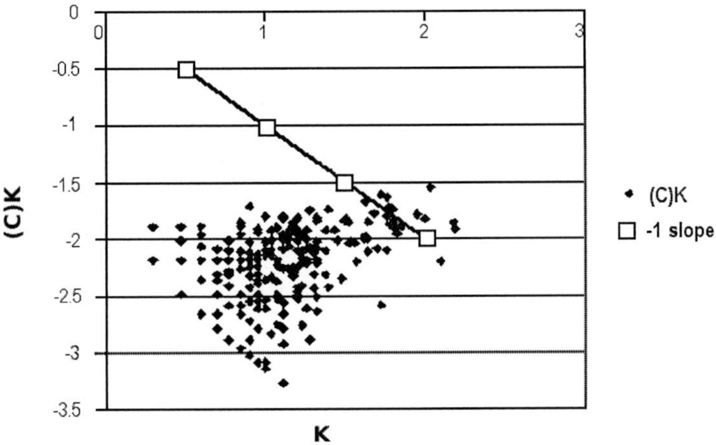

Figure 6.8 Log-Log plot of connectivity (degree (k)) against clustering co-efficient for the "Internet at router level", a non- hierarchical network (Ravasz and Barabási 2003: figure 4a). The line indicates a slope of -1.

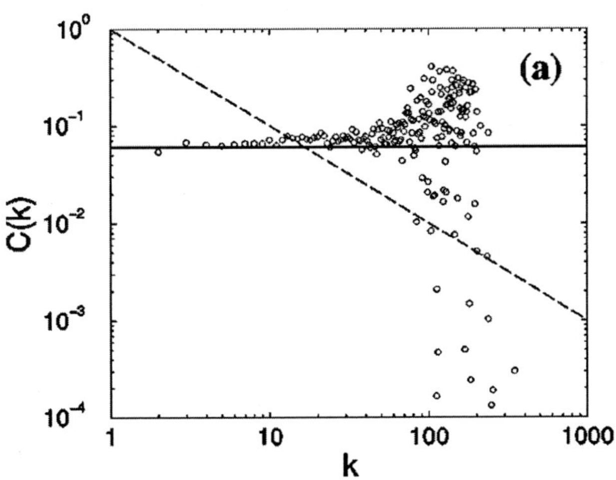

6.6 Assortative versus disassortative mixing

The child network, in agreement with the adult network, displays disassortative mixing. Correlating the count of the first word in the bigram with the second word count, the Pearson co- efficient is -0.21 compared with the adult value of -0.26. A positive value indicates assortative mixing whilst a negative value disassortative (Newman 2002). Therefore, in the adult and the child word co-occurrence network, hubs *tend* not to connect to other highly connected words. This result is in agreement with syntactic dependency networks (Ferrer i Cancho, Solé and Köhler 2004; Corominas-Murtra, Valverde and Solé 2007 a&b).

6.6.1 Disassortative networks: the correlation between degree and betweeness centrality

As noted earlier, betweenness centrality (BC) and degree correlate in disassortative networks (Goh, Kahng and Kim 2003). In the adult text, degree correlated strongly with BC for highly connected nodes. The child's data also shows a correlation, albeit a weaker one than the adult word co-occurrence network (Figure 6.9). Both the child and adult networks do not attest a degree-BC correlation for very poorly connected words.

Figure 6.9 The plot of word degree against betweenness centrality (BC) for the child's network (a) compared with the adult network (b).

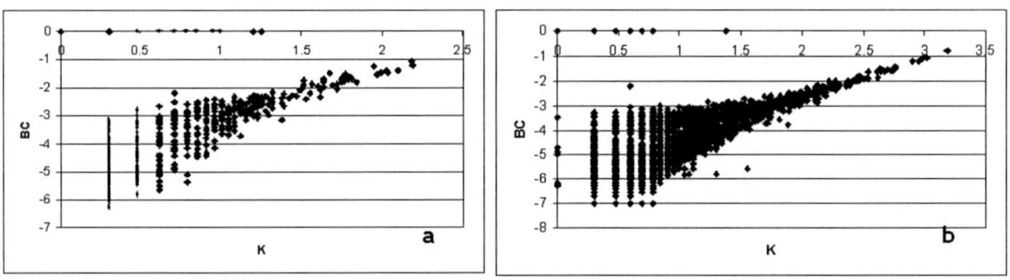

The distribution of BC values for the child text displays two power law regimes (Figure 6.10a), similar to the adult BC plot (shown as Figure 6.10b for convenience). In the first regime, the slope has an exponent of -0.10 compared to the adult -0.15. The BC values in this regime account for 74% of the total number of words (as you recall, 75% in the adult network). For higher values of BC, located in the second regime, the steeper slope has an exponent of -0.78 compared to the adult counterpart -0.88.

Figure 6.10 The distribution of betweenness centrality (BC) for the child network (a) and the adult network (b).

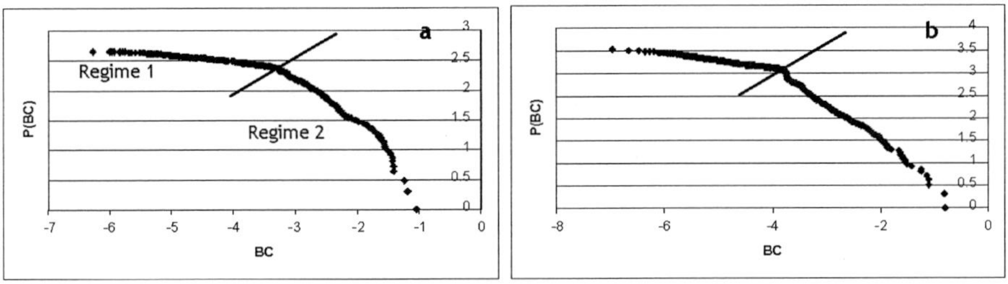

The below table (6.3) shows a comparison of the words which are hubs, the words which are the most frequent and the words which possess the highest BC values. Bolded text indicates that the word is both in the twenty most highly connected hubs and BC list, whereas italics indicate that the word is in the twenty most frequent word list. Words which occur in the frequency list but not included in the twenty largest hubs and BC values are underlined.

The words that occur in the twenty most frequent and connected words that have a high BC are *uh, want, mommy, it, that, you, daddy, this, down, on, i*. It is predicted and observed that these words will be the focus of extensive link patterns within the child's speech (see Table 8.2, 8.6 and 8.12 in chapter eight).

Table 6.3 For the child network, the twenty largest hub words are compared with words with the highest BC value and the twenty most frequent words. A bolded word is both in the twenty most highly connected hubs and the BC list. An italicised word is in the twenty most frequent word list. Words which occur in the frequency list but not in the twenty largest hubs and BC list are underlined.

The twenty largest hubs	**child**	*whats*, ***uh***, nomi, im, ***want***, ***mommy***, ***it***, ***that***, ***you***, me, ***daddy***, **is**, to, ***this***, *georgie*, ***down***, ***on***, ***i***, **going**, *go*
	adult	**be**, ***erm***, ***er***, ***on***, ***of***, ***and***, ***you***, *we*, ***was***, ***a***, ***to***, this, ***the***, ***that***, ***in***, ***i***, *have*, ***for***, ***it***, ***is***
The first twenty words with the highest BC	**child**	***it***, *the*, ***on***, a, ***down***, ***i***, *up*, ***uh***, ***you***, my, ***mommy***, ***this***, in, **is**, ***want***, baby, ***daddy***, *there*, **going**, ***that***
	adult	***and***, ***the***, ***of***, ***to***, ***a***, ***that***, ***in***, ***for***, ***it***, or, ***on***, ***you***, ***erm***, ***i***, with, ***is***, ***was***, **er**, were, **be**
The twenty most frequent words	**child**	*i, it, whats, this, that, want, uh, down, the, daddy, you, get, mommy, there, no, on, up, go, georgie, more*
	adult	*the, and, to, i, a, that, of, it, you, we, in, erm, is,* they, *was, on, have, for, it's, er*

6.7 Hubs and authorities

The hubs and authorities (H=A) in the child word co-occurrence network were plotted against the number of words that are both hubs and authorities (H&A). The cut off point for the plot occurs once the quantity of hubs and authorities reach the connectivity of seven, that is, 4.5% of the maximum connectivity i.e. 155. Unlike the adult network, a non-linear relation is observed. When compared to the adult network, the child network has a greater number of words which are both hubs and authorities in the equivalent number of hubs and authorities (after 100 words)

Arguably, this increase in the number of H&A's for the less connected words may be an artifact of earlier patterns within the child language such as pivot structures. Examples of pivot structures are given in 6.6.

Example 6.6 Pivot Grammar (Braine 1963; Gregory's speech within the first four months of producing word combinations). Pivots are bolded.

a) **P1** + X (e.g. **allgone** shoe/vitamins/egg/lettuce/watch)
b) X + **P2** (e.g. do/push/close/buzz/move **it**)
c) X + X (e.g. mommy sleep, milk cup)

Since a 'pivot' occupies a variable position, it is observed in early child speech that a pivot may occur both before and after a given set of words e.g. **allgone** juice; juice **allgone**. As such a 'pivot' word would potentially be both a hub and authority in a word co-occurrence network. Since pivot structures tend to be replaced by syntactically marked constructions such as item specific constructions around the age of two (see Tomasello 2003), the pivot's connectivity gradually diminishes in relation to the other nodes in the network. Indeed, "*it* is rapidly replaced by *a* as the main hub as soon as purely functional words emerge" around this age in a dependency network (Corominas-Murtra, Valverde and Solé 2007a: 6). Hence, the relative abundance of words being both a hub and an authority for less highly connected words. These words are in fact ousted pivots. A more detailed analysis of this issue is unfortunately beyond the scope of this study and a matter for future research.

Figure 6.11 For the child network (a), the amount of words which are both hubs and authorities (H&A) were plotted against the number of hubs chosen to be represented (the amount of hubs chosen equalled the number of authorities also represented: H=A). Figure b shows the equivalent adult network plot.

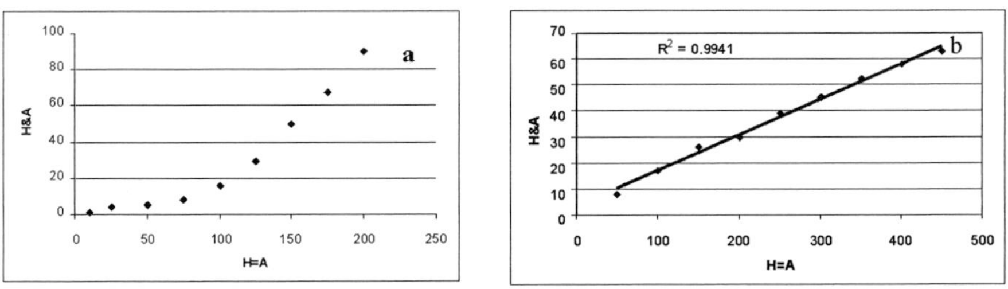

The specific words analysed as hubs and authorities in the child network up to 100, are given below in Table 6.4.

Table 6.4 The breakdown of hubs and authorities in the child co-occurrence network. As the number of husband authorities chosen to be represented in the network increases (first column), so too does the number of words given as both hubs and authorities (fourth column).

Number of H=A	Hubs (H)	Authorities (A)	H & A
10	*Whats, uh, im, want, it, that, you, daddy, georgie, i*	*I, take, man, sleeping, going, can, hug, go, doing,*	i
15	*Whats, uh, im, want, mommy, it, that, you, daddy, this, georgie, down, I, going, go*	*I, going, go, take, man, sleeping, talking, can, standing, boo, hug, doing, slide, afraid, turning*	I, going, go
30	*whats, uh, nomi, my, im, want, mommy, a, get, it, that, you, baby, me, daddy, is, to, thats, this, the, georgie, down, up, dont, jenko, on, I, going, go, can*	*I, going, can, go, holding, take, man, sleeping, talking, give, standing, sneeze, putting, boo, hang, jumping, hug, bump, ah, move, lyinging, love, doing, pick, finish, bring, slide, banging, afraid, turning*	I, going, can. go

60	*whats, uh, no, nomi, draw, there, my, im, oh, gonna, want, mommy, what, a, agra, where, piggie, get, all, did, put, he, it, that, you, baby, doggie, wanna, and, wheres, sit, me, daddy, is, to, cant, more, kitty, moon, thats, sandy, this, the, georgie, here, georgies, see, kangaroo, down, up, dont, its, jenko, on, do, I, going, can, ah, go*	I, going, can, ah, go, kiss, holding, taped, splashing, splashed, take, shaking, said, push, liking, man, kno, kicking, sleeping, falled, doed, crasheded, count, talking, cannot, pushed, give, bumped, standing, sneeze, putting, made, boo, hit, hang, jumping, close, talk, hug, finished, climbing, open, bump, stand, move, rocking, pull, lyinging, love, doing, pick, fooing, eating, finish, bring, slide, banging, afraid, color, turning,	I, going, can, ah, go

100	*whats, uh, no, nomi, lamb, at, coco, there, im, oh, boy, recorder, gonna, want, in, what, girl, horsie, another, agra, where, piggie, um, get, look, hi, all, shoes, kittys, did, truck, put, eh, he, it, that, you, baby, doggie, wanna, her, and, sun, wheres, george, me, uhs, toast, daddy, is, yeah, to, fish, cant, more, kitty, moon, thats, sandy, this, georgie, here, make, like, georgies, see, theres, kangaroo, down, up, smoke, for, dont, its, jenko, on, gammy, penguin, birdie, are, okay, nightnight, yes, toy, i, eat, draw, my, man, mommy, drink, a, going, can, sit, the, ah, go, bring, do*	i, eat, draw, my, man, mommy, drink, a, going, can, sit, the, ah, go, bring, do, hurt, reading, kiss, holding, broke, taped, found, splashing, splashed, feeling, take, shaking, said, cry, push, liking, kno, kicking, sleeping, falled, doed, crasheded, count, talking, cannot, feel, pushed, had, give, bumped, standing, coming, sneeze, snapped, sing, played, sip, putting, change, made, boo, loving, fixing, hit, hang, jumping, close, talk, eated, rest, drawed, hug, finished, climbing, cleaned, open, bump, stand, washing, playing, ate, have, move, rocking, pull, drawing, lyinging, love, singing, doing, pick, fooing, read, finish, eating, finded, slide, lying, banging, coffee, afraid, color, turning, play,	i, eat, draw, my, man, mommy, drink, a, going, can, sit, the, ah, go, bring, do

Table 6.5 below shows a child and adult comparison of hubs and authorities with the ten most frequently occurring words. When comparing the ten largest hubs in the child and adult network, only three words occur in both lists: *that, you* and *I* whereas there is no overlap between the child and adult network for the ten largest authorities. However, words *i, it, that, the* are found in the ten most frequent words for both the child and adult network. Thus, the labels of the hubs and authorities differ quite considerably between the child and adult network with only a partial overlap observed.

Table 6.5 For both the adult and child co-occurrence network, the ten largest hubs and authorities are compared with the ten most frequent words. Bolded text indicates words are found in the ten most frequent words.

The ten largest hubs	**child**	**whats, uh**, im, **want, it, that**, you, **daddy**, Georgie, **i**
	adult	erm, of, and, you, a, to, the, that, in, i
The ten largest authorities	**child**	**I**, take, man, sleeping, going, can, hug, go, doing,
	adult	friends, bin, point, evening, don't, year, yeah, treasurer, it, is
The ten most frequent words	**child**	**i, it, whats**, this, **that, want, uh**, down, the, **daddy**
	adult	the, and, to, i, a, that, of, you, it, we

6.8 Summary

In line with the adult counterpart, the child word co-occurrence network is a hubterranean small world, evidenced not only by the relatively low shortest path and high clustering co-efficient, but also by the (multi-) power law of link length i.e. the storage cost. The power law has four regimes with a truncated tail, in agreement with the adult network counterpart. The truncation is indicative of the network moving away from an optimal point due to infrequent word combinations producing bigrams that have a low cohesion and thus, are costly to store.

Since the power law represents the distribution pattern of cohesion i.e. syntactic constituents within the child's utterances, at a very basic level, the child's syntax has the same distributional properties as those in adult speech. Furthermore, the child network also displays the three power laws: the storage cost power law, the degree power law and the frequency power law which relate to syntagmaticity, paradigmaticity and entrenchment respectively. Thus, for certain aspects of syntax, the network approach in combination with a usage-based account emphasizes the continuous nature of certain primitives in syntax.

In line with a network formed by a cost constraint, the network does not display hierarchical structure. However, it does display disassortative behaviour, in that hubs *tend* not to connect with other hubs. And in line with this type of mixing, there is a correlation between the high values of degree and betweeness centrality, albeit weaker than the correlation observed in the adult network. Hubs and authorities were identified and a non-linear relation was attested between the amount of words being both a hub and an authority. Thus, in the child network more highly connected words form both a hub and an authority compared to the adult network for 100 hubs/authorities and over.

Furthermore, significant hubs were also identified by their high BC, connectivity and frequency values. It will be shown that these hub words will form anchors in the network to which link patterns reiterate; these patterns representing the distributional regularities in the child's speech. The next section will introduce the notion of significant link patterns i.e. motifs and evidence that the most significant motifs found in the adult network will also be the most significant motifs in the child network.

CHAPTER SEVEN

MOTIFS: SIGNIFICANT LINK PATTERNS

7.1 Overview

In chapter five and six, it was evidenced that by optimizing the storage cost and connectivity in a word co-occurrence network, a hubterranean structure is produced with its distinctive multiple link lengths i.e. inverse transitional probabilities. As discussed, this cumulative property is useful in understanding the origins of basic syntactic constituency. From this cumulative property, the study now explores more local properties of the word co-occurrence network and how they relate to syntactic structure (see chapter eight).

Linguists may establish syntactic elements by (functional) distributional analysis as well as identify their distinctive role within the construction. For example, function words aid the identification of (item-specific) constructions. Furthermore, linguists use a particular function word's fairly unique syntactic distribution to identify abstract categories (see section 2.8 and 9.3). Thus, it would be relevant to identify local word distributional properties in the network. Within network theory, local properties are not couched in categories but rather link patterns i.e. subgraphs. Subgraphs consist of any number of nodes and are defined by their unique connective pattern within a given set of nodes.

An example of a 3-node subgraph (a subgraph consisting of three nodes) is given in Figure 7.1a, within a small network (Figure 7.1b) composed of two instances of the subgraph. Nodes *a, b* and *c* form the same subgraph as *X, Y* and *Z*, likewise with nodes *f, d* and *e*. The nodes within a given subgraph can be fully connected or only sparsely. For example, the subgraph in Figure 7.1a, is not fully connected as node *Y* and *Z* are not linked.

Figure 7.1 (a) An example of a subgraph 6. (b) A network in which two examples of the subgraph 6 represented by blue links.

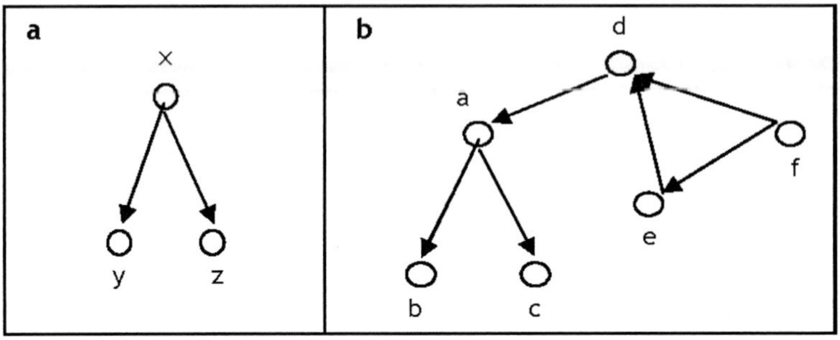

Commonly, there are a few small but significant subgraphs, known as motifs that occur in a given network more often than by chance (Milo, Shen-Orr, Itzkovitz, Kashtan, Chklovskii and Alon 2002). It has been argued that these motifs are pertinent for distinguishing functional designs in systems (Milo, Itzkovitz, Kashtan and Alon 2004). However, this functional (adaptive) approach has been questioned (see Solé and Valverde 2006 for discussion). Motifs may in fact be "evolutionary spandrels" of the network's formation, in that "(i) they are the by-product of building rules; (ii) they have intrinsic, well-defined, non-random features; and (iii) their structure reveals some of the underlying rules of construction." (Solé and Valverde 2006: 421).

On this view, motif formation is resultant of the constraints on the network. In our case, the constraints are (storage) cost and connectivity. Whether motif distributions are functionally driven or a "by-product" of the formation of hubs, it is expected on both views that the child and adult network will display similar motif distributions and indeed this is the case for the triads and for the highly significant 4-node motifs.

It will also be shown (in section 8.2.6.1) that different constructions are potentially represented by the same link pattern within a given network and as such, each motif does not denote a unique function. Indeed, different motifs may capture facets of the same construction (see section 8.3.1) Furthermore, it will be shown in chapter eight that a motif in the adult network may represent a very different (potential) construction than the same motif observed in the child network. Thus, the significant link patterns i.e. motifs pertain to linguistic form and not function. This observation does not evidence a "functional adaptive interpretation" of motifs but rather points to certain motifs being "evolutionary spandrels" (Solé and Valverde 2006: 420-1; see section 8.2.3).

This chapter will briefly cover the technical considerations of analysing motifs in the word co- occurrence network. A useful relation between nodes i.e. automorphicity within a given subgraph is defined in section 7.2 and the extension of a subgraph is discussed in section 7.3. It is argued that subgraphs pick up linguistic form patterns generated by surface generalizations (section 7.4). How to detect significant subgraphs i.e. motifs in a network is the subject of section 7.5. Results of the significant 3-node and 4-node motifs in the adult and child network will then be presented in sections 7.6 and 7.7 respectively. In chapter eight, the results of this chapter will be interpreted using the linguistic perspectives discussed in chapter two.

7.2 Automorphic subgraphs

Certain subgraphs have nodes that share the same relation with respect to *all* the nodes within the subgraph. Such a relation is known as *automorphic equivalence* (Kashtan, Itzkovitz, Milo and Alon 2004). The more formal definition is given below (Kashtan, Itzkovitz, Milo, and Alon 2004:7).

> "Let $S=(V_S, E_S)$ be a subgraph. An automorphism is a one-to-one mapping, τ, from V_S to $[E_S]$, such that $(v_i, v_j) \in E_S$ if and only if $(\tau(v_i), \tau(v_j)) \in E_S$. Two nodes v_i and v_j are automorphically equivalent if and only if there is some automorphism τ that maps one of the nodes to the other [$\tau(v_i)=v_j$]. For each subgraph *S*, we classify its *n* nodes into roles by examining structural equivalence of all possible pairs of the nodes. By the transitivity of automorphic equivalence, one is guaranteed to partition the nodes into distinct roles."

In Figure 7.2, the Y-nodes are automorphically equivalent. This equivalence is easy to spot if the relations between the different nodes are written out (Example 7.1a-c, on the next page), paying attention not only to the number of input and output links the two Y nodes have, but also the direction of the links. Note that node X does not share any similar relations with node Y1 and node Y2: node X's links do not point to the same nodes that Y1 and Y2 nodes do.

Figure 7.2 An illustration of subgraph 6, repeated for convenience.

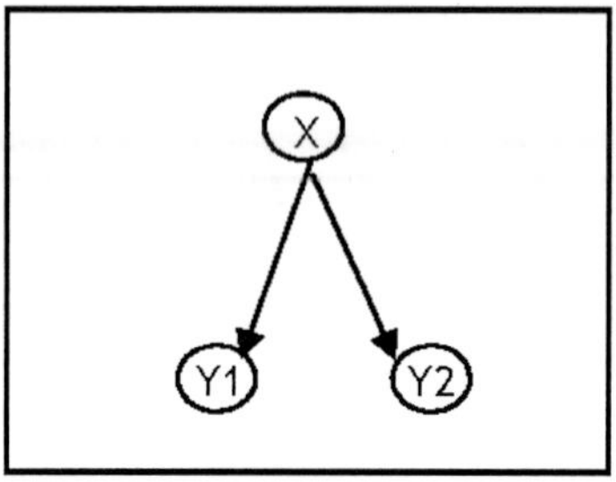

Y1 and Y2 do share the same automorphic equivalence since their output and input links point to the same nodes. Therefore, Y1's relation with node X and node Y2 is analogous with Y2's relation with node X and node Y1. In short, Y1 and Y2 are automorphic.

Example 7.1 Link relations between nodes in subgraph 6 in Figure 7.2

a) Node X: output link to Y1 and output link to Y2
b) Node Y1: input link from X and no link to Y2
c) Node Y2: input link from X and no link to Y1

7.3 Subgraph extensions

Subgraphs of the same pattern do not just occur in lonely instances as shown in Figure 7.1a and Figure 7.2. Clusters of subgraphs may be found throughout a network; the dense connectivity of the hubterranean structure being a favourable environment for such clustering (Vázquez, Dobrin, Sergi, Eckmann, Oltvai and Barabási 2004). Within this structure, subgraphs with the same link pattern tend to group together and reiterate around certain nodes known as anchors forming generalisations or extensions.[1] The anchor thus is the node to which the extension is fixed. In Figure 7.3a, subgraph 6 is depicted with one of its potential extensions (b). The node Y3 is said to extend the motif as all nodes labelled Y share the same relation within the subgraph relative to the anchor node X, the node that fixes the three separate subgraphs.[2]

[1] The term motif extension is preferred since a linguistic generalisation may then be unambiguously differentiated from a motif generalisation.

[2] The three separate subgraphs forming the subgraph generalisation are XY1Y2, XY1Y3 and XY2Y3.

Figure 7.3 (a) An illustration of subgraph 6 and (b) its extension in which the anchor X fixes the automorphic nodes Y1, Y2, and Y3.

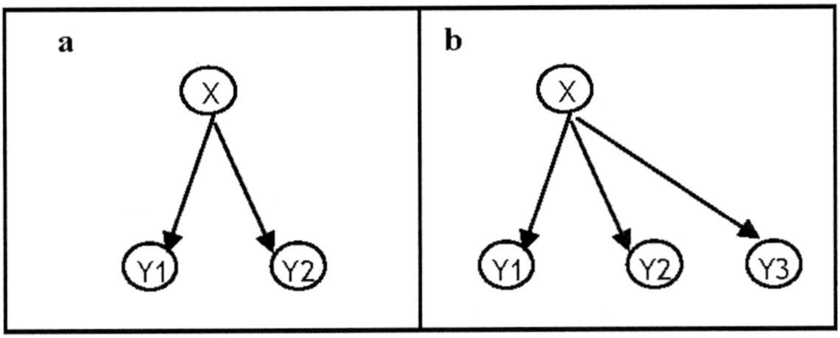

Automorphic extensions i.e. extending automorphic relations as in Figure 7.3, are of interest in respect to early linguistic generalisations since "A PATTERN IS CONSIDERED EXTENDABLE BY LEARNERS ONLY IF THEY HAVE WITNESSED THE PATTERN BEING EXTENDED" (Goldberg 2006: 99, caps are authors own). Automorphism implies that a relation has already been extended. For example, the XÆ Y1 relation in Figure 7.3 is duplicated (see Solé and Valverde 2006: 420, Figure 1) by the addition of XÆ Y2. Thus, Y3 extends these automorphic relations.

7.4 Subgraphs in the word co-occurrence network

Subgraphs have been discussed in a general network sense, so now its time to contextualise the discussion in linguistic terms. In the adult and child networks under study, subgraphs are patterns of word co-occurrences in the network. Recall that link direction denotes the linear order of words in that the preceding word points to the following word e.g. see Figure 7.4 for the sentence marked in red, *I will try to explain* and its corresponding links. Note that orthographic homonyms are considered one single node. Differences in the homonyms' meaning (although not all) are attributable to the construction and context in which the word occurs.

Figure 7.4 A word co-occurrence network. (A) The text is converted into a word co-occurrence network (C) whose links represent precedence relations. The lighter the node's colour, the higher its degree i.e. connectivity (Solé, Corominas Murta, Valverde and Steels 2005: 2, Figure 2A&C).

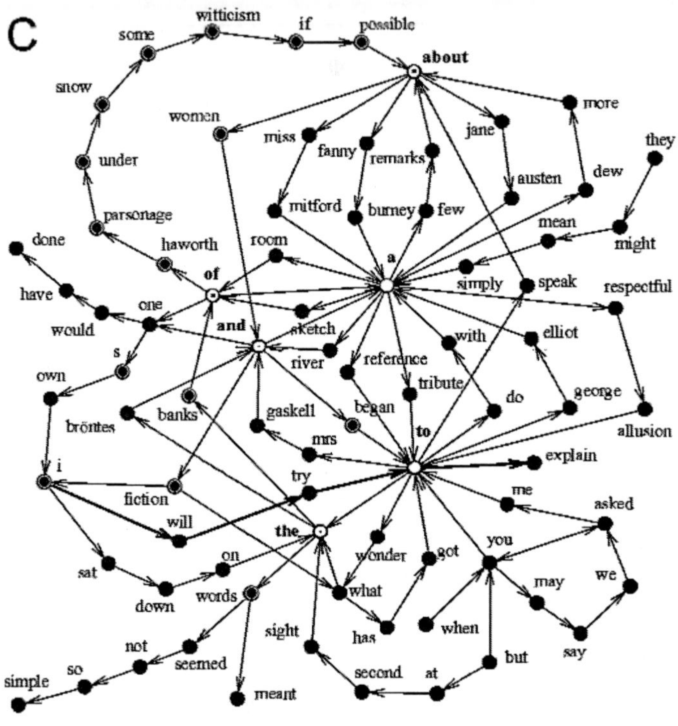

A *"But, you may say, we asked you to speak about women and fiction -what has got to do with a room of one's own? **I will try to explain.** When you asked me to speak about women and fiction I sat down on the banks of a river and began to wonder what the words meant. They might mean simply a few remarks about Fanny Burney; a dew more about Jane Austen; a Tribute to the Bröntes and a Sketch of Haworth Parsonage under snow; some witticism if possible about Miss Mitford; a respectful allusion to George Elliot; a reference to Mrs Gaskell and one would have done. But at second sight the words seemed not so simple."*

- Virginia Wolf, *A Room of One's Own*

It is important to understand that even though the network is based on utterances *produced*, the word co-occurrence network is not a model of language *production* (or comprehension). Rather, the word co-occurrence network may be seen as a mnemonic device: a user friendly history of language usage or an optimised storage configuration for linguistic experience. This representation is useful when the approach adopted is one that emphasises that current language usage is based on and extends previous language experience.

In Figure 7.4, the word co-occurrence network has many subgraph 6 patterns. The word *I* points to *sat* and *will*. That is, the word *I* occurs before the English verbal words *sat* and *will*. The word *a* points to the English nominal words *river* and *reference*. Notice how the traditional verbal and nominal categories are implicated in the subgraph 6 pattern. It will be argued in chapter eight that this type of link pattern will be useful in distinguishing paradigmatic relations i.e. syntactic categories. Indeed, such patterns are used intuitively by linguists using distributional analysis (particularly in SVO languages).

However, there are many types of link patterns representing word distributions. For example, in Figure 7.4, the words “will try to” (in red) and the words “to wonder what” share the same link pattern. The task of the linguist is then to identify and categorise these patterns of form in ways that reflect the functional aspects of the language. That is, to identify constructions and their relations between each other. One step further is to understand how the patterns of form and function are generated and how they become mapped together. We will then understand how novel words are incorporated with ease into speech e.g. a wug, two wugs (Berko 1958) and how unusual sentences never heard before are felicitously produced and comprehended e.g. “She sneezed the foam of the cappuccino” (Goldberg 2006: 100).

The network approach adopted in this study is essentially a (non-functional) distributional analysis in that we are interested in the patterns of form and their origins. In this study, we are interested in surface generalisations of linguistic form. A surface generalisation[3] captures the extension of a construction i.e. a primitive syntactic element (Croft 2001); the extension being motivated by the *intention* to utilise the construction’s communicative roles and/or structure.

---

[3] See Goldberg (2006) for discussion.

Such a process is purposeful as it is a conscious choice by the speaker to take advantage of previously known syntactic structures that canonically express the required communicative function. In other words, interesting surface generalisations do not happen by chance. Thus, as Kilgarriff (2005: 263) points out, "Language is never, ever, ever, random".[4] Indeed, a construction's definition highlights the inextricable functional component of linguistic form: constructions are "LEARNED PAIRINGS OF FORM WITH SEMANTIC OR DISCOURSE FUNCTION" and can be "morphemes, words, idioms, partially lexically filled and fully general phrasal patterns" (Goldberg, 2006: 5).

In the word co-occurrence network, words and their adjacent combinations are represented by means of linked nodes in the network. It is thus likely that basic syntax will emerge from these link patterns in the same way that (functional) distributional analysis identifies constructions through word combination comparisons. Whilst the network is bereft of the functional component, an analysis of the word combinations is a good first pass at identifying potential syntactic elements and their syntactic relations. To capture linguistic patterns from a word co- occurrence network that does not allow for punctuation or does not directly encode meaning, a particular method is required to ensure that non-linguistic patterns are eliminated.

Figure 7.5 An artificial word co-occurrence network drawn by Pajek (de Nooy, Mrvar and Batagelj 2005) from sentences in Example 7.2 (a&b).

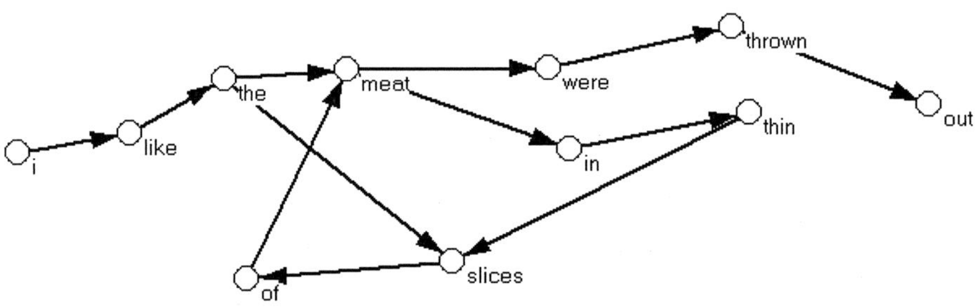

In an artificial word co-occurrence network (Figure 7.5) formed from two utterances (Example 7.2 a&b), non-linguistic patterns that could potentially arise in a network are given in example 7.2 (c&d). The link patterns in Example 7.2 (c&d) stem from combining the link patterns in a and b, resulting in syntactic forms that are not attested (c&d). Suffice to say that

[4] Thanks to Michael Harrington for discussion on this point.

these patterns in c and d would not be used in distributional analysis.

Example 7.2 Grammatical (a, b) and random sentences (c, d)

a) i -> like -> the -> **meat** -> in -> thin -> **slices**
b) the -> **slices** -> of -> **meat** -> were -> thrown -> out
c) * meat -> in -> thin -> **slices** -> of -> meat -> were
d) *i -> like -> the -> **meat** -> were -> thrown -> out

A way of filtering out such random linkage patterns is to compare the link patterns of the word co-occurrence with a randomly generated network. Any patterns that occur above chance will be linguistically interesting and useful for distributional analysis. The method naturally picks up surface generalisations; the significance of surface generalisations being judged in a non- subjective way. That is, no preconceived notions of general categories or constructions are applied to the pattern analysis.

It is pertinent to briefly mention null forms that are often used in paraphrasal accounts (both derivational and representational) to account for mismatches in surface form and meaning. This network model establishes the patterns of any generalisations that have an effect on word distribution - given that the method to establish generalisations is to examine those word distributions. Furthermore, due to an accumulation of language experience recorded by the network, paradigmatic as well as syntagmatic patterns will potentially be discovered.

7.5 Detecting motifs

To establish whether a subgraph is significant that is whether it is a motif of a network, the random null hypostudy is employed (see Milo, Shen-Orr, Itzkovitz, Kashtan, Chklovskii and Alon 2002; Milo, Itzkovitz, Kashtan, Levitt and Alon 2004). This method is adopted in this study. The word co-occurrence networks under study are compared to other networks of the same size that are randomly generated.[5]

[5] It is also important to note that chance i.e. random processes may still create patterns that look as though there is some underlying organisation. With this in mind it is important to compare multiple random controls so that such chance patterns are averaged out. This strategy is employed in the network methodology (see section 6.3.3).

7.5.1 Generating the random network

There are various ways in which a random network is generated. In Milo et al 2002 study, the number of links, nodes and degree (incoming and outgoing links) in the real networks were held constant. In addition, the n-1 node structure was preserved. For example, when evaluating 4-node motifs, the 3-node structure of the real network is replicated in the random networks.

In principle, any number of structure preservations may be enforced on the random networks to achieve specific analyses. For example, Arty-Randrup, *et al* 2004 argue that in a C. elegans neuronal structure, neurons tend to connect to neighbouring neurons rather than to neurons further away (White, Southgate, Thomson and Brenner 1986). Therefore, on Arty-Randrup et al's view, for a null hypostudy to be complete, the randomised networks must also favour local connections. Only then would a common "baseline" between the real and randomised networks be achieved. Not factoring in such a constraint in the random process may cause a "completely random but spatially clustered network" to be misclassified as a "non-random [network]…that has significant network motifs" (Artzy-Randrup *et al* 2004). As such, the distance constraint would be overlooked as the principle in the design, even though the motifs are a by-product of this principle i.e. the local density of connections being a favourable environment for motif formation.

Whilst one is mindful of such matters, Milo, Itzkovitz, Kashtan, Levitt and Alon (2004) argue that when comparing the C. elegans neuronal network to a random network generated by favouring local neuronal connections, although showing similar distributions of SOME motifs (ID 8 and 11) to the real network, the random network does not display similar distributions for *all* motifs that are found in the real c. elegans network. In short, the real and created network have different 'significance profiles' for its motifs (see section 7.6 and 7.7 for significance profiles of the networks under study).

Consequently, Milo et al (2004) suggest that it is the evolution of both the geometric locality constraints *and* connectivity that yields the required "highly designed circuitry" and that, in turn, is represented by motifs. Although a favourable environment for motifs i.e. the hub structure, is formed through the optimisation of a geometric factor (contextualised as storage cost) and node connectivity in the word co-occurrence networks, this study does not make the assumption that the actual motifs themselves are resultant of these two optimised pressures. As such, the geometric and connectivity constraint will not be factored into the formation of random networks.

As Milo et al (2004) propose, "the present approach based on degree-preserving randomized networks is a simple first step for comparing networks and for discovering potentially interesting overrepresented and underrepresented patterns for further analysis", overrepresented meaning significant, whilst underrepresented, insignificant. This study also takes the simplest analysis of motifs formed by word co-occurrence networks and so the simplest null-hypostudy is adopted with no *a priori* assumptions concerning constraints that may be implicated in the generation of such motifs.

The randomised networks in this study has the equivalent number of nodes and arcs (directed links), and the same in/out-degree for each node as the corresponding real network, for both the adult and child network. The triad 3-node structure (N-1 level) was not preserved for 4-node motif evaluation, in the spirit of Sporns and Koetter's (2004) question: "The choice...reflects the specific question we ask about motifs... independent of the level [N], how do the motif number, diversity and composition of real brain networks compare to other network typologies".

Note that the level of a motif in language equates to the amount of words represented in a given pattern e.g. level N-1 (a 3-node motif) represents three words. It is highly unusual to utilise the number of words as a syntactic metric for establishing syntactic elements. A more useful approach would be to use the function of the words. Recall that this study aims to evaluate how network methodology can shed light on the distribution of syntactic form. Given this aim, of interest is the type of motifs found, their composition, and their variety. Therefore, the approach uses the most basic and simple model i.e. there is no N-1 level preservation for random networks.

In the word co-occurrence network, it may be possible to represent item specific constructions as an extended link pattern. A motif may consist of item specific words which act as anchors to the extension. However, without functional considerations, motif analysis will only highlight potential item specific constructions and may not be solely relied upon to ascertain such constructions. An example of an item based construction that may be potentially represented by a motif extension is given below (Figure 7.6).

The subgraph anchors are the items that are specified in the construction (*a* and *the*) and identify the construction. For example, in Figure 7.6, the subgraph 204 (bifan) typifies the distribution of the traditional 'Noun' category i.e. *a/the* + *Noun*. Note that the link pattern and the anchors facilitate the identification of the construction. Also, it is interesting that there are two pairs of automorphic nodes. The anchors {*a, the*} form an automorphic set, as well as the nodes extending the motif {*berlei, kettle, pig, ration, saucepan, sort*}.

Figure 7.6 Pictured in the inset is the bi-fan (204) subgraph. An extension of subgraph 204 is shown, with anchors *a* and *the* fixing the instances of the subgraph. The input nodes (nodes that the arrows point to) are the automorphic nodes: {*berlei, kettle, pig, ration, saucepan, sort*}.

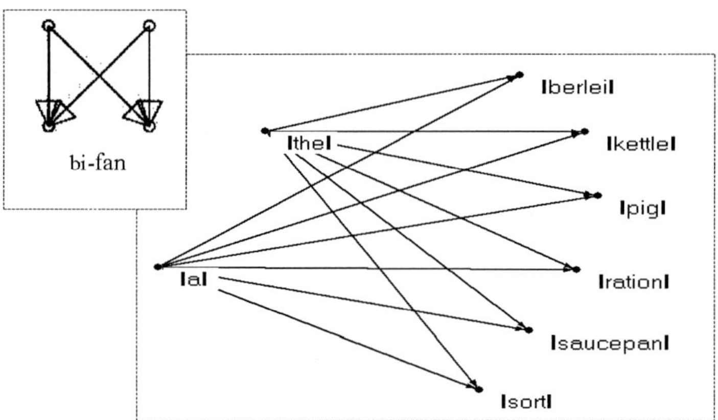

7.5.2 Computational considerations

Mfinder (2002-2005a), a program designed to extract network motifs, was used to identify both 3-node and 4-node motifs. The original adult network used to calculate global properties is much larger than the child network - increasing the computational time of the motif evaluation to an unfeasible level. Consequently, another smaller network was created from one of the public lectures in the BNC ("D8Y - Museum Society meeting. Make-do and mend"). The whole sample contained 6748 words of speech and was recorded in a public context. Although a public lecture, the speech is dialogical, incorporating rhetorical questions, actual questions, answers and discourse markers such as *well*.

Only part of this data was used: specifically 4984 words/items consisting of 1415 types (with a type/token ratio of 28.39%) creating a network with 3542 links (edges). From this smaller BNC network, Mfinder (2002-2005a) computed 4-node motifs in 22.8 hours, whereas the run time for the child network[6] was much shorter i.e. 3.63 hours. The adult network was not decreased further to match the number of types i.e. nodes in the child network (1662 to 1016) as this move would make the adult text studied unacceptably small. Note that the texts sizes do not bear on the process of identifying significant subgraphs. If you recall, the studied network is compared to a randomly generated network of the same size. Consequently, the adult data set size needs to be large enough to provide interesting distributional regularities.

7.5.3 Identifying 3-node motifs

The triad patterns of 100 random networks were compared with the triad patterns in the adult network and the child network. Over-represented patterns in the adult or child network would point to significant 3-node (triad) subgraphs. Mfinder (2002-5a) establishes the top significant subgraph patterns by listing each pattern in respect to a score dependent on the Z-score, and a condition known as "uniqueness", explained below. N.B. only patterns which occurred more than once were evaluated. This decision reflects the parameter setting: mfactor > 1.1.

7.5.4 The definition of Z-Score

The Z-score establishes how significant a subgraph is when compared to other networks of similar structure that are generated randomly. In this way, the subgraphs that occur by chance are normalised, allowing the subgraphs which occur above chance to be identified and their significance judged in respect to the number of subgraphs present. As noted previously, significant subgraphs indicate underlying regularity or organisation.

---

[6] If you recall the child network has 1016 nodes and 3030 links (edges).

Equation 7.1 The Z-score Z where "*N*real$_i$ is the number of times the subgraph appears in the network and (*N*rand$_i$) and std (*N*rand$_i$) are the mean and standard deviation of its appearances in the randomized network ensemble" (Milo, Itzkovitz, Kashtan, Levitt, Shen-Orr, Ayzenshtat, Sheffer and Alon 2004: 1513).

$$Z_i = (N\text{real}_i - \langle N\text{rand}_i \rangle)/\text{std}(N\text{rand}_i)$$

The Z-score of a subgraph pattern is computed by subtracting the sum total of the subgraph pattern in the studied network by the average sum total of the subgraphs in the random networks. Then dividing this figure by the standard deviation from the sum total of the average random network. The higher the Z-score the more significant the subgraph; subgraphs are significant over a score of 2. It has been noted that Z-scores tend to rise as the network size and motif pattern increase in size (Kashtan, Itzkovitz, Milo and Alon 2004:1752; Milo et al 2004; Milo et al 2002).

7.5.5 The definition of uniqueness

The uniqueness criterion "is the number of times a subgraph appears in the network with completely disjoint groups of nodes. The value in the output is not the maximum number of disjoint groups but is a lower bound for the real number" (Mfinder 2002-2005b). The uniqueness bounding value of a motif is taken to be 4 (the default setting on Mfinder). As a practical example to illustrate this uniqueness criterion, the below Figure 7.7 (over page) shows a small network consisting of all the instances of subgraph 110 only. This subgraph has a uniqueness value of three meaning that there are at least three disjoint (non-overlapping) groups of nodes.[7] Three disjoint node groups[8] have been identified and assigned a different colour.

In linguistic terms, the group of disjoint nodes are separate sets of words with the same link pattern. For example, in section 7.4, the subgraph 6 was shown to occur with the word sets {*I, sat, will*} and {*a, river, reference*}. Uniqueness, then, tends to differentiate between potential item specific constructions that are composed of very different components i.e. word types. The higher the uniqueness value, the higher the occurrence of separate sets of the subgraph. As such, more diverse linguistic groupings will be differentiated. Consequently, a significant subgraph has both a high

[7] Note that the subgraph is not significant in any of the two networks under study
[8] This configuration is one of a few possible configurations. However, each possible configuration has in common at least three disjoint groups.

Z-score and a high uniqueness value.[9]

Figure 7.7 (a) This small (undirected) network represents the minimum disjoint instances of subgraph 110 in the adult network (the arrows have been omitted to simplify the scheme, aiding the identification of disjoint groups). (b) The grouping below, represented by the outlined area, is one possible representation of three disjoint node groups; there are other alternatives but each one still produces at least three disjoint groups of nodes.

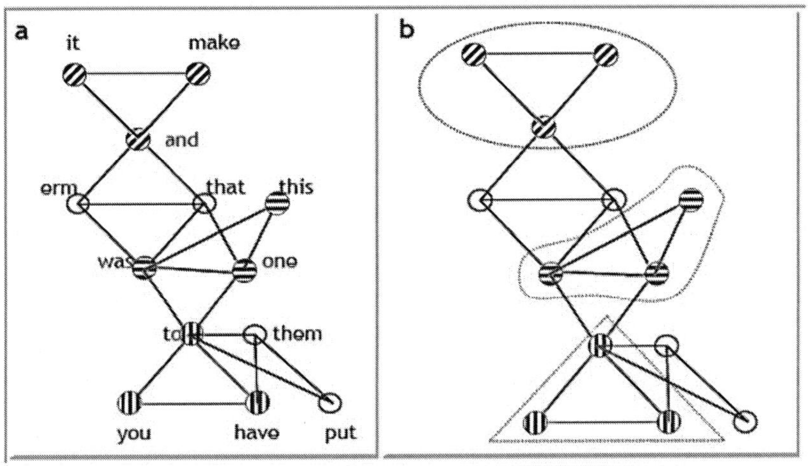

7.6 The Z-scores results for 3-node subgraphs

The Z-scores for the child and adult network are given below in Table 7.1. If you recall, Z-scores over the value of 2 are significant subgraphs i.e. motifs in the network. The motifs in the adult network are triads 6, 12, 14, 36, 74 and 78. The child network shares four of the adult network motifs, namely triads 6, 12, 36, 74. The other two adult network motifs, triads 14 and 78, are not motifs in the child network.

Recall that word co-occurrence networks at a global network level display a tendency for complete triads i.e. they exhibit a high clustering co-

[9] Although the uniqueness criterion captures diversity in the motif instances, the criterion is not necessary requirement for the distributional analysis of a text. The identification of a productive pattern in linguistic form does not rely on disjoint groupings. The uniqueness criterion is retained in this analysis to allow comparison of the language data with other networks.

efficient (see section 5.4). It is interesting that the complete triads are insignificant individually and that a feedback/forward function is not favoured. Since the patterns in the networks are created by extending constructions, it is likely that the explanation for the insignificant of complete triads lies in the nature of the linguistic generalisations. As will be shown in the next chapter, motif 6 (and perhaps motif 36) is implicated in generating syntactic categories, in that the automorphic nodes reduce to a category. As such a paradigmatic relation is formed and an abstract level of syntactic representation is established. See section 9.3 for further discussion.

Table 7.1 Z-scores of the triad subgraphs for the adult and child network. Z-scores over 2 identifying network motifs are italicised. The Z-score of subgraph 46 for the child network, which is borderline significant, is bolded.

Subgraph ID	Adult Z-score	Child Z-score	Subgraph ID	Adult Z-score	Child Z-score
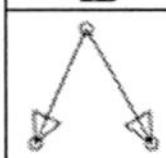 6	*6.09*	*5.76*	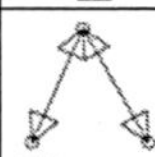 78	*3.86*	0.33
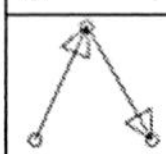 12	*5.76*	*4.78*	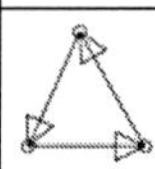 98	-3.75	-1.23
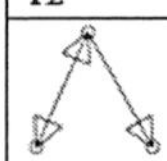 14	*4.15*	-0.69	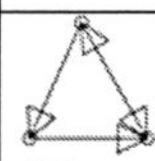 102	-5.02	-1.58
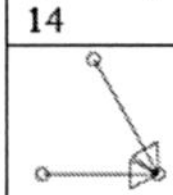 36	*5.95*	*5.23*	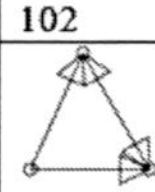 108	-3.57	-2.08
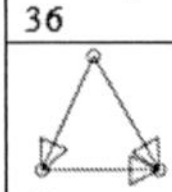 38	-5.77	-5.72	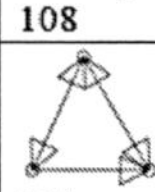 110	-3.59	-0.28
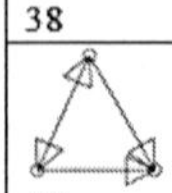 46	-1.45	**1.86**	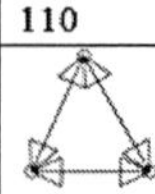 238	-1.12	-0.2
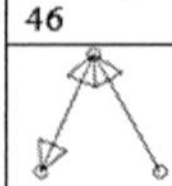 74	*5.46*	*2.47*			

7.6.1 Significance profiles for 3-node subgraphs

This section compares the motifs of the adult and child network. As previously noted in section 7.5.1, the significance profile of a network is used when comparing Z-scores. When comparing Z- scores of different networks, the Z-scores are normalised since "the normalization emphasizes the relative significance of subgraphs, rather than the absolute significance. This is important for the comparison of networks of different sizes, because motifs (subgraphs that occur much more often than expected at random) in large networks tend to display higher *Z* scores than motifs in small networks." (Milo et al 2004: 1539; Milo et al 2002). Z-scores are normalised using the below Equation 7.2, where SP_i stands for the significance profile value plotted on the graph and Z_i, the Z-score of a given subgraph.

Equation 7.2 The significance profile value (SP_i) is calculated by dividing each of the Z-score Z_i numbers for a given subgraph by the square rooted sum of the squared Z-scores for all subgraphs in the profile (Milo, Itzkovitz, Kashtan, Levitt, Shen-Orr, Ayzenshtat, Sheffer and Alon 2004: 1513).

$$SP_i = Z_i / (\Sigma Z_i^2)^{1/2}$$

In a recent study, the types of 3-node subgraphs observed in English, French, Spanish and Japanese texts were compared (Milo, Itzkovitz, Kashtan, Levitt, Shen-Orr, Ayzenshtat, Sheffer and Alon 2004).[10] The figure 7.8 below shows the normalised Z-scores of the 13 possible subgraphs for 3-node link patterns in these languages (note the non-standard labelling of the subgraphs from 1-13).

All the languages share similar significance profiles for the 3-node subgraphs. That is, the same motifs i.e. significant subgraphs, are observed in all the four languages. All of these motifs are incomplete triads – they are not fully connected whereas the fully connected triads are not found in significant numbers i.e. there is an under-representation (Milo, Itzkovitz, Kashtan, Levitt, Shen-Orr, Ayzenshtat, Sheffer and Alon 2004).

---

[10] The bipartite network was a non language network, formed from two groups with node size 10 and 1000 respectively – see Milo *et al* (2004) for further details.

Figure 7.8 Subgraphs within separate networks composed of English, French, Spanish and Japanese sentences as well as an artificial bipartite network. The X-axis numbers 1-13 correspond, respectively, to the following binary notated motifs: 6, 36, 12, 74, 14, 78, 38, 98, 108, 46, 102, 110, 238 (Milo, Itzkovitz, Kashtan, Levitt, Shen-Orr, Ayzenshtat, Sheffer and Alon 2004: 1539).

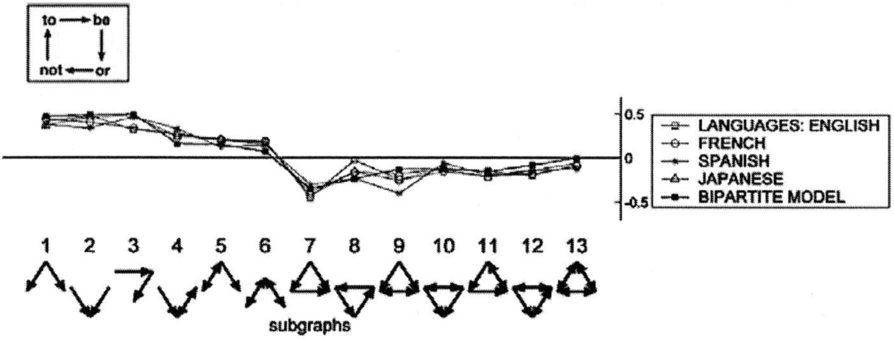

The significance profiles for the adult and child network is given in Figure 7.9. The pink squares represent the 3-node subgraph scores for the adult network and the blue diamonds for the child network. The adult network, under study in this study, has a similar significance profile (SP) to the texts analysed in Milo et al's (2004) study,[11] confirming that reducing the size of the adult network for the subgraph analysis is acceptable in that it does not alter the subgraph results detrimentally.

In addition, the child network SP is similar to the adult network SP, with the exception of subgraphs 14 and 46 scores (and perhaps subgraph 78). That is, all incomplete subgraphs are significant whereas complete subgraphs are not. All incomplete child subgraphs apart from subgraph 14 (number 5 in Figures 7.8 and 7.9) are significant. Only one complete child subgraph is significant – subgraph 46 (number 10 in Figure 7.8 and 7.9).

[11] The number of nodes (N) and edges i.e. connections (E) in the individual text networks for the Milo *et al* (2004) study are as follows: English (N=7724, E= 46281); French (N= 9424, E= 24295); Spanish (N= 12642, E=45129); Japanese (N= 3177, E= 8300).

Figure 7.9 3-node Z-scores for the adult and child network. The X-axis numbers 1-13 correspond, respectively, to the following binary notated motifs: 6, 36, 12, 74, 14, 78, 38, 98, 108, 46, 102, 110, 238. Pictures are from the motif dictionary (Kashtan, Itzkovitz, Milo and Alon 2005c).

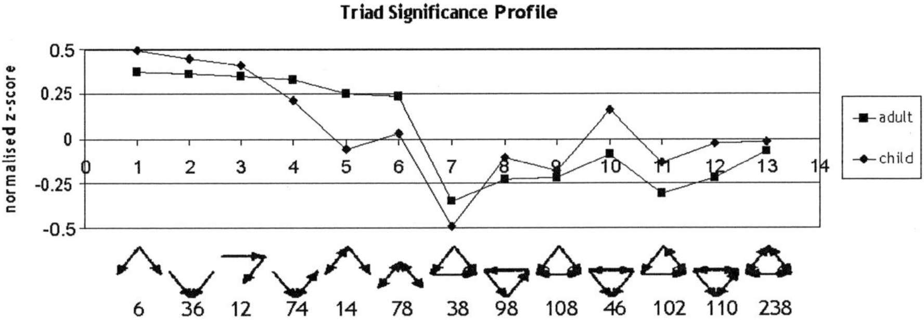

Notice that two of the most highly significant motifs are the link patterns that, when extended, denote a hub structure: motifs 6 and 36 (see chapter eight and nine for discussion). This finding is in keeping with the idea of the hub structure being important in capturing syntactic structure within a word co-occurrence network. Motif 12 is also important as it represents linear word order. In addition, motif 74 represents "recursive" structures such as the *and*-construction and the *of*-construction. These motifs as well as other significant subgraphs will be briefly discussed in chapter eight.

7.6.2 Uniqueness values for 3-node subgraphs

The uniqueness values of the triads for the adult and child network are given in Table 7.2. All subgraphs with a Z-score of two and over (shaded) also have a uniqueness value over 4 and so are highly significant.

Table 7.2 Uniqueness values of the triad subgraphs for the adult and child network. Z-scores over 2 identifying network motifs are italicised - all motifs have a uniqueness value of four and over. The bolded numeral of subgraph 46 indicates a borderline 1.86 Z-score value in the child network. Pictures are from the motif dictionary (Kashtan, Itzkovitz, Milo and Alon 2005c).

Subgraph ID	Adult uniqueness	Child uniqueness	Subgraph ID	Adult uniqueness	Child uniqueness
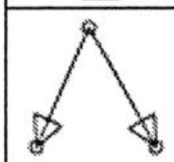 6	*126*	*95*	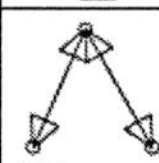 78	*10*	9
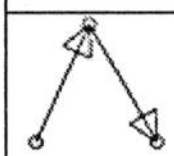 12	*171*	*89*	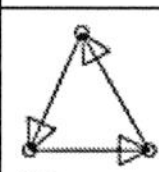 98	21	17
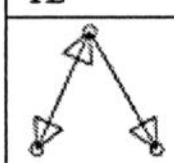 14	*17*	24	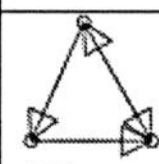 102	10	6
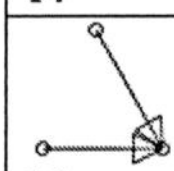 36	*130*	*77*	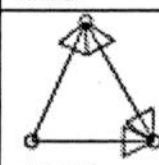 108	7	10
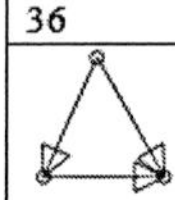 38	27	30	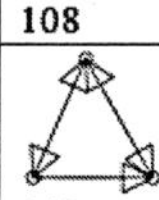 110	3	5
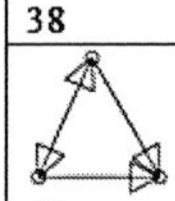 46	6	**9**	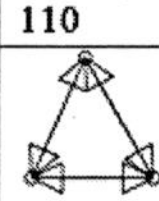 238	0	1
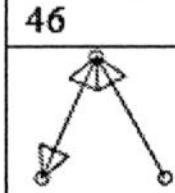 74	*17*	*24*			

Three of the most highly significant triad motifs also obtain the highest uniqueness value: namely motifs 6, 36 and 12. Large extensions of motif 6 and 36 link patterns form hub structures, and indeed, the hubterranean structure represents a high percentage of patterns and unique words composing those patterns. Thus a high percentage of functional regularity occurs in the hubterranean structure substantiating hubs as important in expressing syntactic structure within a word co-occurrence network. The high uniqueness value of motif 12 in both the adult and child network relates to variations in linear word combinations.

It is easy to observe that the adult network has many more unique words associated with motifs 6, 12 and 36. However, the adult network had roughly 40% more nodes i.e. type words, than the child network. Therefore, the values are normalised as per the Z-scores – see Equation 7.2 above: Figure 7.10 shows the normalised graph.

Figure 7.10 3-node uniqueness scores normalised for the adult and child network. Equation 7.2 gives the normalisation procedure.

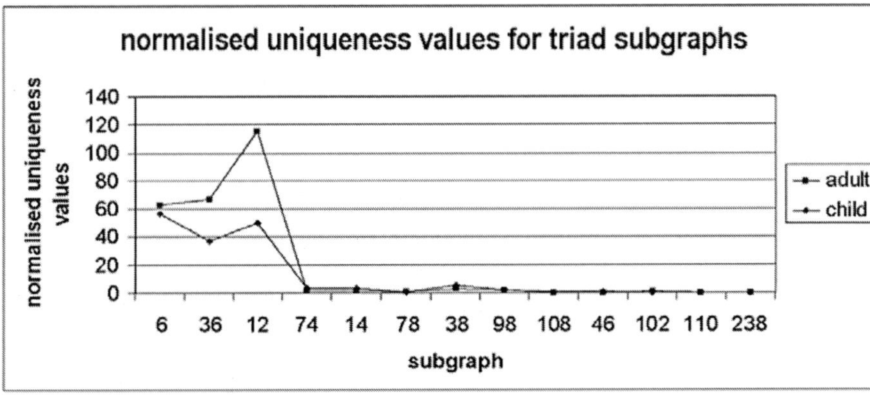

Even after normalisation, it is interesting to observe that the adult network has many more unique words associated with motifs 12 and 36, than the child network. In linguistic terms, there are many different words occurring in the motif 12 and 36 link configuration, in that the link patterns do not overlap as much when compared to motif 6. The higher uniqueness score of the motif 12 for the adult network may be related to the adult vocabulary being more specialised than the child's, and may be used more selectively in conversation.

Moving on to motif 36, the difference between motif 36 and motif 6 is the direction of the links. The automorphic nodes in motif 6 are input nodes (links point towards the nodes) whereas in motif 36, the automorphic nodes are the output nodes (from which the links point). The two largest extensions for the adult motif 36 represent the *and-construction* and the *of-construction*, respectively. These structures are heavily recursive and utilise many different nominals and, in respect to the *and-construction*, many different types of words. Interestingly, these structures are virtually absent from the child's motif 36 largest extensions (see chapter eight). The discussion will now move on to motifs which consist of four nodes.

7.7 Significance profiles for 4-node subgraphs

Identifying significant network 4-node motifs in the adult network requires a comparison of randomised networks which have the equivalent number of nodes and arcs, whilst retaining the same in/out-degree for each node. As noted above, triadic patterns found in the adult and child networks were not transferred to the randomised networks.

7.7.1 Z-scores and uniqueness values

The significance profiles of the 4-node subgraphs for the adult and child network are shown in figure 7.11(a-d). Due to tendency for the 4-node Z-scores to depend on the network size, the significance profile does not show the Z-scores, but rather "the abundance of each subgraph *i* relative to random networks" calculated with the equation below (Milo et al 2004: 1541).

Equation 7.3 The equation establishes the significance profile value Δ_i given "the abundance of each subgraph *i* relative to random networks" (Milo et al 2004: 1541) since the Z-scores of 4-node subgraphs tend to depend on the network size. Where $N\text{real}_i$ and $N\text{rand}_i$ is the absolute number of times the subgraph occurs in the real network i.e. the network under study and the random network respectively and where ε is set to four to compensate for the large "misleading" value of Δ_i that occurs when subgraphs appear only a few times in the real and random network (Milo et al 2004: 1541).

$$\Delta_i = \frac{N\text{real}_i - \langle N\text{rand}_i \rangle}{N\text{real}_i + \langle N\text{rand}_i \rangle + \varepsilon}$$

It is interesting to calculate the significance profiles for 4-node subgraphs, since networks showing similar triad significance profiles can display different 4-node (tetrad) profiles (Milo et al 2004).

Figure 7.11a The significance profile of the first forty-eight 4-node subgraphs (14 to 862. See motif dictionary Kashtan, Itzkovitz, Milo and Alon 2005).

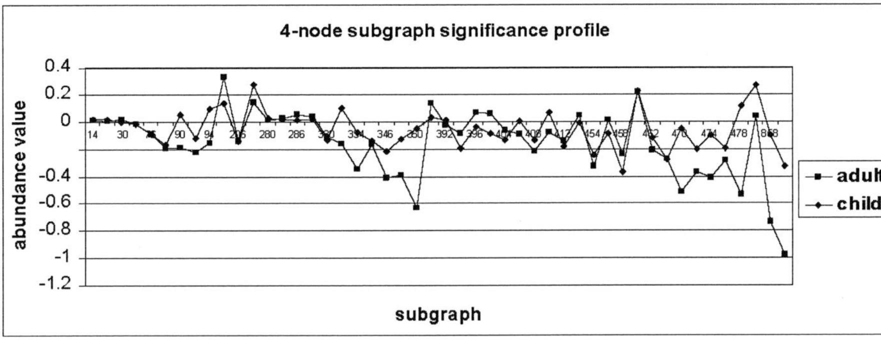

Figure 7.11b The significance profile of the subgraphs 904 to 4564.

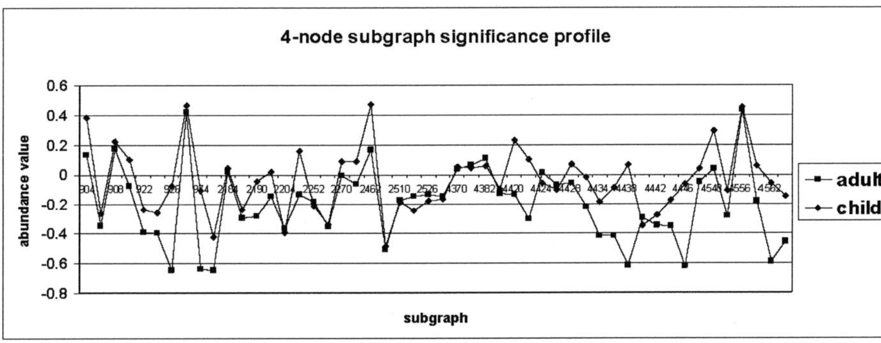

Figure 7.11c The significance profile of the subgraphs 4566 to 5084.

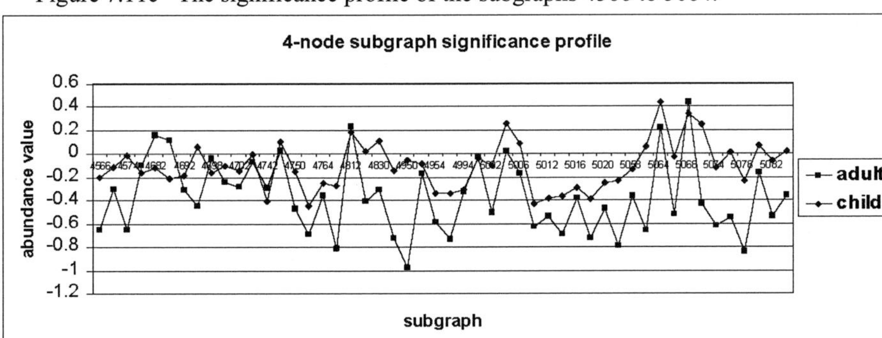

Figure 7.11d The significance profile of the subgraphs 5086 to 15326 (there were no 31710 subgraphs in either the studied or the random networks generated)

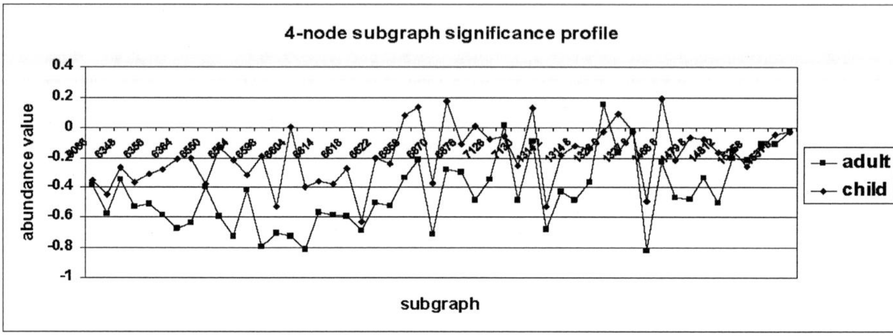

The uniqueness scores of the 4-node subgraphs, normalised as per Equation 7.2 have been profiled in Figure 7.12 a-d. Motifs, along with a Z-score over two, have a uniqueness score of four and over.

Figure 7.12a The normalised uniqueness profile of the first forty eight 4-node subgraphs (14 to 862) for the adult and child network.

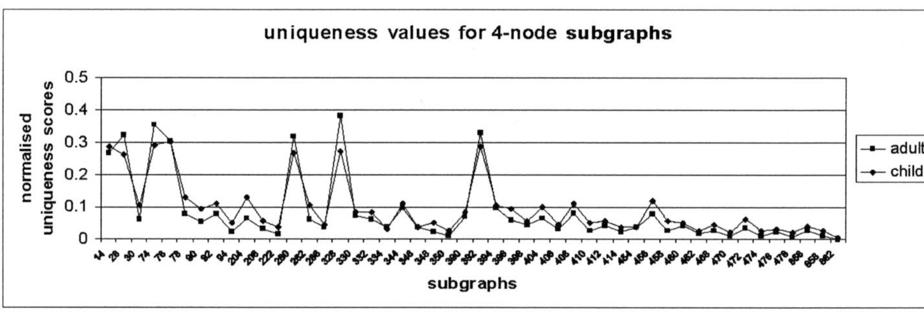

Figure 7.12b The normalised uniqueness profile of the subgraphs 904 to 4564 for the adult and child network.

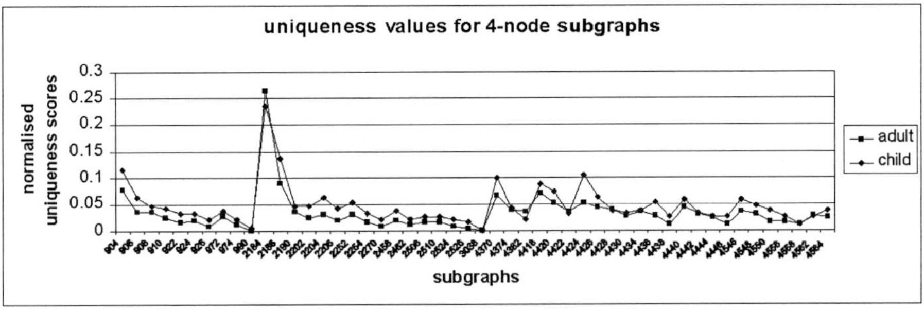

Figure 7.12c The normalised uniqueness profile of the subgraphs 4566 to 5084 for the adult and child network.

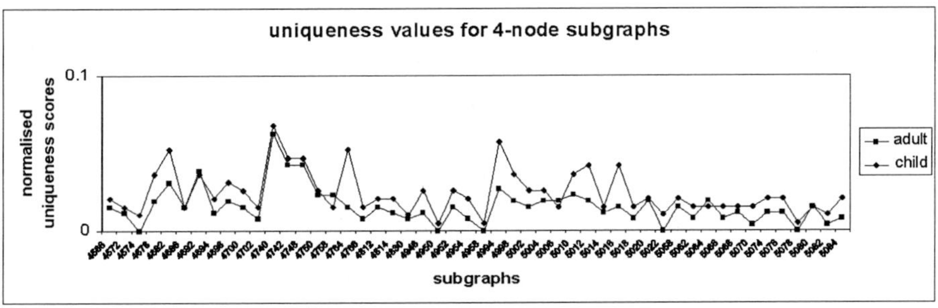

Figure 7.12d The normalised uniqueness profile of the subgraphs 5086 to 31710 for the adult and child network.

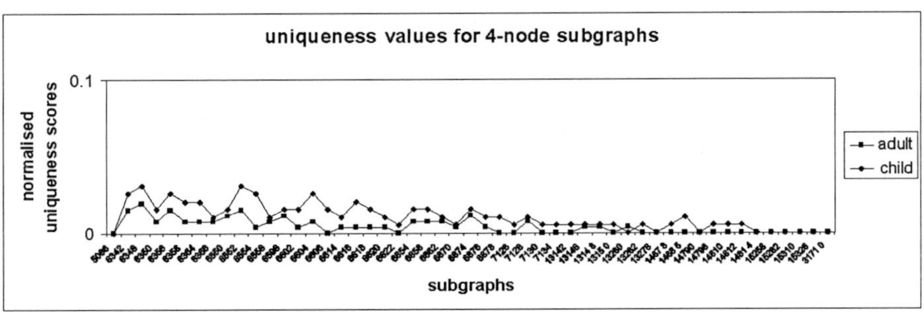

There is some variation between the adult and child abundance scores, and less so for the uniqueness values. Due to the sheer number of subgraphs and time constraints on the study, it is impractical to explain individual subgraph divergences. Instead, a more realistic goal would be to focus on the most highly significant motifs in both the adult and child networks. If you recall, subgraphs with Z-score>2 are classified as a motif and a high uniqueness score coupled with a high Z-score gives a highly significant motif.

By this method, seven highly significant motifs were found in the adult and child networks (Mfinder 2002-5a). Of the six most significant 4-node motifs in the adult network, the same six motifs are also the most significant in the child network, shown here in Table 7.3 as 204 (bi-fan), 460, 904 (bi-parallel), 908, 972, 4556. The child network had an extra pattern (4548) which was not highly significant in the adult network. The Table 7.3 gives the Z-scores and uniqueness values of these motifs for the adult and child network.

It is interesting to note that the bi-fan (204 motif) and the bi-parallel (904 motif) are also significant in functionally similar networks i.e. information processing systems such as the neuronal structure of the worm C. elegans and electronic circuitry i.e. forward logic chips - S15850, etc (Milo et al 2002). Motifs that are shared between networks of a certain kind are known as 'consensus motifs' (Milo et al 2002:825).[12] However, not all the motifs found in C.elegans also occur in word co-occurrence networks. For example, motif 456 is significant in former network (Itzkovitz and Alon 2005) but not in the latter networks.

The significance profiles of the highly significant motifs for both the child network and the adult network are compared in Figure 7.13. If you recall, Z-scores are dependent on network size and so the abundance value is used (Equation 7.3). Clearly in the child network, there is a distinct over-representation of the 904 motif whilst the 204 motif is under-represented in relation to the adult significance profile.

[12] It has been argued that such consensus motifs may "define broad classes of networks, each with specific types of elementary structures" (Milo et al 2002), suggesting perhaps the bi-fan and bi-parallel motifs characterize systems that process information.

Table 7.3 A comparison of the Z-score and the uniqueness value for each of the most significant motifs in the adult and child network. The bracketed Z-score indicates the motif is not found in significant numbers in the network. Pictures are from the motif dictionary (Kashtan, Itzkovitz, Milo and Alon 2005c).

motif reference	adult Z-score	child Z-score	adult uniqueness	child uniqueness
204 	9.35	4.02	17	25
460 	4.11	5.90	11	10
904 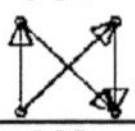	3.46	19.16	20	22
908 	2.70	4.84	9	9
972 	4.16	7.06	7	7
4548 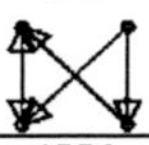	(0.49)	7.51	8	9
4556 	4.51	6.78	4	5

That the child favours the 904 link pattern over the 204 pattern may be due to the 204 having two sets of automorphic nodes (whilst 904 has one automorphic set). As such the 204 motif expresses a more complex set of relations within the four nodes than the 904 motif. Thus, the 904 motif may be easier to construct and learn. Furthermore, in the child network, the 4548 pattern has the second most significant Z-score and is highly abundant and yet it is practically insignificant and sparse in the adult network (see chapter eight for discussion).

Figure 7.13 The abundance value of the most highly significant motifs represented in Table 7.3 above for child (diamond) and adult (square) data. Pictures are from the motif dictionary (Kashtan, Itzkovitz, Milo and Alon 2005c).

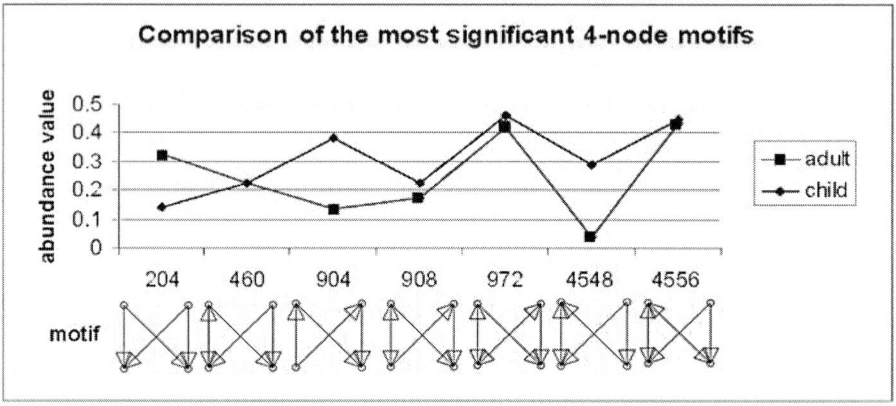

For comparison, the uniqueness scores for the adult and the child network have been normalised by Equation 7.2. Most of the uniqueness values for the adult and child motifs are very similar: the exception perhaps is the 204 motif. The adult network has less isolated instances of this motif than the child network (Figure 7.14).

Figure 7.14 The normalised uniqueness scores for the most significant 4-node motifs in the adult and child network.

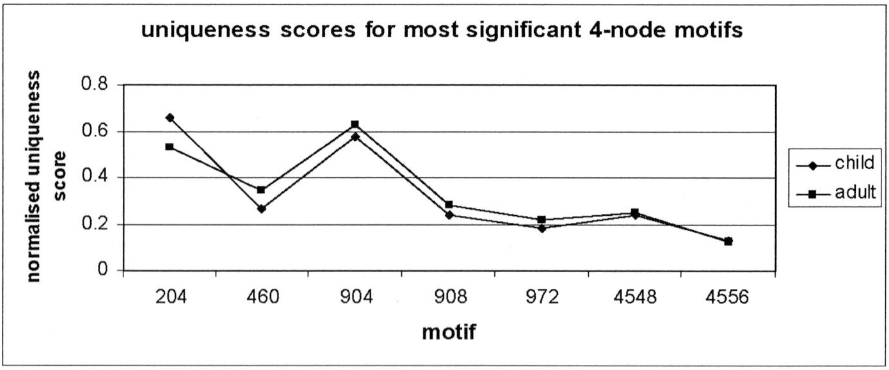

Thus the adult network uniqueness scores and Z-scores for the 3-node and 4-node most highly significant subgraphs resemble those of the child network. Chapter eight will elaborate on these results.

7.8 Summary

The notion of subgraph and significant subgraphs i.e. motif were discussed. A general method of finding motifs was presented. Word co-occurrence patterns that appear frequently in the studied networks reflect distributional patterns that are used by linguists to ascertain syntactic elements and structures. Regularities in word co-occurrences and thus in the link patterns are not formed by chance but reflect the extension of constructions. Thus, any pattern within the network that does not conform to an actual utterance (since the network does not distinguish sentential boundaries) will be random. Thus, the word co-occurrence network is compared to networks that are generated randomly: the random null hypostudy which factors out such randomness (see Milo, Shen-Orr, Itzkovitz, Kashtan, Chklovskii and Alon 2002; Milo, Itzkovitz, Kashtan, Levitt and Alon 2004). Naturally, this method does not reflect how children pick up regularities in language: children look for functional regularities matched to word co-occurrences – however the random null hypostudy does account for link patterns representing constructions that have been extended functionally in an intentional and non-chance way.

Both the two word co-occurrence networks were compared to their respective randomly generated networks and the significance profiles of 3-node and 4 node motifs for the adult and the child network were presented. In regard to the 3-node significance profiles of the adult and child network Z-scores, only two of the thirteen subgraph results diverge. The most significant 4-node motifs in the adult network are also the most significant in the child network. Moreover, motifs found in a neuronal structure of a worm and electronic circuitry, are also found in the adult and child networks (the bi-fan (204) and the bi-parallel (904)). In conclusion, the most highly significant patterns in the child network closely resemble the highly significant motifs in the adult network. The differences in the significance scores between the two networks will be picked up in the next chapter.

CHAPTER EIGHT

INVESTIGATING MOTIFS WITHIN THE ADULT AND THE CHILD NETWORK

8.1 Overview

This chapter focuses on the most highly significant motifs found in the adult and child network and discusses their relevance to possible syntactic structures represented within the word co- occurrence network. Due to the accumulation of language experience recorded by the network, paradigmatic as well as syntagmatic patterns will potentially be represented by the motifs.

Motifs may be split into two groups: automorphic motifs whose nodes share automorphic relations and non-automorphic motifs whose nodes do not (see section 8.2 and 8.3). Informally, automorphic nodes are those nodes that form the same relations in respect to all the other nodes within the motif (for a more formal definition, see section 7.2). Of the highly significant triad motifs, motif 6, 36 and 78 have two automorphically equivalent nodes; whilst motifs 12, 14 and 74 do not possess automorphic nodes. In respect to the six most significant 4-node motifs in the adult network, four motifs have automorphic roles e.g. motif 204, 904, 972, 4556. Motif 204 has two pairs of automorphic nodes, the other motifs have one pair. The two 4-node motifs that do not have any automorphically equivalent roles are motifs 460 and 908.

Sections 8.2.1 and 8.2.2 discuss the triad motif 6 and its ten largest extensions within the adult and child network respectively. The majority of the automorphic nodes anchored around each extension of motif 6 may be represented by the traditional category noun or verb. However, network analysis proves evidence against general categories in line with Croft (2001). In section 8.2.1.1, the topic of redundancy and degeneracy, a property pervasive in complex biological systems, is briefly discussed in respect to verb conjugations e.g. *was* and *were* and their representations as hubs. The anchors and automorphic nodes of the largest extensions of motif 36 in adult and child network are presented in sections 8.2.3, 8.2.4 and the adult motif 78 in section 8.2.5. Section 8.2.6 details the 4-node automorphic motifs.

The most highly significant non-automorphic motifs are discussed (section 8.3). Two examples are given to highlight how this type of motif captures asymmetry inherent in word combinations. The first example utilises the *and-/of-* constructions to show that a gap in usage may be resultant of an unrealised communicative intent i.e. there has not been any conceptual motivation to use a certain word combination (section 8.3.1 and 8.3.2 for the adult and child network respectively). The second example, a fragment of the Covariational Conditional, illustrates that elements within a construction are constrained semantically causing a gap in usage (section 8.3.3). Variation in the significance between the adult and child motifs is broached in section 8.3.4.

The first section of this chapter will briefly discuss the anchors and automorphic nodes that are represented by the largest extensions of the most highly significant automorphic motifs in the adult word co-occurrence network.

8.2 Automorphic motifs

8.2.1 Adult motif 6

An automorphic motif is composed of an anchor (or a set of anchors) which is a specific word (or words) and a set of automorphic nodes (see chapter 7 section 7.2 on the automorphic relation). The anchors of the ten largest extensions for motif 6, pictured in Figure 8.1, are given in Table 8.1.

The traditional general category represented by the majority of automorphic nodes is also given in Table 8.1 with the respective anchor. The anchors tend to fix the traditional major categories of verb and noun. For example, the word *the* in Figure 8.1 anchors two automorphic nodes – the traditional nouns *age* and *country*. Notice how all the anchors in Table 8.1 are traditional function words. This is not surprising since function words have been argued to have a high connectivity i.e. a high type frequency (see chapter two for discussion). As such, the automorphic nodes would constitute lexical words as their connectivity would typically be low considering the network is disassortative i.e. hub words tend not to connect

Figure 8.1 (a) Motif 6 from Mfinder motif dictionary (Kashtan, Itzkovitz, Milo and Alon 2005c). (b) In the adult network, the word *the* anchors motif 6. The motif was drawn by PAJEK (de Nooy, Mrvar and Batagelj 2005).

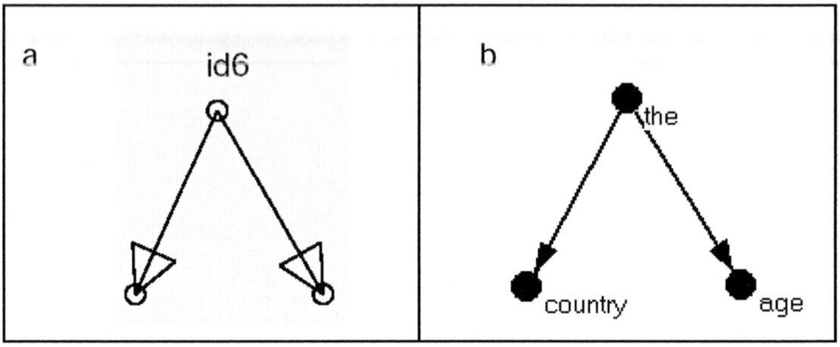

Table 8.1 For the ten largest extensions, the anchor word is given as well as the category assigned to the majority of the automorphic nodes. The nominal category is bolded. The anchors *the, and, of* fix the most automorphic nodes.

anchor	the traditional category assigned to the majority of automorphic nodes
the	**noun (79%)**
and	**noun (34%)**
of	**noun (60%)**
a	**noun (68%)**
to	verb (69%)
was	verb (42%)
in	**noun (43%)**
were	verb (52%)
for	**noun (42%)**
that	verb (23%)

to each other. As such, function words may be seen as an anchor of the motif 6 pattern with lexical words constituting the automorphic nodes e.g. *age* and *country* in Figure 8.1. Since both the words *age* and *country* share the same equivalence relations within the motif, one may want to capture this regularity by positing a category.

Previously, category accuracy measurements have been used as a value judgment on the categorial status of automorphic nodes found in productive linguistic environments, whether these environments are bigram formations (Cartwright and Brent 1997; Redington et al. 1998) or frames (Mintz 2002). This analysis steers clear of this measurement as it presupposes that a distinct *general* word category exists and is used to capture extensions across constructions.

In support of the evidence presented by Croft (2001), there is reason to reject these general syntactic categories when employing network theory. Recall that the word co-occurrence network is a richly detailed distributional analysis of word patterns, in respect to linguistic form (only). As such, significant patterns in the network represent the word distributional patterns that may be utilised by linguists when making a first pass at establishing the syntax of the language. The motif 6 or 36 pattern captures the notion of a syntactic category. The automorphic nodes, within the motif, form the members of the syntactic category which is anchored by the associated function word. For example, the function words *the* and *a* anchor the traditional noun category as noted above.

Furthermore, the integration of two motif 6 patterns is represented by the motif pattern 204 as shown in Figure 8.2 and as evidenced in the word co-occurrence network (Table 8.10). Since the nodes representing *the* and *a* are automorphic, it is tempting to posit a category for these words, as indeed is traditionally done e.g. a determiner or an article. However, the 204 motif anchored by *a* and *the* only expresses the overlap in the members of the two motif 6 patterns. The 204 does not encompass the distributional facts of all the automorphic nodes anchored by *the* and *a* in the motif 6 primarily because the two motifs do not possess an identical set of members e.g. compare **a rice* and **a scissors* to *the rice* and *the scissors*.[1]

[1] An asterisk indicates ungrammatical or unconventional usage.

Figure 8.2 The 204 motif. The words *the* and *a*, which are traditionally classed as determiners or articles, anchor the set of words {N1, N2, N3, N4, N5, N6}, all of which are traditional nouns.

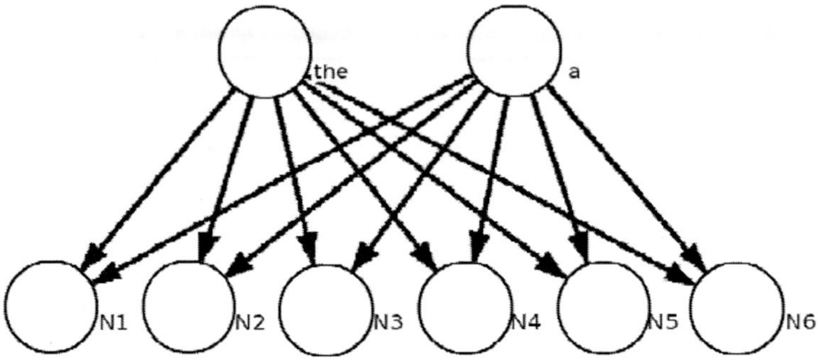

Thus, if one hypothesizes a determiner category for the two anchors *a* and *the*, the category is not associated with the nouns *rice* and *scissors* and so the determiner category is not a consistent syntactic environment for a nominal.[2] To represent the whole distribution of the noun, all the syntactic environments where nouns occur must be included in the motif pattern. That is, all the function words that are associated with this major category must be included in the 204 motif pattern. Thus, the 204 motif must - at the very least - include the function words {*the, and, a, of, in, for*} as shown in Figure 8.3.[3]

Whilst this allows a general category to be formed by the in-degree automorphic nodes, there is the question of the out-degree automorphic nodes that represent the environments in which the general category occurs i.e. the function words. Since these function words are denoted by automorphic nodes, they too must logically form a category.[4] However,

---

[2] Even if one were to connect the word *rice* to the node representing *the* in 204 network shown in Figure 8.3, *rice* would still not be classed as a noun as it would not share an automorphic relation with the node set {N1, N2, N3, N4, N5, N6}. In that, *rice* does not connect to the node representing the word *a*.

[3] Naturally, morphology also provides an environment for identifying a noun but this analysis is beyond the scope of the study.

[4] Even if one were to create an ad-hoc process whereby only the {*N1, N2, N3, N4, N5, N6*} set of nodes were represented by a discrete element, it is unclear how such a move would be facilitatory. Generalisations across item-specific constructions may be accounted for by the overlapping links within the 204 motif (see section 9.4.1.1) without recourse to the general category. Indeed, the general category

the words {*the, and, a, of, in, for*} do not felicitously combine into a single homologous class. In fact, it is not clear how such a category would help in capturing any existing regularities with its eclectic membership. Instead, such words are traditionally classified as determiner/article, conjunction and preposition.

Figure 8.3 The 204 motif captures a distribution environment associated with nominals. The function words {*the, a, and, of, in, for*} are represented by the out-degree[5] nodes and a potential nominal set {*N1, N2, N3, N4, N5, N6*} by the in-degree nodes. For two general categories to be posited, both sets of automorphic nodes e.g. {*the, a, and, of, in, for*} and {*N1, N2, N3, N4, N5, N6*}, must reduce. Thus, the two syntactic categories are mapped one-to-one.

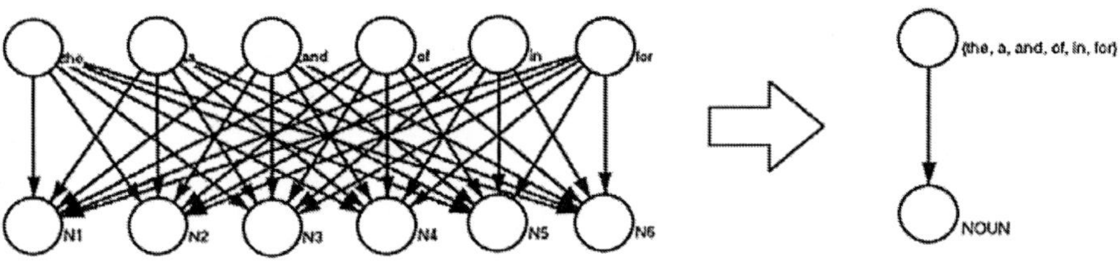

Note that recourse to the label functional category[6] is not admissible. In the same way that a noun is a subclass of a lexical category, so too must the category that captures the member set {*the, a, and, of, in, for*} be a subclass of a functional category. This argument extends to the general verb category too. If such a category were posited then the function words *to, was, were, that,* etc would be required to form an (inadmissible) syntactic category.

Fortunately, the rejection of general categories does not equate to the reject of all syntactic categories. A syntactic category may be represented by the automorphic nodes in motif 6 and 36 and therefore such a category

---

functions in exactly the same way as a specific category. Thus, positing an ad-hoc stipulation to justify a general category is only motivated by the desire to be consistent with previous traditional accounts of syntactic categories. This motivation has no objective basis.

[5] Out-degree and in-degree refers to connections pointing away from the node and to the node respectively.

[6] Since "speakers analyse low-frequency items more than high-frequency items" (Bybee 1985: 133), it is highly unlikely that a functional category is cognitively viable, particularly considering the entrenched nature of function words.

is strongly associated with the function word anchoring the automorphic nodes. Since a function word facilitates in identifying a given (item-specific) construction, then the syntactic category associated with the function word may potentially be associated with its respective construction. For example, the construction frames would be 'the + N_{the}', 'a + N_{a}', 'and + N_{and}', etc where the variables N_{the}, N_{a} and N_{and}' are specified in form and meaning for each construction instance. Hence, construction-specific categories are evidenced in the word co-occurrence network.

Whilst some linguists do recognise the failings of general syntactic categories, the general category's usefulness lies in its convenience. However, the set of construction-specific categories within a given language need not be cumbersome since extracting motif 6 link patterns from the data set allows potential syntactic categories to be identified. Therefore, the network theory will potentially allow easier analysis of large quantities of construction-specific categories.

8.2.1.1 Redundancy and degeneracy

It is interesting to note the occurrence of *was* and *were* in the ten most highly connected nodes in motif 6. The two hubs arguably generate redundancy in the network, since both forms predicate for the same construction (in conjunction with relevant morphosyntax), and express number agreement. Certain computational linguistic theories seek to reduce redundancy from linguistic representation e.g. the Minimalist Program, particularly in respect to the storage of linguistic representations (Croft 2009). However, as Croft (2009: 13) points out, this assumption is "unwarranted" since redundancy in a given analysis may assist in parsimonious computation.[7]

On a network approach, since *was* and *were* are hubs i.e. they are represented by highly connected nodes within the network, the redundancy endowed by the two forms assists in reducing the fragility of the network, a well-known vulnerability of hub structure (see section 4.2.4). Since a certain function word may identify a particular construction e.g. the use of

---

[7] Croft (2009: 12-14) compares the complicated derivational rules in Minimalism and the storage of specific construction instances in Construction Grammar. Croft (2009: 12-14) claims that the former theory is computationally non-parsimonious, whilst the latter is more costly in terms of storing representations. However, on a network approach, the storage of elements is optimised, in that chunking occurs for highly frequent word co-occurrences, lowering the storage cost.

was and *were* in the passive construction in combination with the /–ed/ verbal inflection, if one of the copula forms become cognitively corrupted i.e. inaccessible to production mechanisms, the other form would still facilitate the identification and usage of the construction. On this view, verb conjugation establishes functional redundancy.

In fact, as Croft (2008:122) asserts, a non-redundant analysis would in general create an odd account of certain constructions as his following examples illustrate (Example 8.1). In a non- redundant account of example (a), to recognise the plurality in the inflection /-s/, one would then need to discount the plurality of the word *three*. The subject-verb agreement, in (b) likewise negates the 3rd person singular meaning of *she* in a non-redundant account. In example (c), the morpheme /-ed/ (in this context) denotes the past tense, and so from a non-redundant perspective the past temporal meaning of *yesterday* would need to be rejected. To reduce redundancy in example (d) when acknowledging the locative meaning of *into*, one would need to reject the locative meaning inherent in the verb *insert.* The last example (e) concerns the "transfer of possession" meaning of the ditransitive construction. A non-redundant account would then deny this "transfer of possession" meaning inherent in *give.* By admitting redundancy into the interpretation of each sentence, such odd accounts are avoided.

Example 8.1 Examples illustrating redundancy in utterances (Croft 2001: 122). Bold text is author's own.

a) **Three** book**s**
b) **She** live**s** here
c) **Yesterday** I bik**ed** to work
d) I **insert**ed the floppy **into** the drive
e) I **gave** her a book

Furthermore, examples 8.1 (a-e) may be better described by the notion of degeneracy, since "unlike redundancy, which occurs when the same function is performed by identical elements, degeneracy, which involves structurally different elements, may yield the same or different functions depending on the context in which it is expressed" (Edelman and Gally 2001:13763). In each of the above examples, the 'redundancy' involves elements whose syntactic structure are different e.g. there is a word/morpheme contrast in examples 8.1 (a-d), and in example 8.1 (e), the verb contrasts with the construction in which it occurs. In languages with an extensive verbal paradigm, the inflectional redundancy would generate much degeneracy within a syntagmatic context i.e. with an utterance. This observation is in line with the idea that "high redundancy might be needed in order to properly build degeneracy into the system" (Macía and Solé 2008: 6).[8]

Returning to the two forms *was* and *were*, their paradigm generates redundancy in that they share the same syntactic structure i.e. both are function words, and each one expresses the same function i.e. facilitating the identification of the same construction. Interestingly, in the syntagmatic context of an utterance (Example 8.2), the noun and the copula promote degeneracy. For example, even though the elements are structurally different e.g. a lexical word versus a function word, both elements express number either singular (Example 8.2a) or plural (Example 8.2b).

Example 8.2 Degeneracy may be expressed through subject-verb agreement. The singular and plural meaning intrinsic to *was* and *were* is also part of the nominal meaning.

a) **The lady was** dancing
b) **The couple were** dancing

In fact, it is the obligatory nature of the number agreement that causes the degeneracy since the lexical word imposes a more specific constraint on the context of the interpretation. In that, the subject of the utterance is specifically identified whereas the copula only serves to constrain the interpretative context of the subject in a very general way, in respect to the number feature of the subject.[9] On this view, a major source of degeneracy in language is obligatory inflection. As such, inflectional morphology will facilitate the generation of degeneracy.[10]

---

[8] Note that in respect to linguistic form, affixation is not necessary for degeneracy as examples 8.1 (d-e) illustrate. Indeed, certain non-affixal languages such as Chinese lack the redundancy of the verb conjugations, but exhibit degeneracy in respect to verbs and the constructions they occur in (in line with example 8.1 e.). As such, it is likely that degeneracy may be found in varying degrees cross-linguistically. Thanks to Randy LaPolla for discussion on this point, although our views may differ.

[9] See LaPolla (2003: 114) for a discussion on how "language … disambiguates the context of interpretation".

[10] It is interesting to note that in respect to dependency grammar, the head of the dependency relation is usually obligatory and identifies which element is to be

The property of degeneracy is interesting as it pervades biological systems, from the immune system to synaptic plasticity (Edelman and Gally 2001). Indeed, "the processes of evolution and natural selection… are accompanied by degeneracy" (Edelman and Gally 2001:13766). Furthermore, Edelman and Gally (2001) point out that biological systems displaying degeneracy tend to exhibit complexity; complexity being defined as the "interplay between functional specialization and functional integration" (2001: 13767).

It is not unexpected that syntactic structures also exhibit complexity with functional specialisation being representative of the separate elements of syntax: the storage component and the computational component (see chapter nine). Functional integration is then realised by the integration of the two functions. As such, complexity in syntax reflects how syntactic properties are derived from the storage and/or the computation of linguistic form.

8.2.2 Child motif 6

The ten largest extensions for motif 6 are anchored by hub words. In this case, most of the anchors of the largest extensions also form the ten largest hubs in the child network e.g. {*I, uh, the, a, it, you, my, is*}, with the minor exceptions of two anchors *I'm* and *want*, whose degree (connectivity) rankings are nevertheless still very high: 19 and 12, respectively (see Table 8.2).[11] Since function and lexical words are defined by their hubterranean qualities as well as being anchors for the automorphic nodes within motif 6, the words *I, uh, the, a, it, you, my, I'm, is* and *want* would constitute the 'function words' of the child's vocabulary.

Indeed, note how the most highly connected hubs in the child network (*a, the, it, you, is*) correlate with the counterpart hubs in the adult network. However, the hub overlap is not substantial. The child's hubs represent

---

marked by inflection (Zwicky 1985: 13&16). Furthermore, the dependent further specifies the (semantics of the) head (Fraser 1994). Although dependency relations are "at the same level in the sentence" (Fraser 1994: 860), it would be interesting to investigate whether degeneracy relations underlie dependency relations.

[11] Note the degree rankings are for the in- and out- degree combined. When only the out-degree of a word is calculated for the child network, the ten highest ranked words are: *I, uh, the, a, it, you, my, is, I'm, want.* As expected, all the anchors for the ten largest extensions of motif 6 are also the ten most highly connected hub words.

many words (with their clitics) that would be viewed as slot-filler structures such as the clitics *I'm* and *where's* (see Lieven, Pine and Baldwin 1997) as well as traditional lexical words such as *want* and *go* (see Tomasello 1992). Names like *mommy* are also highly connected nodes, correlating with the experimental data on familiar names facilitating adjacent word recognition (Bortfeld et al 2005).

Table 8.2 The extensions of motif 6 that have more than ten automorphic nodes with their corresponding anchor. Degree (connectivity) and frequency rankings are given respectively. The ten largest extensions are bolded.

anchor	no. of automorphic nodes	anchor	no. of automorphic nodes	anchor	no. of automorphic nodes
I (3,1)	115	that's (36,40)	28	oh (41,57)	17
uh (4,7)	100	and (29,46)	27	uhs (58,82)	17
the (1,9)	98	to (23,42)	26	can (44,48)	16
a (5,26)	68	get (26,12)	24	don't (44,34)	16
it (2,2)	57	more (30,20)	24	what (70,53)	16
you (6,11)	55	where (38,51)	24	he (52,96)	15
my (7,29)	54	Georgie's (31,74)	23	another (44,53)	14
I'm (19,28)	50	its (41,68)	23	draw (44,60)	14
is (10,21)	48	what's (41,3)	23	find (64,96)	14
want (12,6)	48	Georgie (17,19)	22	gonna (55,113)	14
where's (26,22)	39	in (12,26)	22	have (52,150)	14
on (8,16)	38	Nomi (28,31)	22	up (14,16)	13

anchor	no. of automorphic nodes	anchor	no. of automorphic nodes	anchor	no. of automorphic nodes
going (15,24)	34	at (39,30)	19	down (22,8)	12
that (9,5)	31	Daddy (17,10)	19	little (86,131)	12
go (19,18)	29	put (39,48)	19	all (70,103)	11
this (11,4)	29	there's (58,70)	18	got (58,79)	11
me (21,44)	28	baby (23,23)	17	no (55,15)	11
see (23,36)	28	Mommy (16,13)	17	there (32,14)	11

Table 8.3 For the ten largest extensions of motif 6, the anchor word is given as well as the traditional syntactic category assigned to the majority of the automorphic nodes.

anchor	the traditional category of the majority of automorphic nodes
I	verb (90%)
uh	**noun** (39%)
the	**noun** (95%)
a	**noun** (84%)
it	**noun** (28%)
you	verb (56%)
my	**noun** (87%)
I'm	verb (86%)
is	**noun** (42%)
want	**noun** (56%)

Of the ten largest extensions, it is found that the majority of automorphic nodes are either nominal or verbal types, in agreement with the adult network findings (Table 8.3). As noted in the last chapter, this finding arguably reflects the tendency of mainstream linguistics to posit two general linguistic categories of noun and verb. However, as noted above, conflating the link patterns of all traditional nominal distributions results in an anchor set that is very eclectic e.g. {uh, the, a, it, my, is, want}. Thus, it is preferable to propose that syntactic categories are construction-specific. The extension of the construction is facilitated by the semantic similarities of the automorphic nodes which are connected to a specific anchor in a given link pattern.

8.2.3 Adult motif 36

Motif 36 has two nodes that share automorphic equivalence (see Figure 8.4). Interestingly, the anchors for the ten largest extensions of motif 36 and motif 6 are exactly the same (in the adult network). In fact, of the forty-three anchors that fix the motif 36 extensions with eleven automorphic nodes and over, only four did not overlap with the anchors of the corresponding motif 6 extensions e.g. {*women, war, people, nineteen*}.

Like motif 6, the majority of automorphic nodes fixed around the ten largest extensions for motif 36 may be assigned the traditional nominal or verbal category (Table 8.4). Essentially motif 6 and 36 represent hub nodes, and so the largest hubs are the loci for (traditional) verbs and nouns. However, as noted earlier, reducing the motif patterns to general syntactic categories is not feasible.

Figure 8.4 Motif 36 with automorphic nodes and the corresponding anchor shown (Kashtan, Itzkovitz, Milo and Alon. 2002-2005c).

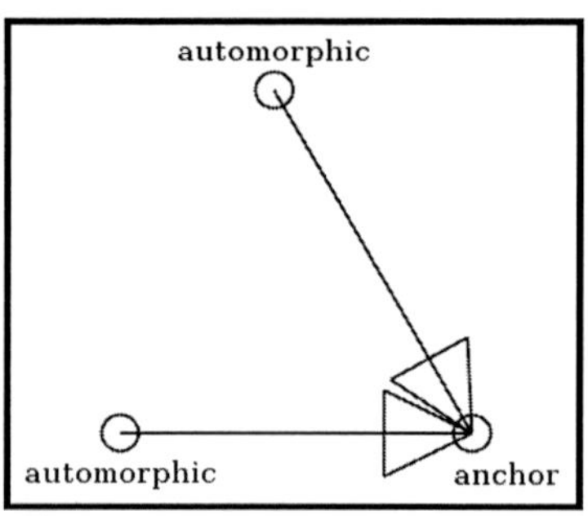

Table 8.4 For the ten largest extensions, the anchor word is given as well as the category assigned to the majority of the automorphic nodes. The nominal category is bolded.

anchor	**the traditional category assigned to the majority of automorphic nodes**
and	**noun (72%)**
of	**noun (79%)**
to	verb (54%)
the	verb (42%)
in	verb (42%)
a	verb (35%)
that	**noun (36%)**
was	**noun (75%)**
for	**noun (53%)**
were	**noun (81%)**

Comparing the automorphic nodes from largest extensions for both motif 6 and 36 (Table 8.5), the anchors of the largest extensions co-occur mostly (a) with nouns e.g. the anchors {*and*, *of*, *for*}; (b) with verbs e.g. the anchor *to*. There are then anchors {*the*, *a*, *in*} which tend to follow a verb but precede a noun whilst the reverse is true for other anchors {*that*, *was*, *were*). Again it would not be prudent to lump together the distributional facts in the first three rows of the table since one would not assume {*and*, *of*, *for*} represent a single abstract syntactic category. Likewise with the anchor sets {*the*, *a*, *in*} and {*that*, *was*, *were*}.

Table 8.5 The ten largest extensions of motif 6 and 36 from the adult network. The most predominant traditional syntactic category of automorphic nodes is given.

anchors from top 10 extensions	***category with majority of automorphic nodes***	
	motif 36	***motif 6***
and	**noun**	***noun***
of	**Noun**	***noun***
for	**Noun**	***noun***
to	**Verb**	***verb***
the	**Verb**	***noun***
a	**Verb**	***noun***
in	**Verb**	***noun***
that	**Noun**	***verb***
was	**Noun**	***verb***
were	***noun***	***verb***

In respect to the network as a whole, the anchors of the ten largest extensions for motif 6 and 36 correlate well with the hub words of the network. The ten most highly connected nodes in the smaller BNC corpora are {*and, the, of, to, a, in, was, that, for, were*}. All of these most highly connected hubs form an anchor for one of the largest extensions of motif 6 and 36. In sum, the hub words tend to be function words that identify potential (non-general) syntactic categories.

Since the optimisation process derives the hubterranean typology, the functional-lexical distinction originates from this process. As such, the distinction is not specific to language; rather counterparts to the distinction will exist in other networks with the same typology. Furthermore, the functional-lexical distinction as well as motifs 6 and 36 may be viewed as "evolutionary spandrels" (Solé and Valverde 2006: 421; see section 7.1) in that they all derive from the hubterranean structure which in turn is generated by an optimisation process.

8.2.4 Child motif 36

Akin to motif 6 in the child network, the anchors of the ten largest extensions are also highly connected words or hub words. Six of these anchors, are also included in the ten most highly connected hub words in the child network. Thus, the words {*it, the, on, my, in, up, down, that, a, daddy*} may be classed as the 'function words' of the child's vocabulary. Again note how the familiar name of *daddy* forms an anchor as well as the traditional verbs *sleeping* and *tired* (Table 8.6).

Like motif 6, the majority of automorphic nodes in the largest extensions of motif 36 tend to be verbal or nominal depending on the given anchor (Table 8.7). Predominately in the largest extensions of motif 36, most of the anchors in an utterance are preceded by traditional verbs whereas in motif 6 many more anchors precede traditional nouns. Unlike the largest extensions of motif 6, there are four locative anchors in the largest extensions motif 36, whilst the four anchors {*it, the, my, a*} are found in both motif 6 and 36 largest extensions.

One advantage of comparing adult and child utterances using network theory is that the theoretical concepts employed in analysing the adult network may also be applied to the child network since they share the same hubterranean typology formed by the same optimisation process. Consequently, it is possible to refer to functional and lexical elements in both networks when these definitions are network-based. Likewise, syntactic constituency is also a by-product of the optimisation process that forms a hubterranean structure and as such is found both in the adult and child utterances as evidenced in chapter five and six.

It is then predicted that any syntactic elements represented by the hub structure will be consistently found both in the child and adult network. The major difference between the two networks is the words represented by the nodes. Although the child's network, as noted before, does not have all the (a)symmetrical (motif) patterns that the adult does. This latter observation is expected considering that the adult's vocabulary and therefore construction inventory far exceeds that of the child's.

Table 8.6 The largest extension of motif 36 with respective anchors with the corresponding number of automorphic nodes. The degree and frequency ranking of the anchor is also given respectively. The ten largest extensions of motif 36 are bolded.

anchor	no. of automorphic nodes	anchor	no. of automorphic nodes	anchor	no. of automorphic nodes
it (2,2)	73	going (15,24)	23	coffee (58,66)	14
the (1,9)	53	you (6,11)	23	fall (70,58)	14
on (8,16)	45	me (21,44)	22	to (23,42)	14
my (7,29)	39	tired (44,82)	20	doggie (58,47)	13
in (12,26)	37	here (35,38)	19	juice (70,32)	13
up (14,16)	37	Nomi (28,31)	19	too (70,143)	13
down (22,8)	34	sleeping (44,53)	19	blanket (70,120)	12
that (9,5)	34	go (19,18)	18	Kitty (44,41)	12
a (5,26)	32	there (32,14)	18	sit (55,45)	12
Daddy (17,10)	32	do (34,33)	17	broken (81,73)	11
Mommy (16,13)	31	eat (37,62)	16	eyes (103,116)	11
this (11,4)	30	flower (64,62)	16	Georgie's (31,74)	11
Georgie (17,19)	29	get (26,12)	16	Jenko (44,68)	11
doing (32,35)	27	off (64,60)	15	sugar (91,143)	11
baby (23,23)	23	book (70,39)	14		

Table 8.7 For the ten largest extensions of motif 36, the anchor word is given as well as the category representing the majority of the automorphic nodes.

anchor	the traditional category represented by the majority of automorphic nodes
it	verb (92%)
the	verb (64%)
on	**noun** (51%)
my	verb (56%)
in	verb (38%)
up	verb (65%)
down	verb (71%)
that	verb (74%)
a	verb (56%)
daddy	**noun** (22%)

8.2.5 Adult motif 78

The highly significant motif 78 has two bi-directional links and two automorphic nodes - shown in Figure 8.5.

Figure 8.5 Motif 78 pictured with the anchor node and two automorphically equivalent nodes N_1 and N_2.

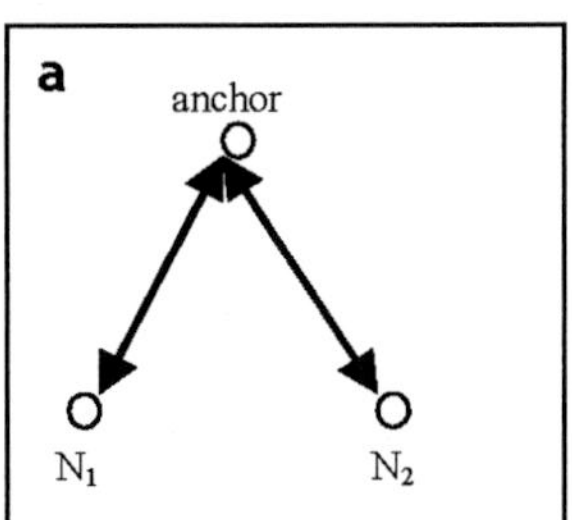

The largest extensions have been collated into Table 8.8. The constructions are given in the right hand column.

It is no surprise that the automorphic nodes anchored by the word *that* are rather eclectic, since *that* has many different functions correlated to various reduced forms (Berkenfield 2001; Ingram 1989). The largest of the extensions represent the recursive structures of the *and*-construction and the *of*-construction, as well as the infinitival *to*. Part of the Covariational Conditional, the Xer part of the 'the Xer the Yer'[12] construction is represented. The anchors for motif 78 are also the most highly connected hub words in the whole network; the anchors being ranked nineteen and higher for degree {*and, the, of, to, a, in, was, that, for, were, you, erm, on, it, with, or, i, be, we, they, had*}.

Whilst motif 78 is significant in the adult network, it is not in the child network. Since motif 78 has two bidirectional links, the reason for the motif's lack of significance in the child network may be that certain word co-occurrences may have an entrenched word order and so uni- directional links, at this stage, are favoured. Admittedly, this point is speculative and requires future research. It will not be discussed further in this study.

[12] For example, *the more the merrier.*

Table 8.8 The largest extensions of motif 78. The constructions represented by the motif are given as are the respective anchors and the number and details of the automorphic nodes.

anchor	no. of N	automorphic nodes (N)	constructions expressed (using traditional general categories)
and	17	charring, clothed, erm, forty, fruit, furniture, it, magazines, mend, newspapers, off, on, pastry, scrubbing, soap, studied, vegetables	and-construction
to	11	come, conform, extend, have, look, me, one, put, speak, take, them	infinitival, to + N, N + to
of	7	one, people, some, supply, those, war, which	of-construction
that	6	erm, for, shows, was, with, women	automorphic nodes are eclectic

8.2.6 4-node automorphic motifs

As stated earlier, there are four highly significant 4-node motifs with automorphic nodes, namely: 204, 904, 972, 4556 (with 204 motif having two pairs of automorphic nodes). Motifs 972 and 4556 do not have large extensions; the largest being composed of three and four automorphic nodes respectively. As such, these motifs are omitted from the analysis.

8.2.6.1 Adult network

Table 8.10 over the page shows the largest extensions for the 204 and 904 motifs. The 52 anchors for the various extensions are made up of 21 unique words {*a, and, be, but, by, erm, for, I, in, my, of, on, she, the, to, was, we, well, were, when, you*}. The count of each anchor found in the largest extensions of 204 and 904 is given in Table 8.9. The anchors in these motifs are all ranked above the fifty most highly connected nodes and apart from *well* and *erm* would arguably be classified as function words.[13]

Table 8.9 The counts of the anchors for the largest extensions of 204 and 904 motifs as shown in Table 8.10 below.

anchor	count	anchor	count
was	6	be	1
and	5	but	1
of	5	by	1
the	5	erm	1
a	4	my	1
you	4	on	1
I	3	she	1

[13] *Erm* may be tentatively classified as either a vocalised pause in speech or a type of placeholder.

anchor	count	anchor	count
in	3	to	1
were	3	we	1
for	2	well	1
when	2		

The anchor analysis reveals potential item-specific constructions although without an accompanying semantic analysis, such observations are tentative (Table 8.10). Potential constructions denoted by motif 904 are the passive construction (with or without the infinitival construction); the copula; the *and*-construction; the *of*-construction. As noted earlier, the 204 anchor pairs have traditionally been assigned the same traditional functional class: pairs such as *was/were, a/the, and/but, for/of, in/of, of/on*; all represented by the multi-output extension of motif 204. Others pairs represented by the multi-input extension of motif 204 are: *a/the, I/you, we/you*. However as noted above, general syntactic categories are not supported by distributional evidence and as such these pairs merely represent an overlap in the distribution of the two syntactic categories anchored by the function words.

Not all pairs conform to traditional functional classes: the different tense conjugations of the anchor pair *be/was* for multi-output 204 extension for example; as well as the pairs *a/you, my/the, a/I, I/the, she/the* for the multi-input 204 extensions. The latter pairs highlight the closeness of traditional pronouns with traditional nouns (M. Laughren p.c.), and how generalisations such as *he-man* and *she-woman* are acceptable (in the spirit of noun compounds such as *student loan*).

Table 8.10 The largest extensions for motif 204 and 904 with respective anchors and automorphic nodes (N).

multi-output 204 motif			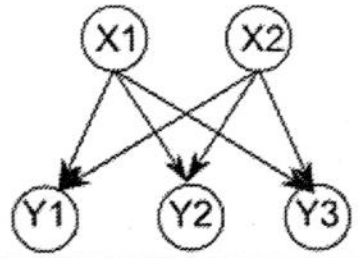	multi-input 204 motif			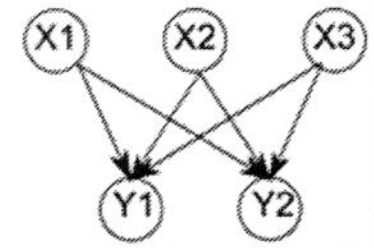	904 extension			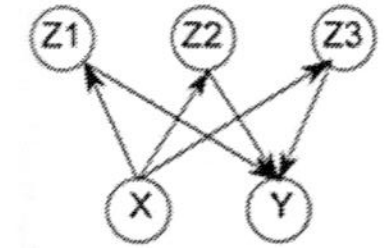
anchor X1	anchor X2	no. of N	automorphic nodes (N)	anchor Y1	anchor Y2	no. of N	automorphic nodes (N)	anchor X	anchor Y	no. of N	automorphic nodes (N)
was	*were*	11y	able, an, banned, considered, going, in, later, not, often, so, very	*a*	*the*	20x	as, that, even, than, and, erm, nine, after, by, for, in, into, of, off, on, to, with, be, have, was	*were*	*to*	12z	able, allowed, asked, considered, expected, going, instructed, issued, later, liable, another, and
and	*was*	9y	a, an, because, in, not, so, the, used, to	*I*	*you*	10x	and, as, because, but, do, erm, so, well, what, when	*of*	*in*	10z	beauty, cosmetics, material, society, water, women, it, you, erm, high
and	*when*	9y	he, I, in, people, she, the, they, we, you	*we*	*you*	9x	and, as, er, erm, mean, so, well, when, that	*and*	*was*	9z	I, she, use, attractive, then, there, most, what, use
a	*the*	8y	berlei, kettle, pig, ration, saucepan, sort, time, very	*a*	*you*	8x	and, as, but, er, erm, of, that, to	*was*	*by*	8z	considered, covered, done, made, produced, retained, used, well

and	***well***	8y	by, I, if, in, I've, we, that, you	***my***	***the***	8x	and, erm, for, from, in, know, of, remember	***was***	***you***	8z	because, so, erm, well, later, the, to, put
erm	***when***	8y	I, in, people, she, the, they, we, you	***a***	***I***	7x	and, as, but, erm, fact, where, yes	***were***	***of***	8z	price, sections, full, more, and, much, most, talking
for	***In***	8y	a, each, many, my, our, the, their, your	***I***	***the***	7x	and, as, colour, do, erm, so, when				
for	***Of***	8y	a, an, beauty, it, life, my, the, their	***she***	***the***	7x	and, as, erm, know, think, when, that				
in	***Of***	8y	a, her, my, nineteen, September, the, their, these								
of	***On***	8y	a, duty, er, how, make, September, the, that								
and	***but***	7y	a, I, I've, she, then, there, you								
be	***was***	7y	a, able, an, in, made, the, very								

8.2.6.2 Child 4-node motifs

Within the child network, there are three 4-node motifs with one pair of automorphically equivalent nodes, namely, 904, 972, 4556 and one motif with two pairs, motif 204. The largest extensions of motifs 972 and 4556 were small – being composed of five and seven automorphic nodes respectively. Akin to the adult motif analysis, these motifs are omitted from the analysis. The 60 anchors for the various extensions are made up of 21 distinct words with close to half the anchors only appearing once: *a, Daddy, Georgie, go, going, gonna, I, I'm, in, is, it, my, Nomi, that, the, this, uh, up, want, what's, you* (see Table 8.11 and Table 8.12).

Table 8.11 The counts of anchors in the largest extensions of 204 and 904 as shown in Table 8.12.

anchor	count	anchor	count
I	11	you	2
the	6	Georgie	1
my	5	go	1
this	5	going	1
uh	5	I'm	1
it	4	in	1
that	4	is	1
a	3	up	1
Daddy	2	want	1
gonna	2	what's	1
Nomi	2		

It is interesting that in both networks, but especially the child network, that there are a few anchors that appear fairly often in the extensions. The count of each anchor found in the largest extensions of 204 and 904 is given in Table 8.11. Table 8.12 below shows the largest extensions for 204 and 904 motifs.

Table 8.12 The largest extensions for motif 204 and 904 with respective anchors and automorphic nodes (N).

multi-output 204 motif

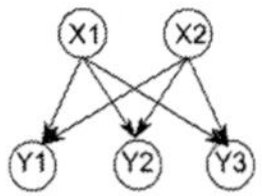

multi-input 204 motif

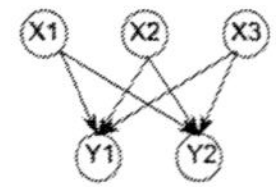

904 extension

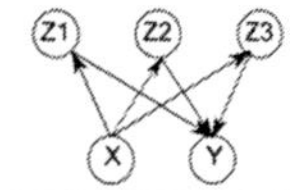

anchor 1	anchor 2	no. of N	automorphic nodes (N)	anchor 1	anchor 2	no. of N	automorphic nodes (N)	anchor 1	anchor 2	no. of N	automorphic nodes (N)
I	*You*	21y	boo, broke, close, cry, did, doing, draw, drink, eat, fall, feel, finished, go, open, put, sit, take, throw, wanna, want, standing	*it*	*that*	18x	at, uh, close, color, did, do, get, hold, jump, like, make, pull, push, pushed, put, take, throw, want,	*I*	*it*	43z	ate, bring, bumped, cleaned, close, color, did, do, doed, doing, draw, drawed, drink, drop, dropped, eated, eating, find, fixing, get, got, jump, kiss, like, move, need, pick, pull, push, pushed, put, putting, read, see, snapped, take, throw, throwed, turning, want, wash, washing, taped
the	*A*	20y	baby, ball, balloon, bee, book, boy, bug, bunny, diaper, flower, girl, goose, hat, kitty, monkey, piece, popsicle, sheep, wall, another	*it*	*this*	17x	at, clean, cleaned, color, do, draw, jump, like, move, need, putting, read, see, take, turn, want, wash	*I*	*the*	21z	boo, broke, change, draw, eat, eating, find, fixing, found, get, hurt, kiss, need, open, put, read, reading, see, want, wash, made
I	*I'm*	19y	afraid, banging, do, draw, drawing, finished, fixing, get, going, jump jumping, lying, lyinging, putting, singing, sleeping, talking, turning, washing	*the*	*a*	14x	for, on, it, that's, there's, uh, eat, find, get, made, read, reading, want, where's	*I*	*my*	20z	ah, bump, cant, color, do, drink, drop, dropped, eat, finish, fooing, go, got, hurt, like, put, putting, see, want, shaking
the	*My*	15y	bed, bib, birdie, book, car, coffee, cookie, daddy, floor, flower,	*the*	*my*	13x	here's, in, on, with, it, that's, uh, eat, hurt, put,	*I*	*that*	20z	broke, close, color, did, do, drawing, eat, found, get,

multi-output 204 motif			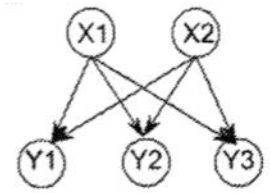	multi-input 204 motif			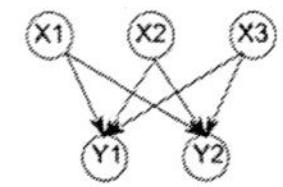	904 extension			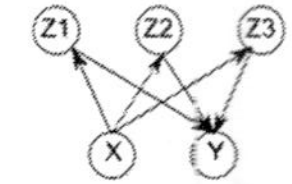
			juice, mouth, sheep, peach; another				see, want, where's				jump, like, love, play, push, pushed, put, take, throw, want, pull
I	*Georgie*	13y	do, sit, bumped, cry, eat, eating, fall, go, going, jump, lying, sleeping, bump	*this*	*the*	13x	at, in, you, draw, moving, need, open, read, see, turn, want, wash, where's	*I*	*this*	17z	color, do, draw, have, jump, like, move, need, open, play, putting, read, see, take, want, wash, cleaned
uh	*is*	13y	fish, kitty, Nomi, spoon, tail, toast, rock; for, on, a, that, Georgie's, going	*daddy*	*Nomi*	11x	that, this, toast, to, that's, uh, uhs, did, what's, where, where's	*uh*	*in*	16z	cookie, down, Georgie, Georgie's, get, go, going, look, play, put, sit, these, throw, toys, you, Daddy's
what's	*uh*	13y	Nomi, baby, daddy, fish, Georgie, Mommy, Agra, down, on, a, that, my, going	*that*	*my*	11x	all, uh, uhs, color, do, eat, like, put, want, what's, where's	*want*	*daddy*	13z	cocoa, cookie, it, juice, my, see, that, the, this, to, toast, water, peach
go	*going*	12y	down, in, on, round, under, up, get, lie, see, sleep, nightnight, slide	*this*	*my*	11x	foo, in, clean, color, do, like, putting, see, want, what's, where's	*I*	*a*	12z	bring, dropped, eat, find, get, had, have, read, reading, take, want, made
I	*gonna*	12y	do, draw, drink, eat, get, go, hurt, jump, kiss, put, throw, rest	*this*	*that*	11x	at, color, do, jump, like, play, take, want, what's, where's, who's	*uh*	*up*	12z	be, come, fish, get, go, going, it, look, me, sit, you, getting
I	*Nomi*	11y	do, doing, drop, eat, eating, going, lying, need, talk, want, putting					*you*	*it*	12z	close, closed, did, doing, draw, drink, put, take, throw, turn, want, knock
uh	*gonna*	11y	do, draw, drink, get, go, jump, kiss, lay, put, throw, donkeys								

As with the adult network, the slot-filler constructions represented by these anchors and motifs are both syntagmatic and paradigmatic in nature.[14] The syntagmatic constructions denoted by motif 904 are what are traditionally known as pronominal frames: fixed subject and object pronominal schemas with a variable verbal element (pronoun + V + pronoun) e.g. I _it , I __the, I _my, I ___that, I __this, I __a, you __it.[15] It has been suggested that pronominal frames "help children learn the transitive pattern" (Sethuraman and Goodman 2004b; Dodson and Tomasello 1998; Childers and Tomasello 2001; McClure, Pine and Lieven).[16]

As will be argued in section 9.4.3, since pronouns are easily identifiable, they facilitate motif extension and so it is no surprise that the child, Naomi, relies heavily on pronominal frames. These results support the idea that "grammatical structure does not revolve exclusively around verbs" (McClure, Pine and Lieven 2006: 716; Akhtar 1999, Childers and Tomasello 2001).

The 904 motif is also anchored by the filler word *uh* - a potential proto-morpheme (Peters 1977; Peters and Menn 1993) or a 'placeholder' (Bloom 1970) - with two separate locative elements *up* and *in*. There is not enough evidence to classify the filler as a proto-morpheme (Peters and Menn 1993). Paradigmatically, the filler word also anchors three 204 multi-output generalisations, pairing with anchors *is*, *what's* and *gonna*.

All three words include an element of (in)finiteness, the first two being finite and the third infinitival. Since the anchor pairs in this case are automorphic, one argument might assume that the filler word might be an infinitival element. However, since there are other automorphic anchors pairs that include the infinitival and the word I (*I/I'm*; *I/gonna*), then *uh*

---

[14] I refer to general syntactic categories since most linguistic discussions utilize such notions. However, I am only using these labels to contextualise the discussion. Such labels are still judged as inaccurate in describing the word distributional patterns and so only construction specific-categories are argued for here in this study.

[15] It would be interesting to observe which constructions the 904 subgraph potentially represents in a language in which the nominative pronoun is optional e.g. Catalan.

[16] Although many transitive utterances are represented by the largest 4-node motif (904), note that not all the links have been formed by the direct use of the transitive.

may be a type of referential element. Although, the prevalence of referential elements as anchors in the largest generalisations of the automorphic 4-node motifs points tentatively to the latter possibility, any hypostudy at present is speculative at best and is a matter for future research.

Other automorphic anchors are similar to those found in the adult network. The following anchor pairs have traditionally been assigned the same functional/lexical class in various linguistic theories: *I/you* (pronouns), *a/the* (determiner), *this/that* (demonstrative), *Daddy/Nomi* (proper nouns). Note that in the child network and in line with the idea of function words being the anchors of the largest extensions, the anchors *Daddy* and *Nomi* are classed as functional elements in the child network contrary to their adult network counterparts but in line with children using familiar names to recognise words (Bortfeld, Morgan, Golinkoff and Rathbun 2005).

As with the adult network, the 204 automorphic anchors represent two different tenses of the same lexeme: in the adult network *be/was*, in the child network *go/going*. In the child network, the automorphic pairs (*I/Nomi*; *I/Georgie*) show how, as in the adult network, the closeness of the traditional pronoun and noun categories. The anchor pairs: *the/my* (204 multi-input); *it/that*, *it/this*, *this/my*, *that/my* (204 multi-output) have adult counterparts: *a/you*, *my/the*, *a/I*, *I/the*, *she/the* (multi-input 204), the difference being that most of the pairs in the adult network are multi-input anchors, whilst most in the child network are multi-output. Whether this difference is significant is a matter of future research.

8.3. Non-automorphic motifs

As noted previously, triads 12, 14 and 74 and 4-node motifs 460, 908, 4548 do not possess automorphic nodes. These non-automorphic motifs highlight asymmetries inherent in language usage. Two examples will illustrate this point: the recursive *and-/of-* construction and a fragment of the Covariational Conditional. The former exemplifies gaps in usage since all possible word combinations may never be completely realized in speech (section 8.3.1), whilst the latter example emphasises the semantic restrictions imposed on elements within the construction, thus constraining word combinations (section 8.3.3).

8.3.1 Adult Network: The *and/of*-constructions

The *and*-construction and the *of*-construction is represented by highly significant motifs, namely motif 78, 12 and 14. Traditionally, recursion is expressed in hierarchical terms but it may be captured by the repeated extension of a motif. For example, motif 78 in Figure 8.6 represents the recursive *and*- construction and *of*-constructions.

Figure 8.6 In motif 78, the nodes N_1 and N_2 are automorphically equivalent. (b) The word *and* anchors the automorphic nodes *on* and *fruit*. (c) The word *of* anchors the automorphic nodes (N) *one* and *some*.

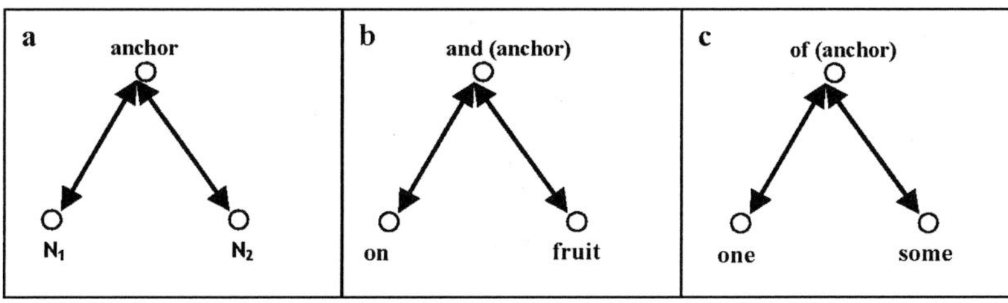

Motif 78 picks up the distributional symmetry that the same words may both precede and follow the word *and* and the word *of* (see Figure 8.6), both in separate word strings and within the same string. All words anchored around *and* and *of* in motif 78 have this same distribution. Obviously, the semantic restrictions placed on the automorphic nodes of these two constructions are very lax, allowing for an eclectic mix of different automorphic nodes (N_1, N_2...).

However, it is not the case that every word precedes and follows the word *and* or *of*. Thus, there is a gap in usage. This gap in usage is reflected in the link patterns of motif 12 (and motif 14), since both are non-automorphic motifs (Figure 8.7). That is, no nodes share automorphic equivalence.

Figure 8.7 Motif 12 is pictured with two non-automorphic nodes N_a and N_b (from the motif dictionary, see Kashtan, Itzkovitz, Milo and Alon 2002-2005c). In (b) the word *and* anchors the different automorphic nodes, as does the word *of* in (c).

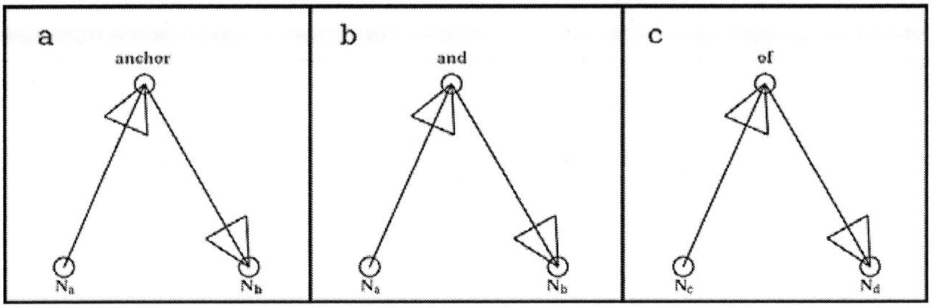

Motif 14 also represents the recursive *and-/of-* construction in Figure 8.8. It shows the middle ground between the purely bi-directional links of motif 78 and the uni-directional ones of motif 12.

Figure 8.8 Motif 14 is pictured with two non-automorphic nodes N_a and N_b (the motif is from the Mfinder dictionary, Kashtan, Itzkovitz, Milo and Alon (2002-2005c). In (b) the word *and* anchors the different automorphic nodes, as does the word *of* in (c).

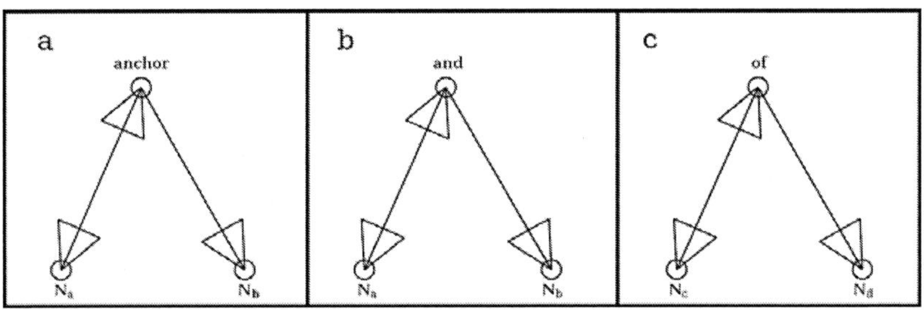

The automorphic (N) words for both the *and/of-* constructions are given in Table 8.13. The anchor words and their bigram count for motif 12 are a subset of those found in motif 14.

Table 8.13 Within a motif 14 extension, pictured in Figure 8.8, the words represented by the automorphic nodes N_a and N_c anchored by the words {*and, of*} respectively.

Automorphic node N_a	anchor	Automorphic nodes N_a	anchor	Automorphic nodes N_c	anchor
erm mend it vegetables charring clothed forty fruit furniture	and	magazines newspapers off on pastry scrubbing soap studied	and	which people supply one some those war	of

Since there is a vast potential for word combinations generated by the functional patterns of constructions, in reality there are usage gaps in the productivity of such constructions and indeed any construction, purely due to the fact that not every potential word combination has yet been motivated by a communicative intention i.e. has occurred in a given person's linguistic experience. This asymmetry in usage may seem obvious. However, the main point is that these usage gaps create certain distinctive connectivity patterns.

Consequently, such asymmetries in usage constitute significant patterns within linguistic form, as suggested by the highly significant motifs that represent them. As such, stipulating that a superior syntactic analysis exhibits symmetry is an "unwarranted assumption" in line with Croft (2009). That the motif analysis of word usage reveals asymmetrical and symmetrical patterns represented by the non-automorphic motifs and the automorphic motifs (that capture construction-specific categories), suggests that the blend of asymmetry and symmetry is fundamental to the patterns of linguistic form.

8.3.2 Child network: The *and*-construction

It is difficult to identify functional limitations in the child's constructions since the child's constructions can diverge from the adult counterparts. The child uses whatever means are available to express the intended message e.g. *bye bye sun* 'bye bye + N_{byebye}' and *all gone sun* 'all gone + $N_{allgone}$' (both examples from Naomi conversation 23, Sachs 1983). On this view, the child's constructions will be viewed as specific to the given child in the light of linguistic knowledge being based on personnel usage and experience. Since a two year child's vocabulary is very small compared to an adult's vocabulary, then certain novel usages will be expected. As such, it is tricky to know whether usage asymmetry of a given construction is actually a limitation or a gap in usage as the intended meaning of the utterance is not open to direct inspection. However, it is possible to take a child's construction and compare how the construction is represented by motifs in the child and adult network. This section briefly examines the *and*- construction.

In the child network, there are only a few examples of elements occurring both pre-*and*/post-*and*: *up and up; Mommy and Daddy; Daddy and Mommy*. Since motif 78 is insignificant in the child network, then the recursive *and*-construction does not occur above chance, being unproductive in the utterances recorded by the transcription.[17] Likewise with motif 46, the motif that also represents the *and*-construction in the adult network. However, the *and*-construction is represented by a significant link pattern, motif 12, in both the child and adult networks.

If you recall, motif 12 is a non-automorphic motif, in that the triad has no nodes that share automorphic equivalence (Figure 8.9). The motif represents the observations that, in the child network, there are a few words that precede the word *and* (and are not found post-*and*) whilst there are many more words that follow *and* but do not precede it i.e. there are many more N_b nodes than there are N_a nodes. It seems that the word *and* occurs more often in a sentential initial position implying a distinct syntactic distribution, perhaps indicating a different functional usage to the adult *and*.

---

[17] Although, it is possible that many *and*-construction utterances were found outside of the data and perhaps would have provided enough evidence for motif 78 to be judged significant. This problem is consistent with intermittent sampling.

Figure 8.9 (a) Motif 12 is pictured with two non-automorphic nodes N_a and N_b (the motif is from the Mfinder dictionary, Kashtan, Itzkovitz, Milo and Alon (2002-2005c). In (b) the word *and* anchors the non-automorphic nodes N_a and N_b which represent the words given in the table.

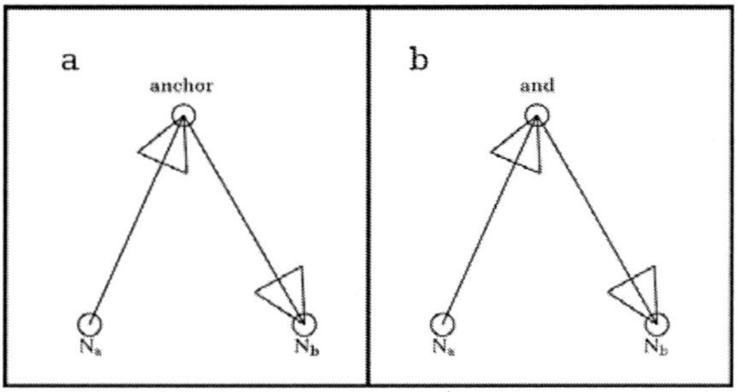

N_a	N_b
laugh, monkey, round, there	a, another, around, back, berries, broken, called, cow, doggie, don't, Georgie, green, it's, Jenko's, juice, Mom, petal, play, read, stars, sugar, telephone, then, wash, we, yellow, you

8.3.3 Adult network: The Covariational Conditional

A fragment of the Covariational Conditional (Goldberg 2006, Fillmore, Kay, and O'Connor 1988; Culicover and Jackendoff 1999) illustrates how functional constraints within a construction can lead to word combinations that are represented by non-automorphic motifs. An example of the Covariational Conditional is given in Example 8.3.

Example 8.3 An example of the Covariational Conditional 'the Xer the Y, the Zer' where X denotes *bigger* , Y *household* and Z, *more they enjoyed it.* Sourced from the adult network (BNC World 2001).

> …from what I gather **the bigger the household** that people were in **the more they enjoyed it** cos the jobs were shared out more.

As Goldberg (2006: 6) points out "The construction is interpreted as involving an independent variable…[the bigger the household that people were in]... and a dependent variable... [the more they enjoyed it]. *The* normally occurs with a head noun but in this construction it requires a comparative phrase. The two major phrases resist classification as either noun phrases or clauses".

Figure 8.10 (a) Motif 14 (b) The independent variable of the Covariational Conditional, the Xer the Y is represented by motif 14 where X denotes *bigger* and Y, *household.*

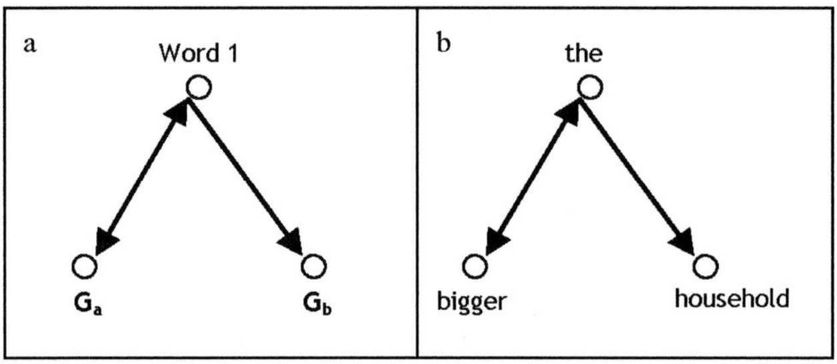

The independent variable part[18] is represented in the distributional pattern of the adult network as the non-automorphic motif 14 e.g. the Xer the Y, where X denotes *bigger* and Y, *household.* (Figure 8.10). The non-automorphic characteristic of the motif reflects the asymmetries between the words that fill the X and Y variables in the construction i.e. the X node may represent a traditional superlative whereas the Y node represents traditional nominals. Thus, semantic restrictions imposed by the construction generate asymmetries in connectivity patterns, that is, the word usage.

8.3.4 Variation in significance between the child and adult motifs

Subgraph 46 (Figure 8.11) is significant in the child network but not in the adult network. Interestingly, the child network has many instances of the subgraph which captures the phonetic homophony of the present tense first person singular conjugation (I **draw**) and the main verb's base form (I can **draw**). Even though the regularity is phonetic, there is still some functional overlap. Naturally the root meaning of capability may be reflected in both the constructions as well as the epistemic truth value of

[18] The motif potentially captures the independent and dependent variable part when these variables are single words e.g. the bigger the better.

the meaning (it is true that I draw), whilst the deontic 'permission' meaning is only suggested in specific contexts.[19]

Figure 8.11 (a) Subgraph 46. Word 1 and word 2 are automorphically equivalent and so may act as anchors or automorphic nodes.[20] In (b) the anchors are *I* and *can*, with the automorphic nodes representing {*do, draw, drink, eat, get, go, jump*}.

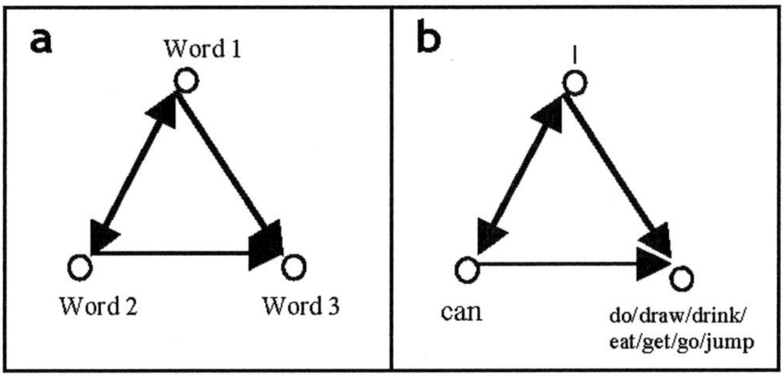

The increased significance of this relation in the child network indicates the frequent usage of the base form of the verb preceded by the pronoun *I*, arguably reflecting the tendency for children to speak often about the here-and-now. In adult speech, this tendency diminishes in relative terms, and consequently, so too does the statistical significance of the present tense-base form relation represented by subgraph 46.

---

[19] For example, having been granted permission to drink something, the child reiterates that she is able to drink "I drink…". Naturally, the habitual meaning of the present tense is not captured by the use of the auxiliary, until an adjunct is present (I can read/I can read every day). Which meaning is intended by the child using the construction is hard to evaluate from the transcription and is open to interpretation.

[20] When word 1 and 2 are automorphic nodes - with word 3 anchoring the subgraph pattern - the two largest extensions in the child and adult network have the same anchors: *the* and *a*, respectively. The word *the* anchors 11 automorphic nodes for both the adult and child network, whilst the word *a* anchors eight automorphic nodes in the adult data and seven in the child data. Given the closeness of the largest extensions for this anchor configuration, the difference in significance between the adult and the child data must lie in another configuration: when word 1 and word 2 are anchors and word 3 takes the automorphic role.

In regard to the 4-node motifs, the Z-score of motif 204 in the child network is lower than the adult network (after normalising to account for the size difference of the network) whereas the 904 motif was found in higher numbers in the child network. In regard to the 204 motif in the adult network, the largest extension of this link pattern was anchored by the traditional copula elements whereas this tendency was not so prevalent in the child network. As such, the lack of inflectional verbal elements (in traditional terms) in the child's speech is captured by the under-representation of the 204 motif.

The largest extensions of the 904 motif are mostly formed by a pronominal schema with a verbal element as the variable elements as discussed in section 8.2.6.2. The largest 904 extensions in the adult network were anchored by an inflectional verbal element and then a verbal element, followed by either the word *to* or *by* or *you*. Furthermore, the second most largest extension involved the word *and* i.e. 'and + N + in' where denotes automorphic nodes. The child's network generally lacks large motif extensions that include the word *and*.

Lastly, subgraph 4548 (Figure 8.12) has the second most significant Z-score and is highly abundant in the child network, and yet it is practically insignificant and sparse in the adult network. Interestingly, the anchors of the largest 4548 subgraph in the child network represent pronominal schemas e.g. 'I + Verbal_1 element + it', 'you + Verbal_1 element + it', 'I'm + Verbal_1 element + it', 'you + Verbal_1 element + I'. The bidirectional link represents the fact that the pronominal occurs in both the subject position and the object position with the given verbal_1 element.

Figure 8.12 The subgraph 4548 representing the pronominal schema 'I + Verbal_2 + it' (from the Mfinder dictionary, Kashtan, Itzkovitz, Milo and Alon 2002-2005c).

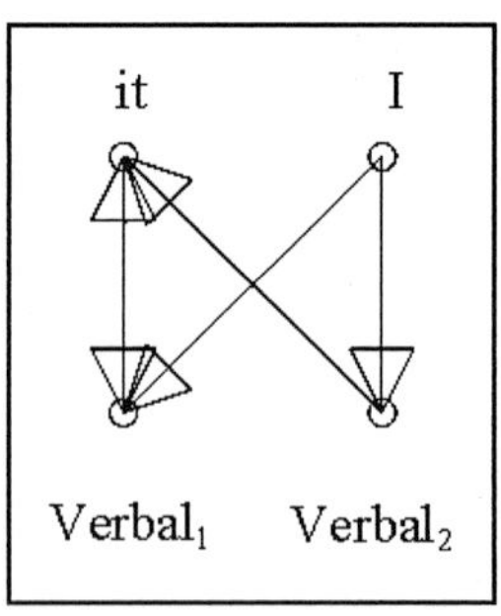

Many researchers have noted that in child speech, the word *it* is favoured more in the object position than the subject position (see section 3.5.3). This asymmetry is represented in the fact that 4548 subgraph has only one bidirectional link connecting the verbal elements and therefore, the subgraph occurs more frequently in the child network compared with the adult network. As such, pronouns tend to anchor large motif extensions (see section 9.4.3).

8.4 Summary

The largest extensions of the most highly significant automorphic triads and 4-node motifs were presented for the adult and child network. In regard to motif 6 and 36 for both networks, the majority of the automorphic nodes anchored around each of the largest motif 6 extensions were classified as either a traditional noun or verb. However, it was demonstrated that motif formation provides evidence against such traditional general categories. Instead construction-specific syntactic categories are preferred in line with Croft (2001).

The anchors of the largest and most highly significant motifs tend to be function words. Since the function words are defined by a high type and token frequency, a function word equates to a hub word. Thus, a function word and indeed the functional-lexical (f-l) distinction is generally derived from the formation of hubs i.e. optimising storage cost and connectivity. Thus, the f-l distinction is an "evolutionary spandrel" as well as the motif 6 link pattern that represents the distinction.

The property of redundancy was also exemplified by the network hubs in the form of verbal conjugations e.g. *was* and *were* that anchor the large extensions of motif 6. Within a syntagmatic environment, it was argued that this redundancy (and obligatory inflection in general) derives degeneracy, a pervasive biological property.

Unlike the adult network, the recursive structures such as the *and*-construction and the *of*- construction are not significant in the child network, represented only by motif 12 (as oppose to motifs 78, 12 and 46 in the adult network). Thus, the *and*-construction in the child network is an example of a construction represented by a non-automorphic motif. The difference in representation of the *and*-construction in the adult and child network may indicate a difference in functional usage. Differences in the significance of the motifs in the child and network were explained by the lack of inflectional elements in the child's speech, the 'here and now' character of child's speech and homophony.

In regard to the 4-node motifs, the anchors of the largest motif 904 extensions in the child network tend to be pronominal frames compared to the anchors in the adult motifs which correspond potentially to the passive construction, the *and-/of-* constructions, and the copula. The anchor pairs of the 204 in both networks represent the traditional functional categories e.g. determiner {*a, the*} and pronouns {*my, the*}. Whilst the adult anchors also represented traditional prepositional pairs e.g. {*for, in*} and conjugations e.g. {*was, were*}, these types of anchor sets did not occur in the largest extensions of the 204 motif in the child network.

It was argued that the highly significant non-automorphic motifs capture asymmetries in word distributions. Two types of usage asymmetry were discussed in the adult network. Firstly, a gap in usage may be due to an unrealised communicative intent. Secondly, semantic restrictions imposed on the elements within a construction cause gaps in usage. In the child network, since constructions may differ from adult constructions, it is difficult to ascertain whether such asymmetries are due to semantic restrictions on the construction's components or to gaps in usage. Since these asymmetries were represented by highly significant motifs in the word co- occurrence network, asymmetries in usage are just as important as symmetries in characterising syntactic patterns (in line with Croft 2009).

The next chapter sketches how abstract syntax may arise from a more item-specific (usage) basis. Furthermore, the chapter suggests how the hubterranean structure may be present on other theoretical levels within syntax.

CHAPTER NINE

THE HUBTERRANEAN STRUCTURE AND ITS SYNTACTIC IMPLICATIONS

9.1 Overview

This chapter will discuss the theoretical implications of a hubterranean structure representing syntactic structure at various theoretical levels e.g. in respect to both the characteristics and the learning of syntax. In section 9.2, it is argued that the task of segmenting linguistic input involves the child attending to both transitional probabilities and type frequency, in line with the adult's strategies when learning artificial languages. These two properties are related to the two constraints of the optimisation process e.g. storage cost is inversely proportional to transitional probability whilst maximal type frequency relates to the drive towards a highly connected universal hub. It will be proposed that the mechanism that facilitates efficient storage of the maximal amount of connections not only provides language with its characteristic function words and syntactic constituency but the mechanism also segments the input too. Thus, storage and segmentation are one in the same task.

In section 9.3, a sketch[1] is given of how potential construction-specific syntactic categories, anchored by the largest extensions of motif 6, may be identified. It has been demonstrated that function words anchor large motif 6 extensions, whose automorphic nodes may represent a non-general syntactic category (chapter eight). It has already been established in section 2.8 that function words may be identified by their hub characteristics namely, high type and high token frequency and these characteristics are evidenced in the adult word co-occurrence network (section 9.3.1). Since the highest ranking words in the power laws of degree and frequency correlate with function words, the interception of

[1] The mathematical details of this sketch are beyond the scope of this study and are a matter for future research.

these distributions i.e. the two power laws with their truncated tails, potentially identify function words.

Once the function words have been ascertained, the automorphic nodes that are anchored by the respective function word in a motif 6 may be reduced to a discrete (construction-specific) syntactic category (section 9.3.2). In regard to extending constructions, section 9.4 examines how the hub structure of a linguistic network and the constraints that generate this structure explain the factors affecting early verb generalisations. Sethuraman and Goodman (2004a) suggest five factors that promote early verb generalisation: semantic lightness of a verb (section 9.4.2), co-occurrence with a pronoun (section 9.4.3), co-occurrence in a varied syntactic environment (syntactic diversity; section 9.4.4), consistent usage in one environment (syntactic consistency; section 9.4.4) and the (token) frequency of the verb (section 9.4.1) which is said to confound the other four factors. I will argue that all these properties relate to the characteristics or the formation of a hubterranean structure. Since it will be argued that these network properties allow words to be more easily identifiable, then verbs generalised early will be associated with a hubterranean structure.

Since it is likely that the verb-construction network may be resultant of an optimisation process of storage cost and connectivity, very general predictions concerning the characteristics of constructions and their distribution fall out from this hypostudy (section 9.5). These very general predictions are based on the power law distributions of frequency (section 9.5.1), degree (section 9.5.2) and link length (i.e. storage cost; section 9.5.3) that would be expected to occur in an optimised verb-construction network.

In section 9.6, the four psychological principles that underpin the relations between a given construction's meaning-to-form pairing in the construction inventory and the size of the inventory[2] are contextualised within the optimisation process. All four principles relate to reducing storage cost and maximising connectivity within the construction inventory. Furthermore, the principles observed in first language learning e.g. Culicover's (1999) notion of *conservatism* and *attentiveness* as well as Israel's (2002) *local* and *global consistency* are manifestations of the two optimised factors i.e. cost and connectivity.

---

[2] See also Ferrer i Cancho and Solé (2003a) in respect to symbolic mapping between "objects of reference" and signals.

Section 9.7 summarises the linguistic structures that are represented by the optimisation process of cost and connectivity: the word co-occurrence network, verb-construction network, the network formed from the construction's form-to-meaning pairing, the construction inventory as well as the child and adult artificial language learning strategies. As a consequence, hubterranean structure is exhibited on many different theoretical levels.

However, the first linguistic topic under discussion is the segmentation of the input by the child.

9.2 Segmentation

In chapter three, studies evidenced how segmentation can be facilitated by a child attending to the transitional probabilities in the input (see section 3.4.1). Even when the knowledge of transitional probabilities is freely accessible, constraining the two computations fundamental to learning a language - segmentation and linear sequencing - would still benefit the learner (see Yang 2004). It has been noted that there are many solutions to dissecting a language and discriminating the right one is no trivial matter. Since keeping tabs on the transitional probabilities of each bigram in the input is often assumed to be rather 'computationally' costly, it has been questioned as to whether 'computations' are performed early on in language development.

Instead, it has been suggested that segmentation is initially a random affair (Perruchet and Vintner 1998) with experience reinforcing observable patterns, with chance patterns dropping by the mnemonic wayside. Repetition is the key here. Words (or part words) are more likely to be repeated in the input than between-word sequences. Since "perception shapes internal representation", repeated percepts are also re-enforced, so that "lasting internal representations are more likely to match a word or part of a word than a between-word segment" (Perruchet and Vintner 1998: 249).

However, early segmentation need not be an entirely random nor a computationally costly affair. In fact, the child need not even be genetically predisposed to a segmentation strategy; at least at a word-constituent level. This study proposes that the mechanism for storing linguistic constructions - the optimisation of minimising storage cost whilst maximising connectivity - is the same mechanism that creates certain properties in the language input; specifically, the pattern of (inverse) bigram transitional probabilities in the input which follows a power law and thus represents constituency at a word level. Thus, the linguistic signal, when stored, will then be segmented by a power law relation into linear sequences by virtue of it being stored as a hubterranean structure, generating the characteristic syntactic constituency as discussed in chapter five. Consequently, words and basic constituency will have been identified.

On this view, the constraints that form the hubterranean network representing the stored linguistic signal are responsible for the capacity to learn this linguistic form. Language is relatively easy to segment since the same mechanisms that created the patterns in segmentation are the same ones that the child uses to segment the input. In short, the segmentation process is not random but instead the actual process of storing the linguistic signal into chunks in which the transitional probabilities correspond to a power law relation and the connectivity between segments is maximised.

Segmenting linguistic input by applying the transitional probability and connectivity relation provides a good starting point for the child. There is evidence that transitional probabilities presuppose other segmentation cues and pave the developmental way for other segmentation skills such as the learning of stress patterns (see Thiessen and Saffron 2003). Thus, other cues such as prosody, function words (see discussion on anchors section 3.5.1) and morphology (Morgan, Meier and Newport 1987) assist the fine tuning of the 'chunking' process. Of the cues, only prosody is specific to language, whilst the segmentation and linear sequencing skills of transitional probabilities, identifying function words and perhaps the use of morphology may not be particular to language learning but rather domain-general and network-specific. This view is supported by function words being represented as hubs in the word co-occurrence network.

One criticism of the artificial languages used to assess the sensitivity to transitional probabilities is their uniform syllabic patterns (Yang 2004). In the case of certain studies, words in the artificial language were all trisyllabic (Saffran, Aslin and Newport 1996; Saffran 2001). In natural languages, the number of syllables varies greatly. In fact, the most frequent words in a given language, namely the function words, are typically short and often monosyllabic (Morgan, Shi and Allopenia 1996). Yang (2004: 452) using a sample of English child-directed utterances randomly collected from the CHILDES database (MacWhinney 1995), found that "on average, a monosyllabic word is followed by another monosyllabic word 85% of the time". Unfortunately, monosyllabic bigrams do not provide a distinction between the probabilities of phonemes occurring within a word and between words.

To test the accuracy of segmentation, bigram transitional probabilities were calculated for the random sample of utterances and these were used to identify word boundaries, thus segmenting the utterances (Yang 2004). However, the results were poorer than expected:

> *41.6% precision rating* determined by the number of correct words divided by "the number of all words extracted by the segmentation process" and *23.3% recall rating* relating to "the number of words correctly extracted by the segmentation process divided by the number of actual words in the text" (Yang 2004: 452).[3]

The network theory investigated in the study offers an explanation as to why monosyllabic words are not a problem for the network approach to segmentation. The segmentation process is actually the optimised process of storing the linguistic stimuli as a hubterranean network. As such, the process not only attends to transitional probabilities but also maximises connectivity to generate the highly connected and frequent hub nodes that represent function words. Indeed, the next section will discuss how such function words are identified in the network. Thus, the segmentation process attends to word connectivity too and involves the formation of hub words. Since monosyllabic words tend to be highly frequent and highly connected syllables i.e. have a high type frequency, it would be expected that monosyllables would be analysed as syntactically independent and given hub word status.

For example, in the child text examined in this study, hub words tended to be monosyllabic words, with the exceptions of familiar names that were highly frequent in the child's conversation (frequency ranking: *Daddy*(10), *Mommy*(13), *Georgie*(19) - also see Bortfeld, Morgan, Golinkoff and Rathbun 2005) and the words *going* (ranked 24th for frequency) e.g. *the, it, i, uh, a, you, my, on, that, is, this, want, in, up,* ***going, Mommy, Georgie, Daddy, I'm,*** *go*. Attention to the frequency of words allows potential hub words i.e. function words, to be identified. Consequently, the segmentation

---

[3] After a mechanism is built in to distinguish words by primary stress i.e. words only have one primary stress, then the accuracy increases to "precision of 73.5% and recall of 71.2%" (Yang 2004: 452). However, this strategy does not help identify function words since they are usually unstressed (Yang 2004).

of monosyllabic words is accounted for by this approach.

On this view both types of knowledge, both transitional probabilities and type frequency, are required for segmentation - that is, both are required to form a hubterranean word co- occurrence network to optimally store the linguistic form of utterances. These two factors also correlate with the two methods of extracting words from an artificial language.[4] If you recall from chapter two and three, type frequency i.e. connectivity of words, and transitional probabilities were utilized as two learning strategies for identifying words (Gómez 2006).

Once the segmentation of specific words and phrases is underway, to extend his/her linguistic knowledge, the child will then need to extract regularities from this linguistic experience creating syntactic abstract representations. The next section will examine how the distribution of connectivity i.e. degree and frequency assists in establishing a non-general syntactic category.

---

[4] By assuming cross-linguistically (in the spirit of Dryer 1997 – see section 1.3) that certain syntactic properties derive from the optimisation process discussed in this study, this theoretical move has interesting consequences for establishing what a word is when semantic information is not available. On this view, a word would not be defined singularly - in respect to its linguistic form - but as a collective distribution pattern. Three such collective patterns are the power laws of frequency, degree and storage cost values i.e. inverse transitional probabilities.

This definition is especially useful if the smallest units of the communication are not known and so appealing to high and low transitional probabilities of these pairs to segment into words is not possible. In this case, to identify a word, one would then chunk the input by establishing a power law function of frequency and connectivity (degree). Once these power law distributions are established in any piece of text or communication that is derived by the above optimisation process, the components of these distributions will represent the smallest unit of syntax i.e. words or possibly morphemes in languages whose relations are morphosyntactic. Indeed, the interception of the frequency and degree distributions would indicate function words. This method may prove fruitful as a first parse when the semantics of a given text is not available as in the case of animal communication. Naturally this proposal rests on the argument (with no lack of controversy) that animal communication is based on the same principles as human communication (see Ferrer i Cancho and Lusseau (2009: 23) in regard to frequency and "dolphin surface behavioral patterns"). There is enough that we do not know about non-human communication to warrant stating that the jury is out on this debate, at least for the present time.

9.3 Identifying (construction-specific) syntactic categories

Identifying a construction-specific category requires a distributional analysis of linguistic form. A *functional* distributional analysis is preferred since linguistic form is only one half of the form-to-meaning pairing of a construction. However, as a first pass at identifying syntactic categories, it would be pertinent to examine word co-occurrence patterns.

The motif analysis described above and in chapter seven, is a very richly detailed distributional analysis. As such, motifs capture major patterns in the links of the network and consequently represent productive distribution patterns in linguistic form. As detailed in section 8.2.1-4, the automorphic nodes of motif 6 and 36 generally constitute members of the traditional nominal and verbal categories. Therefore, these two automorphic motifs are potential candidates for identifying a syntactic category (aside from semantics considerations, section 2.8.6).[5]

The word co-occurrence network, however, exhibits many link patterns and so a mechanism is required that can cut thorough this complexity and identify these automorphic nodes. One way of recognising large extensions of the motifs 6 and 36 is by their anchor, which tends to be a function word. In turn, function words are recognised by their high type frequency i.e. a high degree and high token frequency as discussed in chapter two. This correlation is evidenced in the word co-occurrence network and is discussed in the next section.

9.3.1 Type and token frequency of a function word

For the smaller adult word co-occurrence network, it is found that words with the highest degree are the traditional function words. The most highly connected word is *and*, co- occurring with 286 different words, followed by the word *the* co-occurring with 219 different words. It is only until the connectivity drops to twenty four, that the first lexical word appears in the list: *women* is ranked 37 out of 1217. Those twenty four connections equate to eleven percent of the links of the most highly connected word *and*. The percentage of function words and lexical words in the ranked connectivity list are shown for the smaller BNC text (Figure 9.1).

[5] For languages in which verbal elements occur pre-nominally, the automorphic generalisers of motif 36 would potentially form a syntactic category.

Figure 9.1 The distribution of function and lexical words in the list ranked by connectivity (in the adult network).

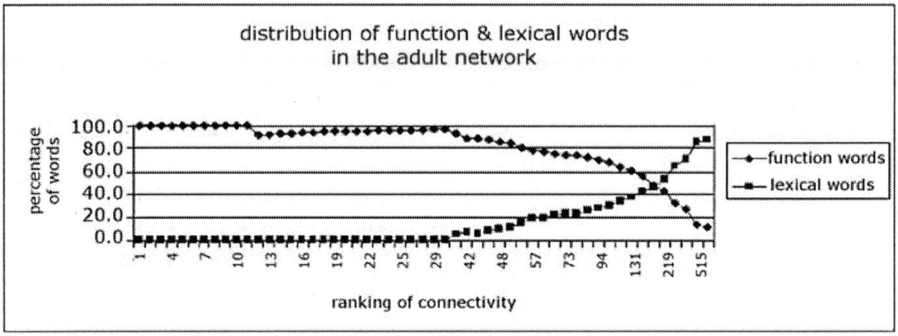

Function words also tend to be highly frequent in speech. Figure 9.2 shows the percentage of function words and lexical words in the frequency ranking list for the smaller BNC text. The function words clearly have a high ranking for frequency.

Figure 9.2 The distribution of function and lexical words in the list ranked by frequency (in the adult network).

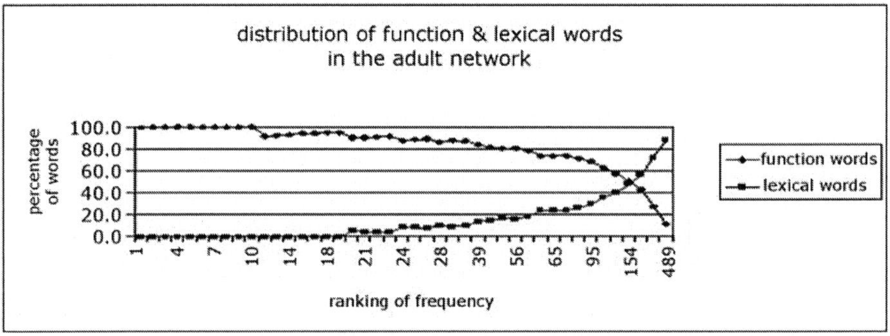

It is pertinent to note that the eight most frequently used words are ranked identical to the eight most highly connected words. The most frequently occurring word in the smaller BNC text is the word *and,* with 219 appearances, followed by the word *the* occurring 204 times. The first lexical word has a token frequency of twenty eight which is thirteen percent of the highest frequency value.

Indeed, this high frequency and degree correlation for function words can be seen in Figure 9.3. The figure shows the fifty most highly connected words with their frequency and degree ranking i.e. the token and type frequency ranking. The words after these fifty are connected to less than twenty other words. The rankings match well, particularly for the first twenty-five words, considering that the lowest connectivity ranking is 1217, and lowest frequency ranking, 489. The Pearson correlation coefficient (0.95) denotes a strong correlation for the first twenty five degree and frequency rankings and for the fifty highest rankings, the co-efficient is a high 0.89. Most words are function words, the exceptions being *erm/er*, *women*, *very*, *know*, *well*, *people*, *got*, and *time*. However, as noted before, these words have eighty two percent less connections than the most highly connected word.

Figure 9.3 The connectivity and frequency ranking for the fifty most highly connected words in the adult network.

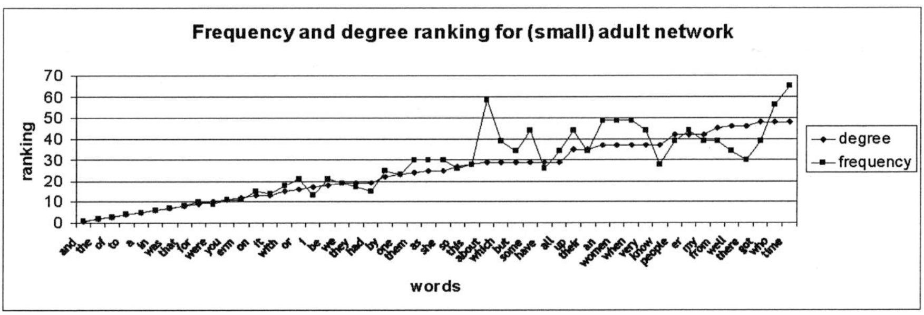

Consequently, function words are hub words in that they are generally identifiable by their high connectivity and high frequency.[6] Thus, when these two characteristics coincide, it is likely (although not guaranteed) that the word will be a function word. Since, the degree and frequency distributions (characterised by power laws with truncated tails) coincide at a high ranking, this characteristic may be used to identify the function words within the interception.[7] On this view, it is proposed that hub

[6] That function words are hubs and anchors of the largest extensions of 3- and 4-node motifs provides an explanation as to why neologisms for function words develop slowly over time. The potential function word would need to build up links within the network, becoming an anchor for the largest n-node extensions.

[7] There is evidence that power laws do interact in a meaningful way. The exponents of the power laws that represent "the probability of a discrete magnitude

characteristics would then aid the detection of these words in a network.

9.3.2 Generating a new syntactic category

Function words tend to anchor large extensions of motif 6 whose automorphic nodes are potential members of a non-general syntactic category (as discussed in chapter eight). Thus, to create a syntactic category, the automorphic nodes within motif 6 that share semantic regularities, are ascertained. Automorphic nodes with these regularities are then represented as a set and denoted by a discrete abstract element i.e. the syntactic category (see Figure 9.4). An advantage of these Boolean type categories is their computational efficiency. Not only are the set of automorphic nodes more easily identified but also the set is more easily manipulated in syntactic computations. For example, the syntactic category assists the analogical process that mediates novel usages.[8] By grouping exemplars of previous usages, if the new word is consistent by analogy with these exemplars (in respect to semantic considerations), then it may then be used in the construction

Figure 9.4 Motif 6 (left) is reduced to an isomorphic mapping (right) of the given function word and the syntactic category which represents the automorphic nodes $\{N_1, N_2, N_3, N_4\}$.

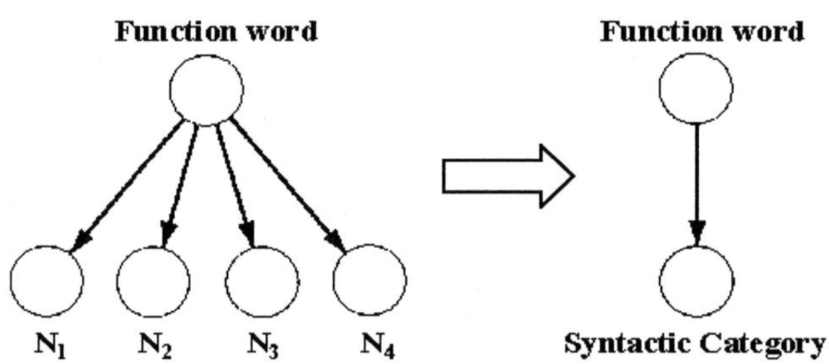

To form a syntactic category, the set of automorphic nodes $\{N_1, N_2, N_3, N_4\}$ are collapsed into a discrete element and a (near-) isomorphic relation is formed with the respective function word (see Figure 9.5A). For this

x" and "the normalized value of x versus rank" coincide with a value defined as the golden number (see Ferrer i Cancho and Hernández Fernández 2008: 2).

[8] See Bod (2009) for a treatment of language learning that felicitously generates phrasal tree structures using the principles of structural analogy and least effort.

step to occur, certain parts of the network must move away from the optimal or critical point, and computationally change state to a more isomorphic form.[9] As such, movement away from the optimal point of the network will be motivated by the increase in the *computational* efficiency of representing the automorphic nodes as a discrete element.

Naturally, displacement from the optimal point within the restricted part of the network is temporary. Once the syntactic category is established, there is movement away from the isomorphic state towards the (near-) optimal point of the network. Movement towards the universal hub increases automorphicity in the network since the universal hub represents the maximal automorphicity possible in the network (when the connections to the universal hub are uniformly directed).

Consequently, the relation between the syntactic category and the function word becomes automorphic. Since the category is automorphically equivalent to the function word, the category links to the automorphic set of nodes $\{N_1 ... N_4\}$. Thus, the motif 6, anchored by the function word, is duplicated. The resulting 204 motif is composed of a function word and a syntactic category which both anchor the same set of nodes $\{N_1 ... N_4\}$ (see Figure 9.5B). In other words, the relations between the function word and the set $\{N_1 ... N_4\}$ are analogous to the relations between the syntactic category and the set $\{N_1 ... N_4\}$.

The only difference is the *specific* relations the links signify. The links from the function word signify precedence relations courtesy of the temporal linearity of speech. The links from the syntactic category represent positional or spatial information, since the nodes in the set $\{N_1 ... N_4\}$ all occur in the same position relative to the function word. Note, that the generated syntactic category shows signs of a 'tinkered' evolutionary origin[10] (see Jacob 1977; Solé and Valverde 2006, Figure 1).

---

[9] Interestingly, in Ferrer i Cancho and Solé's (2003a: 788) model of a bipartite network composed of signals and "objects of reference", there is a sharp transition between a one-signal-to-all-objects language represented by a universal hub type structure and an isomorphic signal-to-object mapping. Within this transition of state, the frequency power law i.e. Zipf's law, is found.

[10] Since tinkering motivates the duplication of structure, automorphism in form underlies this process. Conversely, the biological property of 'degeneracy' i.e. two different structures serving the same function, underpins automorphism of function (see section 8.2.1.1). Degeneracy is also exhibited in the generation of a syntactic category, when attending to the nodes in the 204 motif. The function word and the

Tinkering is a process that reuses the same structural form for different functions (see Jacob 1977). In this context, the duplicated structural form is the link configuration of motif 6 whose function expresses either temporal or spatial relations.

Figure 9.5 (A) The isomorphic relation between the function word (f-word) and the syntactic category (s- cat) is represented by the bold dotted line. (B) As the system moves towards a more automorphic state, the relation between the f-word and the s-cat becomes automorphic (dotted line). (C&D) The links form a 204 subgraph that straddles two levels of linguistic representation.

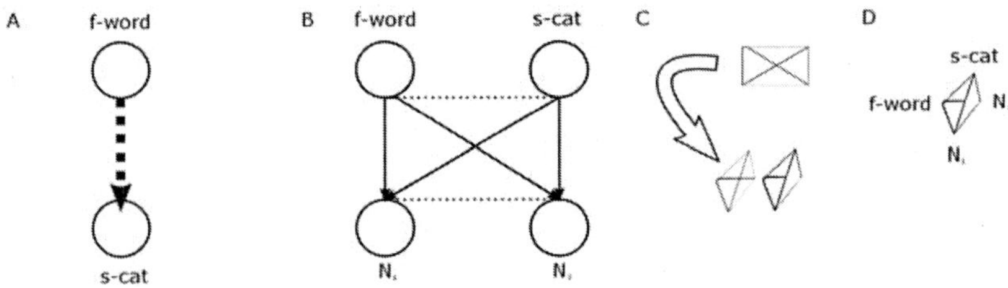

In Figure 9.5, notice how the 204 motif forms a tetrahedron when the automorphic relations and links are viewed as geometric dimensions (see Figure 9.5 C&D; a smaller set of automorphic nodes are used to simplify the figure). The automorphic relations produce a new syntactic level of analysis in which the syntactic category is established. Note also how disassortative mixing (see chapter five) aids this level formation, since high frequency elements are easily identified and "speakers analyze low-frequency items more than high- frequency items" (Bybee 1985: 133). Thus, the frequency asymmetry inherent in the motif connections (due to disassortative mixing) facilitates the development of syntactic categories.

---

syntactic category inhabit different structural levels and yet both identify the same set of nodes i.e. they perform the same function. Thus, tinkering and degeneracy correlate closely (in line with Solé, Ferrer i Cancho, Montoya and Valverde 2003: 23). Furthermore, in respect to syntactic category formation at least, it seems that tinkering derives degeneracy, since it is the duplication of motif 6 that establishes the different syntactic level. On this view, tinkering and degeneracy would be inherent properties of a system in which automorphic relations are encouraged or maximised.

Figure 9.6 Escher's 'up and down' (1947) lithograph (© Cordon Art - Baarn - the Netherlands), illustrating the representation of the same object (the buildings and the people) but from two different perspectives. In the same way that the ceiling of the eaves and the floor intercept, both the specific function word and the abstract syntactic category form part of the 204 motif.

In the same way that Escher's 'up and down' lithograph illustrates the idea of opposite perceptions of the same scene, likewise, specific and abstract (syntactic) elements reflect the same linguistic knowledge but from a different perspective. The 204 motif that creates the new perspective is likened to the mid point of the drawing which sees the ceiling of the eaves being interpreted as the bottom of thc floor. Thus, within the interception of these 'up-and-down' views resides the contrary characteristics of the object. In the same way, the 204 motif is composed of both specific and abstract elements.

Note, too, how the syntactic category is neither atomic nor primitive (Croft 2001). As Croft (2001: 47) states "ATOMIC units are those that cannot be broken down into smaller parts in the theory. PRIMITIVE units are those whose structure and behavior can not be defined in terms of other units in the theory". A syntactic category is non-atomic in that it is composed of a set of automorphic nodes and it is not a primitive in that it is defined by its respective function word.

In sum, movement away from the (near-) optimal point in the network establishes isomorphism between a set of automorphic nodes and the function word. This movement is temporary and the system returns to the optimal point inducing automorphic relations between the function word and the syntactic category. The motif 6 duplicates to form a 204 subgraph.

A new dimension is established in the network namely, the paradigmatic/spatial dimension, separating the function word and the syntactic category.[11] Consequently, a new level of (abstract) syntactic structure is born.

The new syntactic category restricts the usage of words within a given construction, constraining the addition of new connections and the growth of the network in a way that is consistent with the network's functional patterns i.e. the constructions in the network. Consequently, the network maintains its (reasonably) optimal state. For example, productive patterns whose semantic considerations are eclectic will accept new members more readily due to the lack of semantic restrictions. A case in point is the *and*-construction. This already highly productive construction will then become more productive.

---

[11] A general question is whether, in other systems, the genesis of spatial dimensions derives from temporal automorphic relations.

However, partially productive patterns with a member set whose semantics are highly restricted, will tend to retain their partial productivity when new words enter the network. As Goldberg (2006: 51) observes, "the main verb in the VVingPP construction is not very productive" and provides the following examples *Bill went singing down the street* and **Bill raced/walked whistling down the street.* The semantics of the main verb is restricted to the verbs of movement e.g. "{go, come, run, take off}" (Goldberg 2006: 52).

On this view, the connectivity of the node representing the syntactic category will remain consistent with respect to the other nodes in the system. That is, already productive categories will grow quickly whereas non-productive categories will remain so. As such the characteristics of preferential attachment may theoretically be exhibited i.e. the global statistics of the network is maintained (as discussed in chapter four, section 4.4.1). Once the majority of the constructions present in the input are established within the adult's linguistic network, the addition of new words will follow the pre-established functional patterns and therefore re-enforce the specific patterns in the network. Consequently, identifying preferential attachment within the network confirms stabilisation of construction learning, either through the native-like attainment of the given language or alternatively through the language learning reaching a developmental plateau.

On this view, the process of optimizing storage cost and connectivity, and the computation process form a cycle. The former process derives the power laws with truncated tails that facilitate the identification of syntactic categories. Syntactic categories then constrain which new connections are constructed, thus confining the word co-occurrence network to a reasonably optimal state. Thus, the frequency and degree distribution as well as the syntactic categories are all derived from and re-enforce the optimisation of the network.[12]

Since there are three truncated power laws in the network, it is interesting to note that the interception of the frequency and storage cost distributions identifies strong collocations or "frequent utterance schemas" (Tomasello 2006). For example, *I think*, *I mean*, and *I don't.* The word pairs have a very low storage cost and yet the first word in the bigram is highly

[12] That the frequency power law or Zipfs Law "is required by symbolic systems" (Ferrer i Cancho and Solé 2003a: 791) suggests that symbolic mapping supports the optimal state of the word co-occurrence network.

frequent. These collocations may be viewed as very basic expressions on which to build more complex utterances using complementation (in line with Kidd, Lieven and Tomasello 2004: 57). As such, the distribution of frequency interacts separately with the distribution of the two contrary factors that are optimised in the network i.e. storage cost and connectivity. Their interception identifies syntagmatic and paradigmatic structure respectively i.e. "frequent utterance schemas" (Tomasello 2006) and abstract syntactic categories.

In respect to language learning, the child's inventory of constructions differs in size and complexity from the adult's. With every new construction learnt, new functional patterns emerge. Consequently, certain properties of the child's word co-occurrence network will change quite extensively until an adult-like mastery of the language is obtained. It seems very likely that the power laws and the truncated tails of frequency and degree are retained (as evidenced in chapter 6).

Thus, the mechanism for generating syntactic categories which maintains the consistency of the functional patterns is also in place in the child's (near-) optimised network (in line with Culicover 1999 and Israel 2002 as discussed in section 3.3). Since the correlation between degree and frequency is weaker in the child network than the adult network (see Figures 9.3 and 9.5), generalisations will be more difficult for the child as the areas in the network that are potential syntactic categories will be harder to pinpoint (in line with Boyd and Goldberg forthcoming; see section 9.4).

In fact, since the power laws mediate the identification of a syntactic category, it is plausible that other abstract syntactic elements and relations only develop once the network is optimised. Indeed, children between 18-20 months utilize item-specific constructions (i.e. construction islands) before moving on to abstract constructions (for discussion, see Tomasello 2006). Thus, "abstraction…increases with age"[13] (Borensztajn, Zuidema and Bod 2009: 186). However, a more detailed examination of this issue is a matter for future research.

---

[13] Abstraction is "defined as the relative number of nonterminal leaves in multiword constructions", using tree structures (for discussion, see Borensztajn, Zuidema and Bod 2009). Note that this analysis does not adopt construction-specific categories.

Another difference between the adult and the child network is the role of the external input. Whilst the functional patterns in the external environmental input closely match the linguistic knowledge within the adult network, for a young child this external input differs extensively from the functional patterns within his or her network. Although the generation of abstract syntax preserves the consistency of the patterns in the child's network, the child also learns new constructions from the linguistic environment e.g. certain specific word combinations may be learnt by the partial repetition of an adult utterance. On this view, the adult network is self-sufficient and self-consistent, whereas the child network, although self-consistent, is dependent on the environment for linguistic knowledge i.e. the child joins the linguistic network as proposed by Ninio (2006a).

The next section will develop the idea of extending constructions with the addition of new verbs from a network perspective. It will be shown that hub formation and therefore, the optimisation process is implicated in the factors encouraging the extension of a verb to a new construal. Note that the term verb is used in the tradition sense to be consistent with the language literature.

9.4 The five factors constraining early verb generalisation

How do child shift from the construal of a verb in a specific construction to generalising i.e. extending the verb to a new syntactic environment? As noted in chapter three, Sethuraman and Goodman (2004a) identify five key factors that constrain which verbs are generalised early: frequency, semantic lightness, pronouns, syntactic consistency and syntactic diversity - with the first factor, frequency, confounding the other four factors. Whilst each factor is not sufficient alone to explain verb generalisation, when taken together, the interaction of the five factors provides a convincing account of early verb generalisation.

Naturally, it is interesting to find out why these factors in particular are the key to early verb generalisations. This issue is where the network approach can shed light. Relating these factors to the hubterranean network provides an explanation of why these particular factors are implicated. In section 9.3, it was proposed that hubterranean structure facilitates the identification of potential syntactic categories and thus aids the formation of a linguistic generalisation. It will be argued in this section that the verbs generalised early are more easily detected (see Boyd and Goldberg forthcoming).

9.4.1 Frequency

Verbs that occur frequently in both the adult and child speech tend to be generalised early (for discussion see Theakston, Lieven, Pine and Rowland 2004; see also Theakston, Lieven, Pine and Rowland 2006; Sethuraman and Goodman 2004a.). A significant characteristic of a hubterranean structure in a small world network is the frequency distribution of its elements, captured by the power law function. This function was observed in the frequency distribution of words in the child network (see chapter six) – as well as in the adult network (chapter five). As will be shown in the next section, verbs that occur frequently also tend to be highly connected. That is, the verbs are hub words which have a high degree or type frequency.[14] As argued in section 9.3, high type and token frequency facilitate the identification of an element. Since hub words have these characteristics, verbs that are hubs are generalised early as they are easy to find in the network.

9.4.1.1 Correlating the token and type frequency of a verb

It is observed that highly connected verbs tend also to be verbs that are used highly frequently. In Figure 9.7 below, the most frequent verbs are shown on the x-axis with their respective ranking (in respect to other words within the network). In the graph, the same verbs with their respective degree i.e. connectivity rankings are shown (as a pink box). Noticeably, all the verbs tend to have a high degree ranking.[15] Even the lowest degree ranked verb *look*, which has a degree ranking of 124, is still highly ranked, in that it figures in the top twenty percent of the degree ranking values (considering the lowest degree ranking is 657).

Another advantage of being a hub verb is that these verbs tend to have more 'chances' to form new generalisations. To understand why, an example of a highly connected verb *draw* is given. The verb is depicted in Figure 9.8 with its outward links pointing at the words: {*this, it, man's, on,*

---

[14] As you recall, type frequency has previously been implicated in the measure of a construction's productivity (see Goldberg 1995: 134-6; see Bybee 1985: 133 for morphological rules). It is important to recognise that productivity and early generalisations of verbs are two very different linguistic perspectives. Whilst there may be an overlap in the perspective, the two notions pose different questions.

[15] The Pearson co-efficient for the verb's degree and frequency rankings given in Figure 9.7, apart from *look*, is 0.78. Including the verb *look,* the co-efficient drops to 0.5.

cloud, them, eye, eyes, man, e, smoke, flower, donkeys}. These links represent the co-occurrences: *draw this, draw it, draw man's* and so on. The inward links point from the words: {*uh, the, I, you, gonna, going, I'm, can*} and represent the co-occurrences of *uh draw, the draw, I draw,* etc.

Figure 9.7 The comparison of degree and frequency rankings of the most frequent verbs in the child word co-occurrence network.

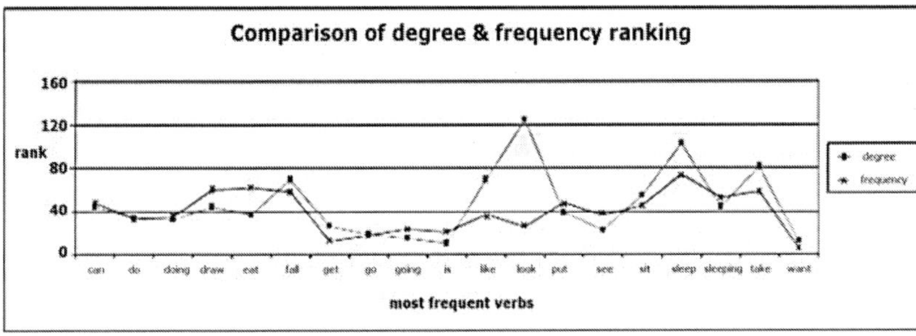

As shown by the Figure 9.8, the word *draw* does not co-occur in the child speech with the word *don't* as there is no link between the two words. But it easy to see that generalising *draw* to the new syntactic environment representing negation, is very possible since the node representing *don't* anchors a reasonably sized 904 motif generalisation shown in Figure 9.9 below.

The dotted arrow in Figure 9.9 represents a potential link between *don't* and *draw* i.e. a potential co-occurrence of *don't draw* in a future utterance. Notice that the potential link creates a larger extension of the 904 pattern with just one link. The important point is that *don't* is a highly connected word itself and so with *draw* being highly connected, there is more of a possibility that *draw* will be connected to an element that is part of a construction anchored by *don't.*

Figure 9.8 The connectivity of the word *draw* in the child's word co-occurrence network. The picture is generated by PAJEK (de Nooy, Mrvar and Batagelj 2005).

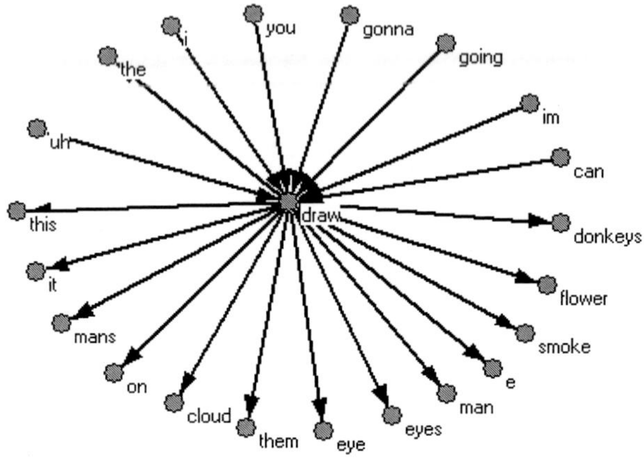

Figure 9.9 The 904 motif anchored by *don't* and *it* from the child network. The dotted links represents the potential formation of a new 904.

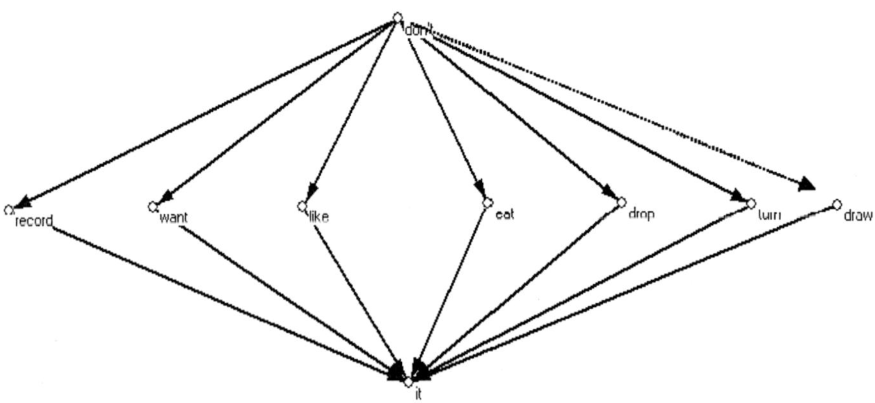

When such a scenario is realised, a verb generalisation can occur easily with very little link cost i.e. storage cost, in that the generalisation is made by forming just one new bigram relation i.e. *don'tÆ draw*. In this case, the word *it* is the element that facilitates the generalisation of draw in the 'don't + N + it' construction, where N represents the automorphic nodes. On this view, generalisations - barring semantic mismatches with existing automorphic nodes and statistical pre-emption (Goldberg 2006) – will occur more favourably when one potential link is able to extend a motif pattern.[16]

Exceptions that may occur - in that there are hub words that are not generalised early - are explained by appealing to the drive behind the creation of generalisations. A communicative context is required to motivate a generalisation. If a context is not there then there is no reason to create a generalisation with a given verb. Whilst it is not likely that poorly connected words will often be generalised, there may exist cases in which this happens if the communicative intent desires so.

In summary, verbs that are generalised early tend to be highly frequent verbs. High frequency verbs in turn tend to be highly connected words. As such the verb is highly identifiable within the network and the high number of connections ensures that the word is indirectly connected to many other motif patterns, allowing new extensions to be formed. Although frequency is cited as a key property of early verb generalisation (Sethuraman and Goodman 2004a, Theakston, Lieven, Pine and Rowland 2004), this study emphasises that the hub structure is the crucial feature; frequency being one of the significant properties distinguishing a hub word. Therefore, verbs that are more likely to be generalised are represented as hubs in the child's word co-occurrence network. On this view, early generalisations reinforce the growth of a hubterranean structure.

[16] If *draw* was not connected to the word *it*, and the word *don't* does not anchor any other patterns, then the motivation for generalising *draw* to *don't draw* would be less since connecting these two words would be more costly and the words would not help expand or enlarge an existing motif generalisation. Thus, early linguistic generalisations and generalisations in general, expand the motif generalisations in the word co-occurrence network. However, there is another factor in extending generalisations. If the communicative intent requires that the two words be used, if no other known construction provides the intended meaning then the verb will still be extended regardless of the existing connectivity. On this view, words felicitously connect when the interpretation necessitates the combination. A felicitious context can in fact vindicate an improbable word combination e.g. *He sneezed his tooth right across town* (Robert Munsch, Andrew's Loose Tooth, Scholastic Canada Ltd., 2002, cited in Goldberg 2006: 6; see section 9.5.3 of this study). Thus, words in the network are as close as the imagination allows (in the spirit of Solé 2008).

9.4.2 Semantic lightness[17]

That light verbs have "general meanings applicable to many situations", suggests that light verbs can co- occur with many different words and indeed they do: highly connected verbs in a word co-occurrence network tend to be light verbs[18] as will be shown shortly. Early generalised verbs, sometimes known as pathbreaking verbs (Ninio 1996, 1999a), tend to be grammatical markers cross-linguistically due to their "extremely general meanings; their semantics consists of little more than pure "argument-structure"" (Ninio 1996: 16; Ninio 1999a). As such, these verbs are suitable candidates for semantic bleaching that occurs in the later stages of grammaticalisation, potentially forming auxiliaries (Heine 1993). Ninio (1996: 14, Table 6) lists the verbs in Table 9.1 as examples of pathbreaking verbs.

Table 9.1 Pathbreaking Verbs (Ninio 1996: Table 6)

intransitives	movement verbs	transitives	other
be	move	get	where is?
stay	come	do	want
sleep	go	make	
	enter	hit	
	get-out	have	
	fall	eat	
		drink	
		pull	

In the child word co-occurrence network, of the fifty most highly ranked words for connectivity (degree), only one out of the nineteen verbs (the verb *draw*) does not figure as a pathbreaking verb: {*is, want, going, I'm, go, see, where's, get, doing, do, that's, eat, put, what's, it's, sleeping, draw, don't, can*}. The words *put* and *see* are typical closed-classed verbs in North-Australian languages (Ninio 1996).[19] Therefore, highly connected

[17] For discussion see Sethuraman and Goodman 2004a and section 3.6.3.

[18] Although the reverse is not strictly true. Light verbs may also have a low connectivity in the network as the verb 'stay' exemplifies. The verb's degree ranking is 461 (out of a possible 657).

[19] Note that in Sethuraman and Goodman (2004a), auxiliaries are excluded from their study of light verbs. I have included auxiliaries as well as lexical verbs as I am interested in how all types of verbal forms are represented in the word co-occurrence network. Copula contractions such as *where's* are also included since it

verbs, in the network tend to be pathbreaking verbs.

Indeed, all the nineteen most highly connected verbs in the child's network anchor the largest extensions of motif 6 and 36, two of the most highly significant (automorphic) motifs. Table 9.2 shows the verbs with the number of automorphic nodes each verb anchors in the respective motif pattern. The highest number of automorphic nodes (N) anchored around motif 6 and 36 for all anchors i.e. not just verbs, is 115 and 73 respectively. Virtually all the 'verbal' anchors for the largest extensions of motif 6 and 36 (with N≥14) are the verbs listed below.[20] As such, the interception of frequency and degree (type frequency) allows such verbs to be easily located within the network.

Interestingly, Ninio (1996: 16) suggests that "Children…also use their "pathbreaking verbs" as if they were a kind of "closed-class" of special verbs: by definition, these verbs have combinatory possibilities that other verbs in their vocabularies do not". This observation is not unexpected since the words that are traditionally 'closed class', i.e. function words, tend to be hub words. Thus, pathbreaking verbs are hub words, and as such, are the most easily found verbs in the network, and are potentially generalised early.

However, it is assumed that semantic lightness alone can not account for early verb generalisation (Sethuraman and Goodman 2004a; Theakston et al 2004: 68). Firstly, verbs which are not classed as light verbs e.g. *cut* and *ride* are used early in diverse syntactic constructions (Sethuraman and Goodman 2004a). As noted before, discourse factors may be at play here, in that the communicate intent perhaps requires that these verbs take advantage of the functional meaning of the new syntactic environment. The choice of verbs in line with Ninio's (2006a) proposal would then be functionally motivated rather than by the connectivity of a word. This exception is rather obvious and hardly contentious since language serves a communicative intent through pairing meaning-to-form in a cost effective way.

---

is a type of verbal constituent in the non-traditional probabilistic sense that is discussed in chapter five, and in the spirit of Bybee (2002a:112) in that "items that are used together fuse together " (Bybee 2002a:112).

[20] The only exceptions are gonna/find/have (14 N) and Georgie's (23 N) for motif 6 and fall (14 N) and doing (27 N) for motif 36.

Table 9.2 Verbs with the highest connectivity are shown. All these verbs are anchors to the largest extensions of the two most highly significant motifs (motifs 6 and 36). Columns show how many automorphic nodes (N) each verb anchors in the respective motifs.

anchor verbs	number of N		anchor verbs	number of N	
	motif 6	motif 36		motif 6	motif 36
is	48		that's	28	
want	48		eat		16
going	34	23	put	19	
I'm	50		what's	23	
go	29	18	it's	23	
see	28		sleeping		19
where's	39		draw	14	
get	24	16	don't	16	
doing		27	can	16	
do		17			

The two early verb-generalising factors discussed so far, frequency and semantic lightness, have been correlated with two key properties of hubs: high token frequency and high connectivity (type frequency). Early generalised verbs tend to have a high token frequency and a high connectivity and as such are easier to locate as well as reinforcing the growth of a hubterranean structure when they are generalised early. The next section will look at why verbs co-occurring with pronouns will be generalised early, again relating this factor to the hubterranean structure in a word co-occurrence network.

9.4.3 Pronouns

Not only do hub verbs get generalised first, but also verbs which co-occur with pronouns in their caregiver's speech. For discussion see Childers and Tomasello (2001); Dodson and Tomasello (1998); Sethuraman and Goodman (2004a). Interestingly, pronouns also tend to be hub words. Table 9.3 below illustrates this observation using data from the adult language. The frequency and degree are high considering the lowest ranking for frequency is 2897 and for degree, 5331.

Table 9.3 The rankings for frequency (token) and degree (type frequency) for pronouns in the large BNC adult network (chapter five).

pronoun	frequency ranking	degree ranking	pronoun	frequency ranking	degree ranking
it	8	9	your	116	75
i	4	10	their	132	80
you	9	11	Me	89	85
we	10	17	She	118	104
they	14	20	Us	126	127
them	58	43	Her	167	155
he	64	64	Him	172	160
our	103	65	his	272	177
my	108	74			

As pronouns are classified as function words, it is unsurprising that there is a strong correlation between degree and frequency (with a Pearson co-efficient of 0.95). Indeed, the pronouns examined in the smaller adult network (see chapter eight) anchor extensions of motif 6 in which the majority of the automorphic nodes are traditional verbs. Thus, in the same way that a function word identifies a syntactic category, a pronoun identifies a co- occurring word. On this view, hub words such as pronouns are easily identifiable within the network and so words associated with pronouns are more easily accessible than those that are not. Therefore, verbs that co-occur with pronouns will more likely be extended to other syntactic environments.

The last two factors to be considered – syntactic consistency and diversity – are explained in network terms by appealing not to the property of hubs but to the process of hub formation in a verb-construction network.

9.4.4 Syntactic consistency and syntactic diversity[21]

Syntactic consistency and syntactic diversity are the final factors Sethuraman and Goodman (2004a) list as influencing early verb generalisation. At first glance, these two factors seem to contradict each other. On the one hand, restricting verb usage to one syntactic construction assists early generalisation. On the other hand, it seems that varying the syntactic environment of a verb also helps the child to generalise the verb early. Nevertheless, once we appeal to the optimisation process that produces the hubterranean structure for an explanation of these two factors, the contradiction is resolved. Syntactic consistency will be discussed first, followed by syntactic diversity.

When a caregiver uses a particular verb consistently within a given construction, the repetitious nature of the input may lead the child to integrate the verb's argument structure as a central sense of the argument structure of the construction (see Goldberg 2006). Later, this argument structure and the construction become automised (see Bybee and McClelland (2005: 384) for integration through repetition and Bybee (2002a:112) for automisation through repetition). For example, the frequent use of *give* in the ditransitive construction designates the meaning "X causes Y to receive Z" to the ditransitive - the meaning becoming construed as the central sense of this particular argument structure construction (see Goldberg 1995: 49; Goldberg 2006). Essentially, the distributional skew provides the repetitious input required for the chunking together of the verb and the construction (Goldberg 2006) i.e. the probability that a given verb occurs in the respective construction is high. Consequently, the verb's meaning becomes associated with the construction.

Since the verb and the construction in which it appears form a relation, it is possible to represent these relations as a subpart network. Figure 9.10 illustrates a fragment of the subpart network for *give*. The word occurs (not exhaustively) in both the ditransitive and the transfer- caused motion construction. Both constructions are themselves subparts of the transitive construction which, in turn, forms a subpart of the subject-predicate construction (Goldberg 1995:109, Figure 4.2). Note that the weighting of the link between the specific verb and the construction would represent the consistency of usage i.e. how costly in storage terms the verb-construction pairing is.

[21] See Goldberg, Casenhiser and Sethuraman (2004); Naigles and Hoff-Ginsberg (1998); see Sethuraman and Goodman (2004a) for discussion.

Figure 9.10 An example of part of a subpart network derived from source sentences such as *Sam gave Mary a cake* (ditransitive); *Sam gave the piece of land to his son* (transfer-caused motion). Examples from Goldberg (1995: 111, Examples 20-21).

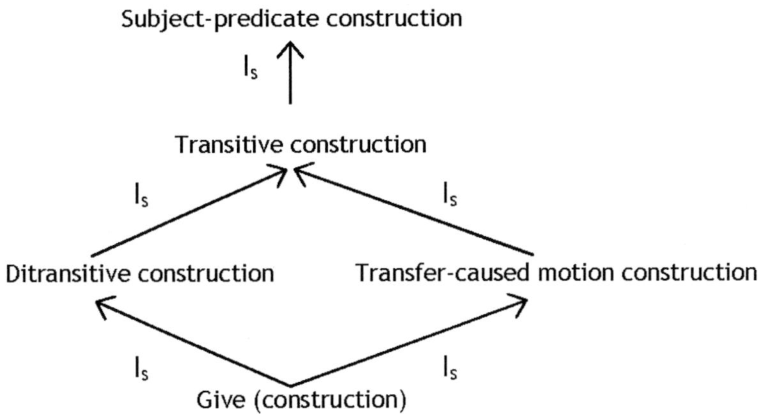

As discussed in chapter two, the chunking of two elements generally reduces the cost of storing their linguistic representations. In the same way that repeating a bigram (two adjacent words) will produce a higher transitional probability and lower the storage cost of the bigram, likewise the repetition of a verb in a specific construction reduces the cost of storing the verb and the construction in a subpart network. Thus, it is likely that the storage cost and connectivity will be optimised in this fragment of the subpart network.

Furthermore, transitional probabilities in word co-occurrences aid the segmentation of an artificial language, in that frequently co-occurring elements are more easily identifiable as a chunk (Gómez 2002, 2006; see chapter two and three). Likewise, it is proposed that a high probability associated with a verb occurring in a given construction facilitates the identification of a verb-construction pairing within the network. Thus, attention is draw to the verb and as such, it is a potential candidate for early generalisation (in line with Boyd and Goldberg forthcoming). Note that such a verb would have a high frequency *within* the respective construction (see section 9.5.1). Therefore, the interception of frequency and low storage cost assists in locating elements to.[22] In addition, the

[22] Since the network optimises the factors of storage cost and connectivity, it is not unexpected that the child is sensitive to forms that facilitate efficient storage. Thus,

repeated use of the verb in a construction would strengthen the verb's functional representation, thus aiding the generalisation of the verb to a new construal.

The factor of syntactic diversity may also be explained in similar terms. In a verb- construction network, a verb that occurs in many different constructions has a high type frequency, suggesting that it has a high token frequency. The interception of these two properties points to a hub word. Recall from section 9.3, hub words are easier to detect. As noted in sections 2.8.1.1 and 3.5.6, high type frequency is important in identifying elements in an artificial language (Gómez 2006). Indeed, the network's optimisation process maximises word connectivity and as such the network attends to high type frequency. Consequently, a high type frequency verb i.e. a hub word in a verb-construction network will be more easily detected within the network. Therefore, a verb heard in many different constructions would constitute a potential target for early generalisation.

9.4.5 Summary

The five factors of early verb generalisation postulated by Sethuraman and Goodman (2004a) are shown in Figure 9.11.

This section argued that these five factors relate to the identification of verbs within a word co-occurrence network and a verb-construction subpart network, in line with Boyd and Goldberg's (forthcoming) study. In section 9.3, it was proposed that hub words i.e. words that are highly frequent and highly connected, are more easily identified within a word co- occurrence network i.e. they facilitate navigation through the network (see Ferrer i Cancho and Solé 2001a: 2264). As such, this section correlated the hub characteristics and the factors that generate the hubterranean structure with verbs that are generalised early.

---

low cost units or chunks are more recognisable within the network than high cost chunks. (See http://www.hjkm.dk/Clusters_OHP.pdf for an analysis of chunks in second language learning). The other factor, maximising high connectivity is reflected in the sensitivity towards highly connected nodes within the network. Attention to elements with a high frequency derives from their entrenched representation.

Figure 9.11 Sethuraman and Goodman (2004a: 86, Figure 3) showing the five factors of early verb generalisation.

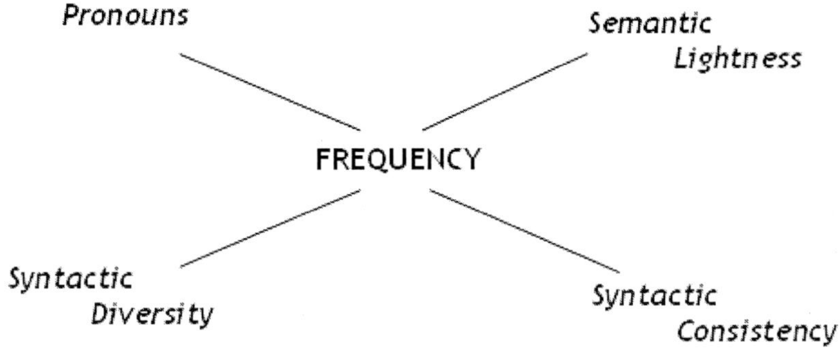

Thus, verbs which are frequent and semantically light i.e. have a high type frequency, exhibit hub characteristics which allow them to be more easily distinguished. Verbs which are associated with pronouns i.e. a hub word, are also recognised by virtue of being connected to this hubterranean structure. The factors of syntactic diversity and consistency correlate with the two artificial language learning strategies detailed in sections 2.8.1.1 and 3.5.6. In that, it was noted how words that have a high type frequency within the artificial language "popped out" (Gómez 2006: 23; see section 2.8.1) and elements that co-occur frequently form a recognizable unit or chunk.

Notice that the two artificial language learning strategies of type frequency i.e. connectivity and attention to co-occurrence frequencies i.e. transitional probabilities relate to the two optimised factors that create a hub structure. Early verb generalisation, then, is influenced by the optimised factors of low cost (high transitional probabilities) and high connectivity (high type frequency) and their interaction with the factor of high frequency. Thus, high frequency words associated with hubterranean characteristics and formation are more easily detected and thus more likely to be generalized early. These factors are schematised below in Figure 9.12.

Figure 9.12 The five factors of early verb generalisation in respect to the hub structure of a word co- occurrence and the verb-construction network with corresponding factors postulated by Sethuraman and Goodman (2004a: 86).

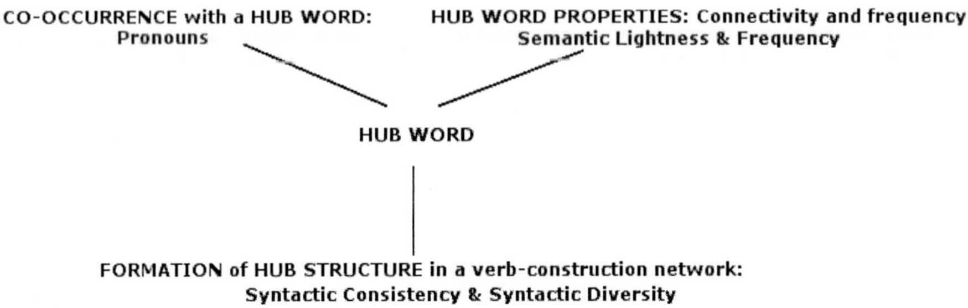

Indeed, this approach sheds light on how constructional islands in early speech are built around frequently used elements (Kidd, Lieven and Tomasello 2004: 57), and also paves the way for an explanation as to how the specific elements and constructions link together within the taxonomic construction inventory. In line with Kidd, Lieven and Tomasello (2004: 57), frequency "play[s] an important facilitative role in the abstraction process". It has been shown above that verbs extended early are those that are more easily identifiable through a high (token/type) frequency. Since the optimisation process is implicated in the verb-construction network, maximal connections between specific verbs and constructions will be established in a cost efficient way. Hence the process will encourage connectivity between constructional islands.

9.5 The verb-construction subpart network

In the last section, it was proposed that the distribution of verb-construction links (co- occurrences) is driven by the formation of a hubterranean structure i.e. the optimisation of connectivity and storage cost. From this view point, it is predicted that there will exist hub constructions with hub properties. The hubs in the word co-occurrence network show a power law for frequency and connectivity, as well as for transitional probability (see chapter 6 for details). Therefore, it is predicted that these characteristics also occur in the relation between verbs, and the constructions they appear in, both for the adult and child networks. I will present some evidence to support this view by briefly discussing each power law in turn, starting with the frequency power law.

9.5.1 The frequency power law

In the Hebrew indirect object construction, verbs in this construction had a frequency distribution captured by a power law both in the child's and mother's speech (2005b&c, 2006a). Moreover, this distribution was represented by two power laws (Ninio 2006b), separating the most highly frequent words, in line with frequency distribution of words within a whole text (Ferrer i Cancho and Solé 2001b).

Ninio (2006b: 4) also notes that this "quantitative difference is also a qualitative one", with generic verbs making up the first power law (the highly frequent word distribution) whilst more specific words are found in the second power law (the less frequent words). Therefore, it is likely that the frequency distribution of verbs within all constructions reflects the frequency distribution of words in a text (at a global level in the network) and a child's verbal learning will display this trend.

9.5.2 The connectivity power law

When a verb occurs in a particular construction, a link is formed between the verb and the construction. In chapter three and the previous section, factors implicated in early verb generalisation point to a verb-construction network being represented as a hubterranean structure. From this observation, it is likely that the connectivity pattern of the construction-verb in the sub-part network will follow a power law in both the child and adult network. That is, it is predicted that a few constructions i.e. hub constructions, will have many verbs co- occurring within and thus connecting to the construction (Figure 9.13 arrow C).

Since type frequency has been associated with the productivity (Bybee 1995; Clausner and Croft 1997; see Goldberg 1995: 134-6), the number of verbs connecting to a construction is related to the construction's productivity. Thus, hub constructions will represent a productive construction (with semantic considerations not withstanding). For example, in the adult network a large extension of motif 6 represents the highly productive 'were + N' construction, with automorphic nodes N representing the traditional verbal category. In the child network, the construction 'I'm + N' is represented by motif 6 with the automorphic nodes N representing a traditional verbal category.

Furthermore, the power law predicts that there will be many constructions with smaller but still fairly large generalisations. These constructions display partial productivity if the verbs display semantic diversity e.g. the ditransitive construction (Figure 9.13, arrow B). Then, there are very many constructions in which only a few verbs appear (Figure 9.13, arrow A). The latter constructions consist of the lexically specific types i.e. formulaic forms such as idioms and in adult language archaic frozen forms such as "be gone with you" and "be that as it may".[23] Included in this type are unproductive sub-classes of verbs.

Figure 9.13 The power law distribution of the type frequency of a construction. The distribution represents the number of verbs which appear in a given construction.

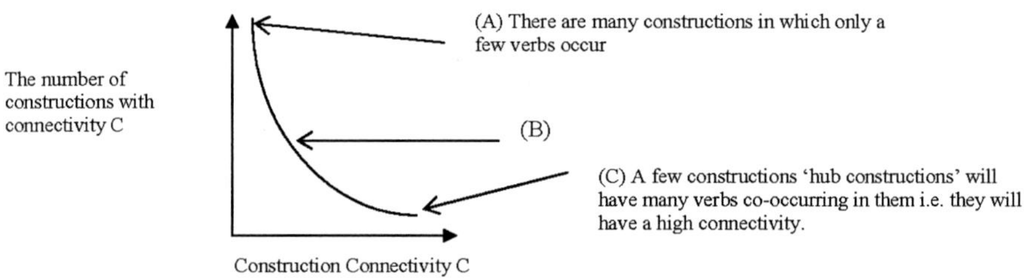

In sum, certain predictions concerning productivity may be proposed when one acknowledges that verb-construction links collectively form a hub structure. These predictions derive from the power law link distribution, in that there will be a few potentially productive constructions and many poorly productive constructions in (the English) language.

[23] The current study is unable to pick out idioms and "formulaic sequences as devices of social interaction" (Wray and Perkins 2000) since the functional part of the construction is not directly employed in the network. Since some frequent collocations imply idioms or formulaic language, even if the meaning was available, a means of establishing the frequency of co-occurrence for elements in a motif as a whole would be required. Currently this is beyond the scope of this study.

9.5.3 The storage cost power law[24]

In a verb-construction link network, the length of the link equates to the storage cost of representing a verb co-occurring in a construction. Storage cost in a linguistic network is represented by inverse probabilities. Thus, the storage cost of a verb appearing in a construction is calculated by dividing the amount of times the verb appears in the construction by the total appearances of the verb and then this number is inversed. The storage cost is then predicted to form a power law distribution in line with Kasturirangan's (1999) observation.

Figure 9.14 The power law distribution of the storage cost of a verb within a construction. The distribution represents the ratio of appearances of the verb within a given construction.

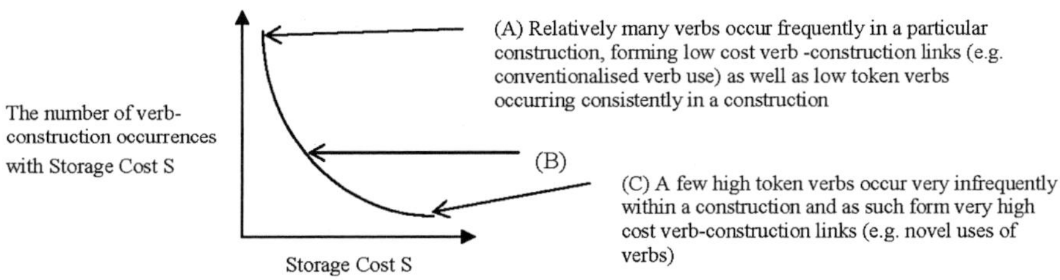

In Figure 9.14, the power law provides predictions for verb-construction pairings. Arrow A points to the low storage cost region of the power law which predicts that there are many verbs that occur consistently within a given construction. That is, the high probability that the verb will occur in the construction equates to a low storage cost value. These low cost values may represent the conventionalized usage of a verb in a construction.

For example, a study has shown that *think* is used more frequently in the construction 'X + think + Y' where Y is a 'sentential complement' in a child's speech (see Kidd, Lieven and Tomasello 2004: 111). A low storage

[24] Empirical proof of this argument relies on calculating the probability that a particular verb will occur in a given construction (using transitional probability, the equation given in chapter 5). It is predicted that the inverse values will form a power law in the same way that values of transitional probabilities of word co-occurrences - once inversed for storage cost – will also display a power law distribution. Unfortunately this calculation is beyond the scope of this study and a matter for future research.

cost is also achieved if the verb does not occur very often in the dataset i.e. it has a low token value, and occurs only in one particular construction. Such verbs would comprise of highly specialised or specific terms not often employed or required in general communication.

Arrow C in Figure 9.14 points to the region in which the verb-construction pairings incur a high cost due to the infrequent use of a (medium to high token) verb in a particular construction. An example of this type is the novel use of a verb. The word *sneeze* is most likely found in an intransitive construction e.g. *he sneezed*. However, the word may be seconded to other constructions for a more novel usage *he sneezed his tooth right across town* (Robert Munsch, Andrew's Loose Tooth, Scholastic Canada Ltd., 2002, cited in Goldberg 2006: 6). The power law distribution predicts that these types of usages are infrequent in relation to other verb uses. Thus, there are only a few high cost verb-construction links in speech.

Arrow (B) in Figure 9.14 finds the middle ground between conventionalized usage and novel usage. The power law distribution shows that there are many verbs that occur fairly often in a construction. These verbs represent the senses that are modified versions of the central sense in a construction. In Example 9.1, the verb *bake* provides a non-central sense in the ditransitive construction.

Example 9.1 The pattern of polysemous senses in the ditransitive (Goldberg 1995:75, shortened version)

a) 'X causes Y to receive Z' (central sense) Example: Joe gave Sally the ball
b) 'X intends to cause Y to receive Z' Example: Joe baked Bob a cake

Whilst all these predictions are very general, the important point is that the distribution of each of the following properties is captured by the power law function: frequency within a construction, co-occurrences of verbs within constructions across the dataset and the storage cost of a verb occurring in a construction. It is predicted that these properties will be found both in the adult and child network, since the factors concerning early verb generalisation point to a hubterranean structure. However, the actual specific constructions in both the adult and child speech will differ. Likewise what constitutes novelty or conventionalized usage is relative to the network at hand.

Naturally, more constructions will need to be investigated to verify that the power law distribution is found in every construction both intra-language and cross linguistically. Note that the power law distributions for connectivity and storage cost do not refer to individual constructions but to all constructions as a whole in the text and empirical work on these distributions is also a matter for future research.[25]

9.6 Optimisation factors

The above predictions arise from viewing the verb-construction subpart network as being driven by two optimisation factors, storage cost and connectivity. When one observes the links between the form-meaning pairings of constructions, these two optimisation factors correlate with four psychological principles that have been suggested in the syntactic literature (quoted from Goldberg 1995). Consequently, the hubterranean analysis would potentially aid the understanding of symbolic mapping of constructions and their relations within the construction inventory.

f **Principle I: "The Principle of Maximized Motivation" (Haiman 1985:20, Lakoff 1987)**

"If construction A is related to construction B syntactically, then the system of construction A is *motivated* to the degree that it is related to construction B semantically… Such motivation is maximized" (Goldberg 1995:67)

f **Principle II: "The Principle of No Synonymy"** "If two constructions are syntactically distinct, they must be semantically or pragmatically distinct "(Bolinger 1977; Haiman 1985:24, Clark 1987; MacWhinney 1989).

f **Principle III: "The Principle of Maximized Expressive Power" (Goldberg 1995: 67).**

"The inventory of constructions is maximized for communicative purposes".

---

[25] It is interesting to note that for another taxonomic system, namely biological species, the frequency of genera with a given number of species exhibits a power law distribution (Burlando 1990, 1993; see Solé and Goodwin 2000: 255-6).

f **Principle IV: "The Principle of Maximized Economy" (Haiman 1985:24-25).** "The number of distinct constructions is minimized as much as possible, given principle III" (Goldberg 1995:67)

Principle II and IV together correlate with the drive to potentially decrease the cost of storing within a construction inventory. That a distinct semantics is related to a distinct form (principle II) produces a chunking of the form and meaning lowering the storage cost of the pairing i.e. construction. Minimising the distinct forms within the inventory (principle IV) aids storage since the inventory will be smaller. Principle I and III encourages connectivity. Principle I promotes semantic and linguistic form connections between the constructions, whilst principle III, by maximising the inventory, the number of potential connections will also be maximised within the inventory.

It is thus predicted that the optimisation of these principles organise the constructions and their symbolic relations (the meaning-to-form mappings) into a form-meaning link distribution represented by a small world hubterranean structure.[26] It is important to note that since the principles may derive from an optimisation process in the mind~brain, these principles not only reflect the nature of adult language but also the nature of child language.

In fact, principles III and IV and both optimisation factors have been captured in Culicover's (1999) notions of *conservative* and *attentiveness* in the learning process. By limiting the number of constructions in a child's inventory, the child is being *conservative* in its syntactic presumptions. However, the child would also like to be as expressive as possible and therefore is *attentive* to new expressions; hence "the inventory of constructions is maximized" (Goldberg 1995: 67).

In addition, Israel (2002) utilizes the notions of local and global consistency in regard to instances and generalisations. As such, this approach represents principle I and III as well as the optimisation factors of reducing storage cost and maximizing connectivity. "Local consistency applies to linguistic units activated online in usage events, and requires these to be as consistent as possible with entrenched utterance types" (Israel 2002: 123). As a consequence, it "favors a massive inventory of

---

[26] For symbolic mapping between "objects of reference" and signals see Ferrer i Cancho & Solé (2003a).

low-scope constructions", that represent "conventional units for every occasion" (Israel 2002: 123). A unit is conventionalized by frequent usage i.e. through the repetition of the form-meaning pairing of the construction. Through repetition, the storage cost of the unit is lowered even though, as stated by principle III, "the inventory of constructions is maximized for communicative purposes" (Goldberg 1995: 67). Thus, conventionalisation of the maximal amount of constructions optimises the construction inventory i.e. the storage cost is minimised.

On the other hand, "global consistency applies to the repertoire of constructions as a whole, and requires that units be represented in ways which maximize their consistency with each other" (Israel 2002: 123), in line with principle I. Thus, connectivity within the construction inventory is promoted. As noted by Israel (2002: 125), although some linguists doubt the significance of abstract representation in grammar, "there is no reason grammars should not include both concrete and abstract representations. Indeed, the consistency principle suggests that the two should in principle be inseparable, and that abstract schemas emerge in tandem with and as a function of an expanding repertoire of item-specific constructions".

This observation is line with the cyclic nature of syntax proposed in this study. Indeed, the generation of construction-specific syntactic categories facilitates local consistency in that, the category specifies the felicitous set of possible usages from which an analogy to a potential usage may be processed. On this view, construction-specific syntactic categories aid local consistency whereas global consistency i.e. knowing the felicitous relations between syntactic categories and between constructions, will allow the child to constrain combinations of constructions in admissible ways.

In sum, Culicover's (1999) conservative and attentive approach, Israel's (2002) local and global consistency model and the principles listed above, all reflect the characteristics of a hubterranean structure, which can be found contextualized in a word co-occurrence network, a verb-construction subpart link network and a construction's form-meaning network.

9.7 Summary

In the first part of the chapter, it was argued that the optimisation process is implicated in the child's task of segmenting linguistic form. Syntax is viewed as a cyclic process whereby the optimisation process of the network generates abstract syntactic elements e.g. the power laws with truncated tails, which mediate the identification of syntactic categories, which in turn restrict the network to its reasonably optimal state. Thus, functional patterns in the network remain self-consistent and the network retains its (reasonably) optimal state. It was argued that this mechanism is found in both the child and adult linguistic network and as such both networks are self-consistent. However, only the adult network is self-sufficient, with the child network relying on the environment for linguistic knowledge.

The remaining sections within the chapter argued that the optimisation process of minimising storage cost and connectivity is found on a number of different theoretical levels. These levels are summarised below specifying the context of the supporting argumentation.

- The word co-occurrence network (see chapter five and six).
  - The interception of degree and frequency distributions mediate abstract syntactic categories.
  - The three factors of frequency, semantic lightness and co-occurrence with a pronoun are implicated in the early extension of a verb to a new construal (see Sethuraman and Goodman 2004a). Thus, degree and frequency interceptions facilitate verb detection.
- Specific verb-construction network.
  - Syntactic diversity and consistency, implicated in the early extension of a given verb to a new construal, result from the sensitivity towards high transitional probabilities i.e. low storage cost and high connectivity. When optimised, these two properties form a hubterranean structure.
- The mapping of the construction's form to meaning.
  - Conventionalised construction meaning i.e. central sense.
  - Polysemous senses within a construction.
  - Psychological principles I and II.
- The size of the construction inventory.[27]
  - Psychological principles III and IV.
- First language learning and adult artificial language learning.
  - Segmentation of input.
  - *Conservatism* and *attentiveness* (Culicover 1999).
  - *Global* and *local consistency* (Israel 2002).

[27] See also Ferrer i Cancho and Solé (2003a) for the mapping of "objects of reference" to signals and the optimal size of a vocabulary repertoire.

The concluding chapter is a short discussion on the advantages and the limitations of the network approach, as well as suggesting future research projects.

CHAPTER TEN

CONCLUDING DISCUSSION

For the last five decades, many linguists have accounted for cross-linguistic similarities in syntax by appealing to genetic endowments. However, a growing number of linguists are discovering that properties thought once to be purely linguistic, actually have an alternative origin. For example, the use of zero and resumptive pronouns has been correlated with their cognitive accessibility (Ariel 1999). This study supports the extension of the possible sources deriving syntactic structure by investigating whether there are certain basic linguistic properties of form, e.g. constituency and the functional-lexical distinction, that are linguistic manifestations of *general* system properties.

The linguistic perspectives adopted in this study i.e. Construction Grammar and the usage- based approach, were presented in Chapter two. One of the advantages of Construction Grammar is that it utilises surface generalisations, as it is a monostratal theory rather than a derivational theory (see Goldberg 2006: 19-44 for a discussion of surface generalisations). The monostratal theory is preferred as there are fewer initial presumptions concerning the types of patterns that will occur in the given language.

Furthermore, the monostratal theory simplifies the analogy between the representation of a particular general system and a given language. Another advantage of the Constructionist and the usage based approaches is the emphasis on the tenet that "languages are learned - that they are CONSTRUCTED on the basis of the input together with general cognitive, pragmatic, and processing constraints" (Goldberg 2006: 3). That these constraints could be domain-general opens up the possibility that the source of the constraints may be extended beyond cognition to a certain type of system.

Importantly, chapter two also highlights how two basic properties of linguistic form, syntactic constituency and the functional-lexical (f-l) distinction, may be characterised in frequency terms that are not specific to the domain of linguistics. The f-l distinction may be couched in type and token frequency terms (see Ferrer i Cancho and Solé 2001a) whilst syntactic constituency at its basic level, may be represented by transitional probability values. This theoretical move facilitates the search for comparable counterpart properties in a general system. Furthermore, studies of adults learning an artificial language has evidenced two learning strategies that employ these frequency measures e.g. transitional probabilities and type frequency learning - with an element in a fixed position (Gómez 2006, Gómez 2002). These findings point to a correlation between language structure and language learning (see Tonkes, Blair, and Wiles 2000; Saffran, Hauser, Seibel, Kapfhamer, Tsao and Cushman 2008; Christiansen and Chater 2008).

The discussion then turned to first language learning (chapter three). Whilst multiple constraints operate on language learning, learners are assisted by a variety of integrated cues. This study focuses on one of those cues: frequency. There is evidence of children being sensitive to word co-occurrences when segmenting their linguistic input (Saffran, Aslin and Newport 1996; Aslin, Saffran and Newport 1998). Highly frequent linguistic chunks or anchors facilitate the extraction of linguistic patterns to form early constructions (Pine, Lieven and Rowland 1998). Through sensitivities to word co-occurrences and attention to these anchors, both continuous and (short) discontinuous dependencies are established (Gómez 2002; Gómez 2006).

By comparing language to other systems in the world that exhibit characteristics comparable to these frequency metrics, we will gain insight into how these properties arise, why linguistic form has certain characteristics and why learners adopt particular learning strategies.

10.1 Networks

One of the simplest ways of representing any system that has entities and relations is a network. The entities correspond to nodes whilst the relations are denoted by links. In chapter four, the discussion focuses on network approaches to language learning. It has already been demonstrated that when language is represented as a network with nodes as words and links as word co-occurrences (up to a window of two words) a small world effect is observed i.e. a small average shortest path and a high clustering co-efficient (Ferrer i Cancho and Solé 2001a).

Furthermore, the network is scale free in that there are a few highly connected nodes (hubs) and many poorly connected nodes (Ferrer i Cancho and Solé 2001a). These hubs arise in a syntactic dependency network around the age of two years (Corominas-Murtra, Valverde and Solé 2007), at the early multi-word stage that is associated with the onset of syntactic structure. The changes in link direction are also observed in syntactic development (Ke and Yao 2008).

In addition, it has been noted that for a network composed of adult speech, hubs tend to represent function words (Ferrer i Cancho and Solé 2001a). Indeed, as noted earlier the f-l distinction may be couched in terms of token and type frequency, the two characteristics of network hubs. Thus, the mechanisms that form hubs will assist in identifying the origins of the intuitive and omnipresent f-l distinction as well as detecting function words in child language.

It has been shown that a hubterranean structure is generated by optimising both the minimal cost of wire linking a new node, and the maximal node connectivity (Mathias and Gopal 2001 a&b). The authors point out that although Euclidean distance is used to establish the wire cost of a link, other non-Euclidean metrics are possible. In addition, Mathias and Gopal (2001 a&b) noted how Kasturirangan (1999) evidenced multiple link lengths in hubterranean networks. When hubs were generated, a signature power law function of wire length was observed. Thus, we have a process and a signature correlating with hub formation (Mathias and Gopal 2001 a&b; Kasturirangan 1999).

Since a word co-occurrence network exhibits a hubterranean structure, this study identifies an equivalent cost metric to wire length which involves transitional probabilities. Transitional probabilities correlate with chunking effects in linguistic form. These effects are observed in the 'short coding' of linguistic form, in response to the storage cost constraint on the form's representation (short coding in the sense of Haspelmath 2008 a&b[1] i.e. forms become shorter and more cohesive).

Since short coding occurs in expressions that occur and co-occur frequently, the transitional probability of two nodes is inversely proportional to the storage cost of representing their linguistic form. This cost metric is

---

[1] Although, Haspelmath (2008 a&b) assumes that short coding originates from predictability rather than cost.

represented in a word co-occurrence network by the link length between the nodes representing the given words. The other factor in the optimisation process, the connectivity of a given node, is associated with the type frequency of the word represented by that node.

In a word co-occurrence network[2] constructed from adult utterances (British National Corpus World 2001), the pattern of storage cost values in a text conforms to a (multi-) power law distribution in line with Kasturirangan (1999) and Mathias and Gopal's (2001 a&b) model. This result provides a possible explanation as to why function words occur in language since hub words tend to be function words. Specifically, function words are represented by hub words that anchor automorphic motif extensions. Function words on this view are a linguistic contextualisation of a general system property and have counterparts in other systems that exhibit (disassortative) hubs. Thus, it seems that function words have not developed solely to serve a particular function but are derivatives of a process by which a hubterranean network is formed. That is, "selection might have taken advantage of the network structure *a posteriori*" and so the f-l distinction may be viewed as an "evolutionary spandrel" (Solé and Valverde 2006: 421).

Likewise, basic syntactic constituency is not a linguistic-specific property, since it may be represented by a power law distribution of storage cost values inversely correlated with the transitional probabilities of words. The storage cost equates to the wire length that is found in multiple lengths in hub structures (Mathias and Gopal 2001a&b; Kasturirangan 1999). Syntactic constituency at a word level is, therefore, derivative of a hubterranean network and like the f-l distinction is an evolutionary spandrel. Moreover, this approach further supports the correlation between the storage of constructions and grammaticalisation (see Rostila 2006).

[2] It is assumed in this explanation that the specific linguistic elements in a speech act are retained and stored in a network format. Furthermore, the word co-occurrence network is a very simple and non-phonological representation of the network representing the storage of linguistic form.

10.2 The optimisation process

In chapter nine, it was argued that the characteristics of hubs and the optimised factors that generate a hubterranean structure are echoed on many different theoretical levels e.g. the size of the construction inventory (for the vocabulary repertoire - see Ferrer i Cancho and Solé 2003a); the verb-construction subpart network; the form-to-meaning pairing of a construction (see Ferrer i Cancho and Solé 2003a for general symbolic mapping);[3] word co-occurrences; segmentation of the input for both children and artificial language learners; early verb generalisations. This observation suggests that the optimisation process is potentially a very general principle. On this view the knowledge of linguistic form not only originates from its usage (as explicated by usage approaches) but its origins are also inextricably tied to the optimisation process of storing linguistic representations and the process of language learning.

As such, syntactic learning and the principles operating on different levels of the syntactic structure are captured by the interplay of two general but opposing factors: minimise cost and maximise connectivity. Isomorphism produces the lowest storage cost of two elements since the transitional probability is the highest i.e. one element always 'primes' or conditions the other. Conversely, a universal hub provides a high level of connectivity between the elements in a network since the diameter of the network is very small (see Figure 10.1). Thus, the optimisation process can be seen as a middle ground or transition state (see Ferrer i Cancho and Solé 2003a) between two extreme structures: isomorphism and a universal hub.

Figure 10.1 The isomorphic forms (left) and the automorphic form i.e. the universal hub (right).

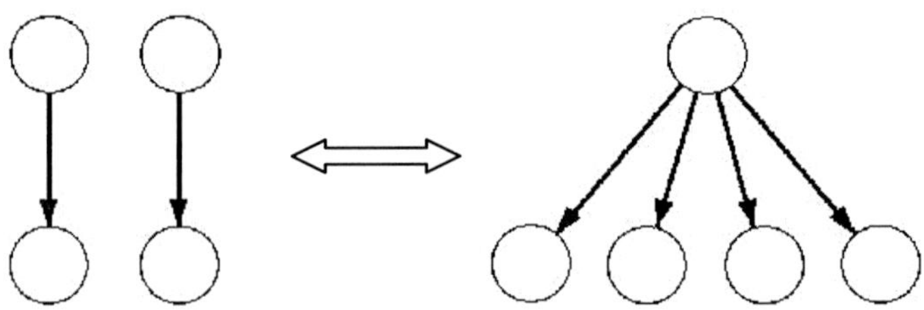

[3] See also Ferrer i Cancho, Riordan and Bollobás 2005 for an account of "protogrammar" and symbolic mapping. See Solé 2005 for a review.

It is potentially likely that power laws mediate the transition between these two extreme states since in actual systems, power laws are found at a phase transition. For example, it has been shown that Zipf's law is found between an isomorphic structure and a universal hub in a model of symbolic mapping of signals and "objects of reference" (Ferrer i Cancho and Solé

2003a). Within a construction in the verb-construction network, a power law distribution of verb frequency has been observed in the development of the in/direct object constructions of Hebrew as well as in indirect object constructions in adult speech (Ninio 2005 b&c; 2006a). As discussed in chapter four, a power law distribution of link length has been found to occur in a generic small world system (Kasturirangan 1999, Mathias and Gopal 2001 a&b) and is evidenced in the word co-occurrence network.

Furthermore, it is likely that the identification of syntactic categories is mediated by the truncated power law distributions (or the interception thereof) within the system. As such, abstract computational elements are created by the same general mechanism pictured in Figure 10.1. Only in this instance, in certain parts of the network there is movement towards a more isomorphic state.

In respect to distributional interceptions, it seems that when words have a high frequency, and high degree (type frequency), or low storage cost, these words are highlighted within the network. Thus, correlations between power laws act as guiding torches, spotlighting certain network areas and locating potential candidates for generalisation e.g. facilitating the formation of syntactic categories or verbal extensions to a new construal.

The frequency power law is also implicated in 'the linguistic cycle' where a periphrastic element may in time become a clitic and may ultimately undergo affixation, with the final stage of the cycle resulting in a null form (see Ninio 2008 for discussion). Ninio (2008) observes that this cycle is an example of self-organised criticality. At the critical point of the system, power law behaviour is exhibited (see Bak, Tang and Wiesenfeld 1987). As such, Ninio (2008) argues that the frequency power law of verb usage is an indication of this power law behaviour (see above and section 9.5.1).

Furthermore, the cohesion of two elements in the context of cliticisation and affixation is relevant to the linguistic cycle. Recall that the storage cost equates to the level of cohesion between two elements. On this view, it is plausible to argue that the storage cost power law would also signal the critical point of the system. In this case, low cost bigrams in a word co-occurrence network that are highly cohesive are of interest since the clitic and the element on which it is dependent, form a highly cohesive relation (see regime five of the storage cost power law found in the word co-occurrence network, section 5.4.3). On this view, it is predicted that the combination of the frequency and the storage cost power laws would drive the linguistic cycle, highlighting potential affixes.

Furthermore, if one views the relation between a word and its morpheme(s) as a bipartite graph (as in Figure 10.1), the linguistic cycle is linked to the process of words and morphemes becoming more highly connected i.e. a universal hub or a more isomorphic form, depending on the part of the cycle the language inhabits (see Ninio 2008; note that Ninio does not use the terms isomorphism or universal hub). It is generally accepted that the characteristics of agglutinative versus fusional[4] and polysynthetic versus isolating are scalar within a given language, although there are languages which typify the extremes. For example, in respect to the latter scalars, Chinese is strongly isolating, exhibiting isomorphism between a word and a morpheme, whereas Navajo verbs are strongly polysynthetic (see Kibrik 2005), in that they contain many morphemes (see Figure 10.2) and may represent a whole sentential event (Example 10.1 below).

[4] The contrast between fusional and agglutinative languages is captured by the difference in the meaning-to-form pairing of the morpheme. Agglutinative languages have an isomorphism between the meaning and the form of a morpheme whereas one morpheme in a highly fusional language may be linked to many different meanings.

Figure 10.2 An illustration of isomorphism (left) between a word (W) and a morpheme (M) and a 'universal hub' (right). Isomorphism is typical in isolating languages such as Chinese. The 'universal hub' word on the right is typical of polysynthetic languages such as Navajo, in which a sentence may correspond to one word (see below example 10.1).

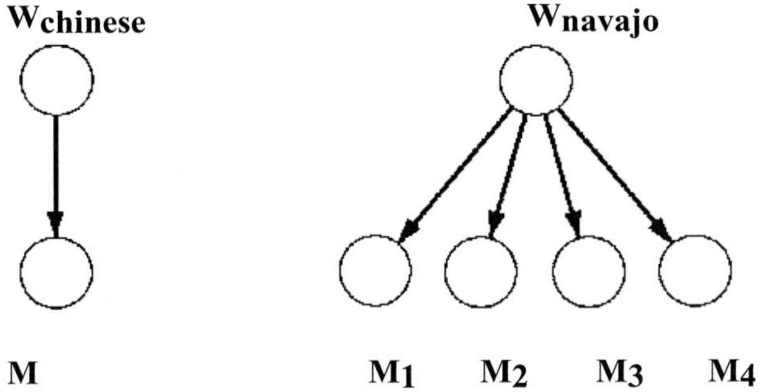

Example 10.1 An example of a word in Navajo (Kibrik 2005:4, example 4). The term *33* indicates that each referential is in the third person; *Acc* denotes Accusative case and a derivational prefix is marked by *Pref*.

> yi- di- lé
> 33.Acc-Pref-handle
> 'He carries it along'

There are many languages that lie within the realm of these two extremes. English singular nouns may be isomorphic e.g. *duck* but certain verb forms are not (*he duck-s*), whilst German words are generally less isomorphic. Thus, the word-to-morpheme ratios vary within these languages. If there exists a transition period between the well connected word–to-morpheme hub and the isomorphic structure, it is predicted that certain languages which reside at this transition point will exhibit power law distributions of word-to-morpheme ratios (in line with Bak, Tang and Wiesenfeld 1987; see Ninio 2008).

As such, it is expected that (a) these languages will have a few words with many morphemes and many words with only one or two morphemes - in line with the degree power law distribution for a word-to-morpheme bipartite network, (b) the frequency of the words would conform to a power law function (see Ninio 2008) (c) the transitional probability of one morpheme co-occurring with another would equate to the storage cost of representing those two morphemes. As such, a power law of storage cost would also be observed. Thus, this type of model may potentially provide an explanation as to why there is a typological contrast between the notions of polysynthetic ~ isolating and agglutinative~fusional as well as the spectrum between these two contrary sets (see Ninio 2008).

For example, these typological sets are created by the process of affixation and elision of morphemes which in turn encourages new paraphrastic constructions to functionally replace the lost morpheme(s). Notice how affixation may be couched in terms of cohesion between morphemes whereas periphrasis is resultant of the drive for expressiveness. Interestingly, cohesion and expressiveness equates to the storage cost and connectivity, respectively, in our model.

These two factors highlight two contrary sets that may be operating in language (Figure 10.3): the contrast~cohesion[5] of elements (see Lindblom 1989, footnote 7) and the connectivity~isomorphism of elements (see Ninio 2008; see Ferrer i Cancho and Solé 2003a for symbolic mapping). The former corresponds to the storage cost scalar and is illustrated by the diametric periphrastic~affixation process in the linguistic cycle. In that, the periphrastic process produces an independent morpheme which lacks cohesion i.e. it is contrastive to other elements. The affixation process on the other hand involves strong cohesion between two elements which lose their contrastive i.e. independent nature.

The connectivity~isomorphism set is indicative of the connectivity factor in a hubterranean network and is reflected in the distinction between polysynthetic~isolating and agglutinative~fusional typological characteristics. Thus, the factors of storage cost and connectivity pertaining to the formation of a hubterranean structure are implicated in very broad typological distinctions, the ‘linguistic cycle’, as well as at many linguistic levels detailed in chapter nine.

---

[5] The symbol ~ denotes contraries.

Figure 10.3 The two contrary sets of properties that are operating in language. Namely, constrast~cohesion (see Lindblom 1989)[6] and connectivity[7]~isomorphism (see Ninio 2008).

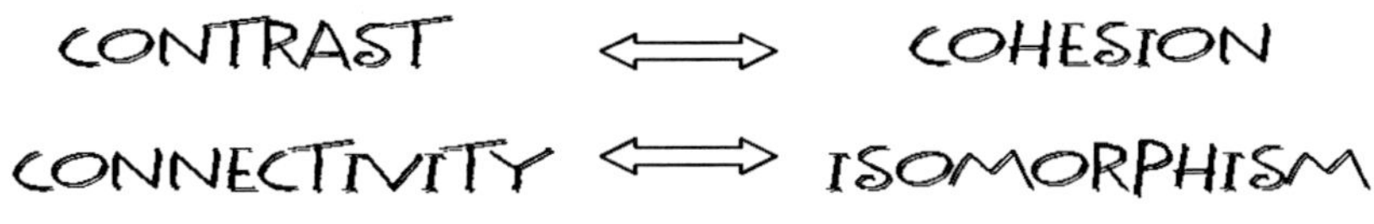

Since power laws are suggestive of a transition between two states,[8] it is interesting to note that power laws are also found outside the grammars of language, being implicated in language diversity within individual countries (and 'population size'; see Gomes, Vasconcelos, Tsang and Tsang 1999 and Solé 2006 for discussion) as well as the number of languages per family (Wichmann 2005). As such, it is suggestive that these power law distributions represent the transition phase between isomorphism and the universal hub e.g. 'one country speaks one language' and 'one country speaks all languages' or 'one family contains one language' and 'one family contains all languages'.[9] Indeed, as an

---

[6] A contrast~cohesion distinction concerns co-articulation, in respect to phonemic perception. Whilst "a motor system tends to default to a low-cost form of behaviour" (Lindblom 1989:413) and so causing co-articulation to occur, phonemes are still shown to be sufficiently distinct and therefore perceivable as separate phonetic entities (Lindblom 1989:425). Such "distinctiveness" is achieved by a competing perceptual condition "sufficient contrast" which curtails signal efficiency (low cost behaviour) when lexical access is hampered. The two competing restrictions (sufficient contrast and low cost behaviour), conditioned by factors independent of the signal such as "word frequency and lexical sound structure", facilitate the production of the short term variations in speech signals (Lindblom 1989: 412). For example, the variations in the forms of 'my' in different types of speech: clear speech, casual speech, fast speech, etc. Thanks to John Ingram for alerting me to this point.

[7] As noted in section 9.3.2, the universal hub may represent the maximal automorphicity between nodes in a network. Consequently, there is a more general contrary set: isomorphism~automorphism.

[8] However, one is mindful that power laws do not exclusively characterise critical states (Solé and Goodwin 2000: 56).

[9] The contrast~cohesion contrary set may be implicated in the dialect continuums e.g. Dutch-German, where two dialects are mutually intelligible when they are close on the continuum scale. Thus, cohesion and contrast would represent mutual intelligibility and unintelligibility of two given languages respectively.

explanation of the language family sizes, Wichmann (2005) proposes the process of self organised criticality which - as noted above - is characterised by a system exhibiting power law behaviour at a phase transition (see Bak, Tang and Wiesenfeld 1987).

10.3 Contrary factors

It is not unreasonable that two contrary factors such as isomorphism and maximal connectivity underline the linguistic behaviour at different theoretical levels even though the two factors may seem diametric. As Engstrøm and Kelso (2008: 121 & 122) note "the human sense of contrariety is ubiquitous" with many "contraries so often interpreted as being mutually exclusive, either/or dichotomies such as whole versus part, self versus other, body versus mind, nature versus nurture".

The idea of complementary factors tending to occur simultaneously is captured by the idea of metastability. "In metastability, the tendency for individual co-ordinating elements to exhibit collective behaviour coexists with the contrary tendency for those elements to express individual independent behaviour" (Engstrøm and Kelso 2008: 123). Thus, each word has its own unique distribution patterns and yet it also forms part of the 'collective behaviour' of non-atomic syntactic primitives such as the power laws of frequency, degree and storage cost.

In describing spatiotemporal behaviour of biological systems, a significant characteristic is the notion of *relative phase*: "Relative phase emerges as a result of nonlinear interactions among coordinating elements, yet reciprocally conditions or "orders" the behaviour of those same elements"(Engstrøm and Kelso 2008: 125). In respect to the word co-occurrence network, the power law of storage cost that is the signature of hub formation, is also a result of non-linear interactions between type and token frequency of word combinations.

The resulting constituency facilitates future decisions on the constituency of new utterances. The power law, thus, is formed by the hubterranean typology of word interactions but at the same time the power law distribution 'restrains' the variation in new word combinations. On this view, basic syntactic constituency is derivative of the network's relative phase; the relative phase in this context referring to the power law distribution of storage cost values formed by the creation of a hubterranean structure.

10.4 Development of abstract syntax

It has been argued that abstract syntactic categories are indirectly derived from the optimisation of cost and connectivity. Furthermore, verbs generalised early to a new construal are characterised by their high frequency and high degree (connectivity) or low storage cost. If the reasonable assumption is made that these categories form the basis of abstract syntactic structure and early generalisations utilise such categories, then it is plausible that a child's knowledge of abstract syntactic structures will only develop once the optimisation process has occurred. Thus, it is hypostudyed that the first signs of the optimisation process will be exhibited in networks based on two-year-old utterances (approximately) since it is around this age that syntactic constructions are observed (as opposed to 'pivot schemas' which have no syntactic roles; see Tomasello (2003:122-143) for discussion).

Indeed, it has been shown that at this age, dependency relations display scale free distribution, perhaps a sign of this optimisation process (Corominas-Murtra, Valverde and Solé 2007 a&b). Before this age, word combinations are motivated partly by communicative intent and partly by the patterns within the linguistic input. Only when the optimisation process occurs does the child have access to the means of extending previously learnt constructions in a consistent way, and hence abstract syntax emerges "for free" (see Solé 2005).[10]

---

[10] Furthermore, a child's vocabulary size is indirectly implicated in the development of a power law distribution simply because such a distribution will require a given number of elements to express the power law function. Obviously the frequency, degree or storage cost of ten words interacting would not strongly fit a power law function. However, the minimal limit of components and connections required to form a power law is an issue that will not be broached here. This issue does however raise the question of whether the optimisation process is operating on the very first words. On this view, it is just a matter of the vocabulary size being large enough to represent a power law distribution. Alternatively, the optimisation process would not operate initially on the first few morphemes learnt. For a small vocabulary, there is less pressure for efficient storage of their representations. However at a certain critical point, resources may be taxed and as such the optimisation process is necessary to reduce the cost of storing the representations. This speculative matter is left for future research. However, note that the segmentation of monosyllabic words e.g. function words such as *the* and *a*, aids the formation of the frequency and degree power law distributions. These function words are highly visible in the network in that they occur within the interception of the frequency and the degree distributions and

Contra to the idea that the usage-based approach describes a discontinuous path of syntactic development, on this view there is continuity between the child and adult's syntactic knowledge. Both the child and the adult word co-occurrence networks have the same word distribution typology i.e. the hubterranean structure that derives syntactic constituency and function words. Furthermore, both the child and adult languagc arc supported by the same mechanisms that generate syntactic categories and extend constructions. Thus, Goldin- Meadow's (2007: 417) critique of usage-based models is unfounded i.e. "Usage-based accounts of language-learning ought to predict that, in the absence of linguistic input, children will not communicate in language-like ways", giving as an example, the invented homesigns of deaf children.

Although the deaf children may lack linguistic input, there is still the task of storing the representations of the signs of their invented language. It is predicted that the storage of these representations will still be optimised for connectivity and cost. Consequently, the deaf children's homesigns will bear the characteristics of the hubterranean network: syntactic constituency, highly connected words and (construction-specific) abstract syntactic categories.[11] On this view, the means of establishing syntactic categories is provided "for free" being "spandrels" of the network (see Corominas-Murtra, Valverde and Solé 2007b: 7; see also Solé 2005; Solé and Valverde 2006: 421) as well as the three non-atomic syntactic primitives i.e. the frequency, degree and storage cost power law distributions.

---

facilitate the formation of abstract syntactic categories. Thus, a near optimal state may only occur in the network when these function words are identified in the input and incorporated into the child's speech.

[11] Support for this hypostudy, is found in the modelling of the growth of a language network, using Zipf's law as the only syntactic prerequisite and a "growing working memory" (Corominas-Murtra, Valverde and Solé 2007b: 9; see section 4.4.2). At the typological transition to a small world network, a power law of degree distribution was observed (Corominas-Murtra, Valverde and Solé 2007b). Indeed, Zipf's law has been derived from modelling the symbolic mapping between signals and "objects of reference" (Ferrer i Cancho and Solé 2003a). As such, frequency metrics are an integral part of the explanation of how a small world is formed and how syntax arises. Indeed, with the above factors facilitating a small world structure, it is not surprising that children learn language with relative ease.

10.5 Limitations

As with all theories and research, there are limitations to an analysis. Some limitations span across the academic domain, with certain questions being out of everyone's reach at present. The combined semantic and syntactic analysis of early child utterances may be one such challenge. However, identifying limitations is a very productive exercise as it opens up opportunities for further development and substantiates potential future research projects. To this end the chief limitations of this study are presented here.

10.5.1 The orthographic word

An obvious limitation of this study topic is the orthographic nature of the study i.e. the lack of phonetic, morphological, semantic and pragmatic information represented by the network, making it difficult to account for allophones, inflections and homonyms. Furthermore, words that are spoken frequently tend to exhibit variation in pronunciation (Bybee 2002a; Bell, Jurafsky, Fosler-Lussier, Girand, Gregory and Gildea 2001).

If meaning and phonetics were directly analysable in the network,[12] there is still the problem of one word being represented by one node. Thus, taking into account meaning and pronunciation requires that individual instances are retained and incorporated into the network – much like Bybee's exemplar approach (Bybee 2002a) and Goldberg's semantic clusters (1995:135). In this way, the meaning of a word in a particular context can have an individual meaning and pronunciation associated with it.

If this meaning and pronunciation is consistent, then the instances reinforce one representation, whereas different meanings and pronunciation produce different nodes each with their own particular sense and phonetics. Thus homonyms and differences in pronunciation due to stress are differentiated in the same way – by retaining instances. Sensitivity to individual instances has already been established since transitional probabilities (which provide the storage cost of representation) are calculated by attending to individual instances.

[12] In what form the meaning would be encoded into the network is beyond the scope of this study. See Steels (2005) and Steels and de Beule (2006) for an introduction to a grounded approach: fluid construction grammar.

Furthermore, in line with Bybee and McClelland (2005), it would be pertinent to represent a whole signal in a network and allowing the signal to be broken into chunks according to semantic information, distributional analysis and transitional probability information between phonemes/ syllables as well as words. For example, in this new network, two whole utterances may are represented separately by two nodes. A chunk that occurs in both of the two utterances (i.e. the interception of the two utterances) may be assigned a node independent of the original utterances - although this new node will still link to the original utterances. In this way the speech signal is broken down. On the other hand, if two chunks occur in an isomorphic configuration then they may be chunked together as one node. Partial chunking will also occur as word co-occurrence is a graded property as we have seen with constituency i.e. the link length between two nodes. Note that such an approach integrates a temporal approach which is lacking in this study.

10.5.2 Motifs and constructions

Constructions are meaning to form pairing and so any linguistic generalisation will require knowledge of this pairing. Consequently, finding constructions in the adult and child network is a tremendous challenge since it requires that semantic information concerning the words and their combinations is known or in the case of the child, is a knowable quality (particularly in respect to text transcriptions).

Functional regularities are *indirectly* indicated by significant motif patterns since regularities in the usage imply that the speaker has intentionally taken advantage of the construction's form to convey a particular meaning. However, it is doubtful that these patterns of form will identify functional patterns. Individual motifs do not map one-to-one onto individual constructions. For example, a motif pattern may represent many different constructions, identified by specific items e.g. motif 6 and anchor *I* and *I'm* identifying two different constructions in the child network.

On this view, motif 'significance' is purely indicative of non-random form and does not have a direct counterpart in the analysis of constructions. 'Significance' may play a more abstract role in linguistic analysis since highly significant motifs seem to be the templates of a rich and detailed distributional analysis that linguists use to capture linguistic form. For example, motif 6 and 36 identify potential paradigms and conjugations i.e. potential (non-general) syntactic categories, whilst motif 904 represents the shortest discontinuous dependency that has an intervening segment e.g. the pronominal frames in the child network 'I + {N 1 , N 2…} + it'. That motifs 6 and 36 may represent the hubs of the network, arguably suggests that these motifs may be derived from the optimisation process. If so, it is unlikely the proliferation of these motifs has a "functional adaptive interpretation" (Solé and Valverde 2006: 420).[13]

Although, general categories are rejected, motif analysis will - as a first pass - aid the identification of construction-specific categories as well as picking out potential productive item-specific constructions. Naturally, small item-specific constructions will be easier to recognise rather than the abstract argument structure constructions due to consistencies of word usage e.g. the *and*- construction is recognized by the specific word *and* whereas the English transitive construction is recognized by its syntactic and semantic structure.

Whether languages share the most highly significant 4-node motifs and differ in the less significant motifs, thus establishing a unique distributional analysis for each language is a matter of empirical research. Interestingly, the most highly significant motifs in the child's network are virtually the same highly significant motifs that are found in an adult's speech in English, suggesting continuity in distributional patterns between the child and the adult language. Again, future projects would need to substantiate that this correlation is observed in other child and adult speech as well as other diverse languages with a large set of individual child and adult speech corpora taken from a wide range of social contexts.

10.5.3 The sequence of learning

In this model, it is not possible to make iron clad predictions as to which words and constructions will be learnt first. A key characteristic of an optimisation process is that although one is aware of the final outcome of the process i.e. the hubterranean structure, the individual steps to this stage are very hard to predict. That is, it is difficult to predict the temporal building of the network – which word/construction will be added next. There are only tendencies and these tendencies may reflect how frequency

---

[13] However, this suggestion is tentative. When modelling the growth of a word network with Zipf's law as the only syntactic prerequisite, motif 6 and motif 36 were slightly overrepresented compared to a real child network (Corominas-Murtra, Valverde and Solé 2007b).

influences the choice of words.

For example, the verb whose meaning has become attributable to a particular construction i.e. the verb *give* in the ditransitive, may be learnt early due to frequency of usage and the fact that its semantics are most strongly associated with that particular construction. This is strictly a tendency as noted in chapter three. When verbs are extended to a new construal, although there are five diverse factors that correlate very strongly with the observable data (see chapter nine), each factor has (without exception!) its own list of exceptions.

Frequency is only one of the motivations in picking a word/construction. The word /construction is required to fulfil the child's communicate intent (Ninio 2006a; see chapter 4.4.3 for discussion). Thus, the child's interest creates a non-predictable factor in learning. This variability is in keeping with the whole notion of using an optimisation process to capture learning.

Other studies indirectly suggest that learning is driven by an optimisation process. For example, the learning of a construction means the child displays a verb frequency distribution that is captured by a power law (Ninio 2005 b&c; 2006a) not that the child has learnt a specific set of words associated with the construction at certain well defined timeframes. Thus, one can predict how the overall structure of the knowledge will manifest but one can not predict the exact development of knowledge i.e. the exact schedule by which certain words will be learnt.

10.6 Second language learning

The results of this study have implications for second language learning given that the learning of any language is based on familiarity of usage and the extension of past linguistic experience. Since a hubterranean network is the most optimised way of representing and storing linguistic form, it stands to reason that the linguistic form of a second language will be learnt more efficiently if it too is stored and represented in the same optimal way. Thus the factors that encourage hubterranean structure will benefit the second language learner.

The practical outcome of this observation is that learners would benefit from having their set of training utterances support the formation of a miniature word co-occurrence network with hubterranean structure. That is, the inverse transitional probabilities of the words in the training utterances would edge towards collectively displaying a power law distribution. In addition, productive constructions are favoured to create hub words with all words adhering to a degree power law distribution.

Furthermore, across various sessions, it would be beneficial for verbs that provide the central sense of the construction to be repeated often within the respective construction, in the spirit of syntactic consistency discussed in chapter nine. Generally, the verb frequencies in each construction presented in multi-sessions would converge on a distribution captured by a power law function in line with Ninio (2005 b&c, 2006a). Also, it would be beneficial for verbs to appear in diverse constructions. That is, certain verbs would display syntactic diversity.

Ultimately, the cumulative utterance set from the multi learning sessions would exhibit these collective properties (if practically possible) with the aim of building a hubterranean network of word co-occurrences within the mind of the second language learner. Thus, the representation and storage of the words' form is optimised, consequently facilitating language development. Even for native language training sets for children, it would be prudent to replicate the global statistics of the word co-occurrence network, in miniature, to maintain a hubterranean representation - thus, reinforcing the child's natural tendency to store linguistic form in this optimised configuration.

10.7 Incorporating natural principles

It is the linguist's aim to express language in the most parsimonious way. And yet, as this study argues, general (natural) principles found in other systems have already established this parsimony: by deriving the most economical way of representing linguistic form without excessively curtailing expressiveness, a hubterranean structure is born. On this view, abstract syntax does not function as the most minimal algorithm describing patterns in linguistic form i.e. it is not just about capturing the shortest description of all possible word combinations. Rather, it is one component of a cyclic syntactic process, derived from the optimisation of storage efficiency and maximal combinatoriness.

Abstract syntax further supports this optimised representation by identifying potential generalisations and constraining how new representations are connected within the network. Thus, abstract syntax creates consistency between past linguistic experience and new usages. On this view, syntax is a navigator, manoeuvring the mind within the ever varying sea of linguistic form and it is a guide, steering the mind to new linguistic experiences.

As argued in chapter nine, abstract syntax will form a separate syntactic level to the storage of word co-occurrences. These two levels will most likely be mapped together by a type of tinkering and degeneracy process (see chapter eight and nine). Note that the abstract elements will also be stored in a cost effective way, maximising intra-relations. Although there are two levels, both specific collocations and abstract syntactic elements will exhibit properties attributable to the storage aspect of syntax i.e. a hubterranean structure.

As such, the constructionist view of a continuum between the lexicon and the (storage aspect of) syntax is retained. On this view, the two syntactic levels i.e. the word co-occurrences and the abstract elements, establish the form component of the constructicon. The other syntactic function – computation - is quite separate from this storage view of syntax. Thus, there is a functional division within syntax: storage versus computation. This division is similar to the formal accounts of a separate lexicon and a computational system. However, this account differs, since it evidences a storage component of syntax which can not be separated from the 'lexicon'. The computational aspect of syntax, though, is a matter for future research.

In regard to the characteristics of syntax, this study rejects the assumptions that syntactic structures must always be symmetrical, non-redundant and capture generalities in opposition to empirical evidence (in line with Croft 2009). As this study argues, it is the *asymmetric* distributions of frequency and connectivity i.e. the non-atomic syntactic primitives that incept and facilitate the generation of syntactic categories and the extension of generalisations.

Redundancy and degeneracy in the form of verbal conjugations (see chapter eight) protect the hubterranean network from its inherent vulnerability - a direct attack on a hub node. *Specific* storage cost values create a distribution that reflects basic syntactic constituency on which the phrasal hierarchy is built (in line with Bybee 2002a). Indeed, the asymmetry of power laws and redundancy (see Wagner 2005 for the latter attribute) are properties prevalent in natural systems.

Consequently, it is important for the linguist to consider that the properties inherent in language may be manifest of more general properties found in the (natural) world. It is a logical step to view language as a natural object in the world and analysis it as such and compare it to other natural systems. It allows linguistic data to contribute to very general questions concerning natural systems and in turn, network theory can reciprocate with potentially comparable properties found in other natural systems. In this way, linguistic theory can contribute to the larger picture of what a natural process is and in return synchronise with other theories of nature.

Since every theory starts with one or more (different) assumptions, the angle of observation imposed on the object under enquiry is always acute and thus limited to some degree. As such, it is implausible that one theory will provide the whole picture. One advantage of network theory is that these initial assumptions are few e.g. entities are represented as nodes and a determined relation is represented by a link between the nodes. Such simplicity in assumptions affords a wide view of observation and enquiry.

By adopting the network perspective with minimal syntactic prerequisites, this study offers a unique account of how basic syntactic characteristics arise "for free" (see Solé 2005). Specifically, the study provides a simple unified explanation as to why linguistic form has certain characteristics and a particular organisation, how abstract (non-general) syntactic categories arise and why certain language learning strategies are observed. This explanation shines a light on the natural mysteries and the shy linguistic creatures that lay hidden within the hubterranean world of syntax.

~~~

It is charmingly appropriate that this study was written in Australia, a beautiful country where Aborigines have long understood the closeness between their language and the land. The idea of calling the landscape into existence through 'naming' may seem poetic, and yet each time our speech breezes across our landscape, we are invoking the very natural patterns that infuse our world. It is not that language happens to echo the harmonies of nature. It is more than language and nature being inseparably entwined. It is simply that the universe is within the verse. Language *is* nature and nature *is* language.
~~~

Appendix A.1

Word Pair Examples with Storage Cost Values from the Adult Word Co-occurrence Network (Section 5.4.3)

Regime 1			Regime 3			Regime 3		
1st word	2nd word	cost	1st word	2nd word	cost	1st word	2nd word	cost
the	beaches	3333	think	it'll	476	we	also	294
and	felixstowe	2000	know	me	476	were	most	294
to	walk	2000	the	lady	476	erm	right	286
i	admire	2000	so	passionately	455	of	each	286
the	unions	1667	do	speak	455	no	surrender	286
a	pudding	1667	what	you'd	455	have	said	286
that	story	1667	if	everyone	455	like	creature	286
of	animals	1429	erm	that's	435	get	food	278

Regime 2			and	he's	435	the	top	278
it	ought	1250	well	designed	417	in	march	270
you	agree	1250	it	depends	417	for	beauty	270
we	extend	1111	the	star	417	a	lady	270
the	kids	1111	to	support	417	that	means	270
in	texture	1111	you	give	400	it's	great	263
and	buttons	1111	a	start	400	from	which	256
to	book	1000	people	tend	400	to	help	256
i	then	1000	i	wonder	400	the	minutes	256
erm	london	833	that	information	400	er	i've	250
the	health	833	at	large	385	i	came	250
a	horse	833	we	hadn't	385	there	really	250
is	enhancing	769	one	obvious	385	very	enjoyable	250
of	artists	714	got	very	385	had	its	250
and	children	714	with	respect	385	it	goes	250
to	listen	667	is	any	370	is	important	250
i	give	667	about	july	370	actually	passed	250
they	usually	667	the	number	370	them	things	244
it	actually	625	as	such	370	this	trust	244
on	average	625	or	send	357	be	helpful	244
you	wish	625	of	your	357	you	mean	244
we	asked	588	and	make	357	yeah	it's	244

have	brought	556	that's	saturday	357	of	her	238
the	morning	556	to	persuade	345	but	i've	238
for	baths	556	all	came	333	think	you've	238
in	that	556	i	sent	333	and	to	238
a	concert	526	was	wondering	333	know	don't	238
and	only	526	are	interesting	323	the	schools	238
that	sounds	526	a	laser	323	right	over	238
it's	obviously	526	not	squeeze	323	we	think	233
to	start	500	that	really	323	so	successful	233
er	question	500	can	practice	313	do	more	233
i	couldn't	500	it	needs	313	a	month	233
there	often	500	on	file	303	up	my	233
this	bit	500	just	wash	303	then	decide	227
be	spending	500	don't	wanna	303	that	you've	227
Regime 3			you	won't	303	if	something	227
yeah	like	476	the	sort	303	what	does	227
of	silk	476	to	buy	294	to	speak	227
but	is	476	mean	they've	294			

Regime 3			Regime 3			Regime 3		
1st word	2nd word	cost	1st word	2nd word	cost	1st word	2nd word	cost
i	am	222	that's	people	179	if	i've	152
could	look	222	say	i'd	175	you	actually	152
he	phoned	222	it's	interesting	175	the	first	149
they	sent	222	the	general	175	time	it's	149
the	thing	222	erm	oh	172	there's	you	149
was	something	217	things	rather	172	time	working	149
in	one	217	to	meet	169	there's	usually	149
erm	some	217	some	faces	169	is	you	149
and	i'll	213	come	more	169	going	shopping	149
been	recommended	213	er	have	169	a	way	147
well	how	213	more	interested	167	me	outside	147
would	also	208	all	through	167	to	find	147
it	down	208	there	own	167	mean	then	147
i'm	studying	208	yes	to	167	we	know	145
of	my	208	we	as	167	quite	right	145
which	obviously	204	said	officers	167	were	expected	145
on	them	204	see	more	167	two	o'clock	145
to	this	204	they	go	167	i	could	145
you	would	204	the	time	164	no	other	143
a	week	204	and	mend	164	any	specific	143

people	feel	200	was	right	164	have	never	141
i	suppose	200	this	lady	164	like	they've	141
that	will	200	be	happening	164	well	they've	141
out	into	200	are	extremely	161	other	matters	141
and	we	196	a	speaker	161	it	might	139
oh	mid	192	yeah	it	161	get	out	139
who	on	192	of	money	161	how	important	137
at	eight	192	not	they	161	for	these	135
one	girl	192	that	we've	161	in	your	135
because	those	192	but	we're	159	by	some	135
with	their	192	think	people	159	you	say	135
i've	asked	189	here	next	159	by	perhaps	135
have	someone	189	can	find	159	a	sort	135
now	one	189	here	yes	159	people	didn't	135
is	something	185	know	is	159	i	had	135
to	keep	185	to	tell	156	go	by	133
about	you	185	it	now	156	and	i'm	133
the	reason	185	in	time	156	an	interest	133
as	every	182	so	why	154	they	can't	133
or	they	182	do	wish	154	the	one	132
for	information	182	just	wanted	154	was	on	132
in	those	182	on	you	154	of	march	132

a	small	179	don't	put	154	it's	still	132
when	that	179	just	talking	154	at	harlow	130
sort	in	179	don't	seem	154	we	all	130
it	will	179	will	interest	154	one	on	128
and	there's	179	and	one	152	got	five	128
that	it's	179	what	he	152	the	meeting	127

Regime 3			Regime 3			Regime 3		
1st word	2nd word	cost	1st word	2nd word	cost	1st word	2nd word	cost
i	got	127	then	you'll	115	to	erm	102
there	must	125	has	already	114	er	we	101
and	if	125	what	time	114	group	except	101
very	happy	125	if	not	114	people	into	101
had	no	125	has	more	114	there	should	101
it	a	125	what	did	114	the	board	100
a	letter	125	it	says	114	theatre	finds	100
is	because	125	have	two	114	put	some	100
erm	that	123	and	what	112	out	now	99
really	twice	123	way	than	112	doing	like	99
actually	look	123	my	first	112	much	a	99
really	i've	123	all	that's	112	this	erm	98
about	our	123	i	did	112	these	soap	98
that	a	123	my	books	112	be	any	98
them	dry	123	could	easily	111	used	you	98
did	reach	123	yes	he	111	mean	there	97
on	with	123	of	their	111	the	town	97
this	up	123	he	just	111	oh	gosh	97
be	quite	122	they	won't	111	who	will	97
the	whole	122	you	will	110	at	one	97

as	soon	122	the	way	110	oh	is	97
you	like	122	harlow	caring	110	at	school	97
or	another	120	want	and	110	we	want	97
meeting	will	120	was	about	110	were	gonna	97
in	any	120	something	basically	109	got	in	96
yeah	that's	120	thing	i	109	because	our	96
of	er	120	for	instance	109	of	this	96
to	write	120	erm	so	109	with	you	96
our	publicity	120	are	concerned	109	it	has	96
but	we	119	are	one	109	last	given	96
think	we're	119	make	good	108	take	care	96
and	there	119	a	number	108	last	two	96
i	hope	119	didn't	even	108	us	with	96
know	an	119	is	quite	106	no	one	95
that's	gonna	118	i	can't	106	but	anyway	95
good	evening	118	been	difficult	106	think	about	95
good	example	118	well	i've	105	a	great	95
cos	this	118	we	might	105	into	their	95
good	but	118	can	assure	105	i've	come	95
should	an	118	your	friend	105	i'll	bring	95
right	right	118	lot	going	105	i've	heard	95
should	only	118	she	really	104	know	you	95

cos	a	118	i'm	very	103	they	actually	94
right	i	118	the	playhouse	103	like	people	94
so	they're	116	which	country	103	that	would	94
do	want	116	just	one	102	now	i'm	94
a	new	115	don't	actually	102	council	do	94
up	as	115	on	seats	102	now	she	94

Regime 3			Regime 4			Regime 4		
1st word	2nd word	cost	1st word	2nd word	cost	1st word	2nd word	cost
was	gonna	93	sorry	about	84	time	now	75
is	in	93	more	this	84	it's	at	75
their	funding	93	all	around	84	i	haven't	75
so	we're	93	sorry	could	84	is	what	75
to	them	93	there	any	84	is	one	75
about	to	93	very	pleased	84	going	what	75
get	erm	93	had	in	83	could	it	74
as	much	91	yes	but	83	they're	having	74
if	anyone's	91	it	doesn't	83	he	hadn't	74
what	we're	91	we	used	83	they	used	74
what	happened	91	still	gotta	83	me	it	74
also	on	91	work	apparently	83	it	could	74
perhaps	erm	91	still	thirty	83	the	theatre	74
where	else	91	said	just	83	the	council	74
also	understand	91	see	him	83	mean	there's	74
only	from	91	isn't	used	83	was	quite	74
year	of	91	is	not	83	as	they	74
or	was	91	is	that	83	as	if	73
for	you	91	is	erm	83	mean	he	73
why	later	90	is	going	83	many	teams	71

a	group	90	they	did	83	use	those	71
when	he's	89	actually	put	83	other	groups	70
sort	it	89	was	this	83	anything	for	69
and	all	89	them	as	83	how	she	68
that	one	89	to	know	82	we	didn't	68
that's	one	88	this	week	82	on	and	68
haven't	looked	88	be	something	82	is	this	68
back	too	88	are	on	82	know	it's	68
i	didn't	88	those	services	82	by	several	68
say	it	88	through	with	81	people	don't	67
Regime 4			have	something	81	all	i	67
it's	really	87	those	few	81	nineteen	er	67
you	got	87	you	had	81	go	at	67
you	just	87	you	get	81	erm	in	67
you	were	87	those	bags	81	yes	we	67
you're	attracting	86	of	us	81	an	amateur	67
things	so	86	not	quite	80	out	from	67
and	that's	85	but	this	80	out	is	67
from	my	85	think	he	79	they	come	66
a	little	85	and	it's	79	do	quite	66
look	lovely	85	money	bringing	79	local	people	66
some	action	85	money	basically	79	you've	then	66

well	i'll	85	here	you	79	wasn't	you	66
come	home	85	can	remember	79	local	advantage	66
er	in	85	money	with	79	does	needs	65
er	is	85	the	same	79	are	in	65
more	information	85	right	well	79	not	too	65
all	in	84	there's	some	75	you	are	64

Regime 4			Regime 4			Regime 4		
1st word	2nd word	cost	1st word	2nd word	cost	1st word	2nd word	cost
question	now	63	very	very	51	years	but	29
had	that	63	that	you	50	if	anybody	28
very	useful	63	these	here	50	harlow	to	28
is	is	63	okay	if	49	harlow	people	27
erm	you	63	one	was	49	here	is	27
actually	have	63	one	and	48	money	is	26
really	it's	62	i've	written	48	they	are	26
them	with	61	i've	always	47	bar	was	25
did	tell	61	into	squares	47	oil	from	24
just	get	61	place	you	47	way	and	23
don't	really	61	place	with	46	we	had	23
on	how	61	only	had	46	them	and	22
i	would	61	year	didn't	45	every	month	21
than	er	61	an	action	45	every	night	20
three	wrote	61	an	interesting	44	campaign	was	19
as	an	61	but	if	44	interested	to	18
meeting	but	61	but	you	43	then	we	17
or	what	61	been	on	43	me	and	16
new	initiative's	60	well	we	42	contact	me	15
her	view	60	see	whether	42	we'll	look	14

done	any	60	still	had	41	point	out	14
new	plans	60	through	me	41	something	about	14
her	resources	60	those	days	40	to	the	14
and	he	60	cos	if	40	Regime 5		
i	said	60	good	time	39	meetings	for	13
know	how	60	just	as	39	ago	and	13
good	and	60	up	on	38	ask	a	13
find	money	59	there's	lot's	38	members	group	13
cos	obviously	59	time	it	38	ago	one	13
find	although	59	there's	going	37	phoned	at	13
say	if	58	was	very	37	looks	quite	13
it's	in	58	as	we	37	we'd	be	13
him	the	58	something	in	37	almost	every	13
always	persuading	58	thing	you	36	ways	there	13
years	though	57	use	this	36	raised	eh	13
might	this	57	use	to	35	club	tonight	13
has	any	57	anything	in	35	worth	getting	13
he	knows	56	a	very	34	worth	looking	13
get	any	56	go	away	34	mrs	will	13
we	need	56	nineteen	eighteen	33	aren't	as	13
see	these	55	does	have	33	aren't	good	13
for	it	55	does	still	32	absolutely	everything	13

erm	we	54	very	well	32	certain	date	13
didn't	he	54	council	takes	31	each	garment	13
as	it	53	than	recycling	31	these	are	12
before	i	52	than	what	30	these	people	12
from	what	52	cos	it's	30	made	by	12
just	like	51	should	come	29	come	and	12

Regime 5			Regime 5			Regime 5		
1st word	2nd word	cost	1st word	2nd word	cost	1st word	2nd word	cost
get	the	12	tell	them	12	any	more	10
is	a	12	different	countries	12	any	other	10
felt	good	12	we'll	have	12	funding	the	9
matter	you	12	already	had	11	everything	else	9
shows	in	12	september	the	11	following	days	9
felt	was	12	problem	in	11	follow	up	9
wouldn't	do	12	might	like	11	necessary	i	9
regular	interest	12	year	and	11	necessary	oh	9
stortford	for	12	keep	them	11	bit	more	9
felt	hats	12	idea	and	11	together	with	9
oil	pollution	12	groups	are	11	black	coffee	9
felt	intimidated	12	green	and	11	black	boot	9
near	it	12	didn't	have	11	worry	about	9
areas	an	12	it's	a	11	together	yeah	9
concert	weren't	12	month	and	11	elected	by	9
saturday	and	12	bring	some	11	form	which	9
areas	extreme	12	found	it's	11	following	up	9
times	smaller	12	us	to	11	elected	to	9
ron	has	12	take	a	11	form	for	9
michael	to	12	you'll	sell	11	elected	as	9

apart	the	12	think	we	11	makes	a	9
gave	me	12	else	is	11	hill	to	9
believe	there's	12	i'll	be	11	february	eighty	9
pound	if	12	into	a	11	herts	guy	9
putting	they're	12	too	bad	11	makes	sense	9
quickly	done	12	providing	them	10	makes	three	9
worked	quite	12	hundred	people	10	makes	you	9
less	shows	12	providing	services	10	pay	a	9
putting	down	12	can't	remember	10	makes	blankets	9
less	interest	12	low	income	10	questions	you've	9
tried	every	12	april	there	10	following	the	9
manager	has	12	low	as	10	held	up	9
expensive	to	12	low	cost	10	held	over	9
space	a	12	april	the	10	weekend	mm	9
expensive	and	12	read	through	10	posters	also	9
less	that	12	april	run	10	greenpeace	they	9
everybody	will	12	low	so	10	greenpeace	sell	9
space	at	12	april	till	10	police	here	9
expensive	luxury	12	raise	something	10	police	officer	9
publicity	in	12	raising	a	10	gone	this	9
publicity	is	12	raise	or	10	gone	really	9
space	not	12	raising	events	10	team	and	9

ever	it	12	jazz	available	10	finished	now	9
be	a	12	fifty	people	10	finished	here	9
work	with	12	he	was	10	finished	managing	9
along	this	12	what's	going	10	gone	that	9
general	meeting	12	they've	been	10	police	superintendent	9
general	manager's	12	last	time	10	popular	was	9

Regime 5			Regime 5			Regime 5		
1st word	2nd word	cost	1st word	2nd word	cost	1st word	2nd word	cost
fine	we	9	persuade	them	7	that'll	do	6
fine	with	9	sound	terribly	7	around	the	6
gonna	have	9	standing	an	7	review	erm	6
pool	who's	9	hand	out	6	consider	this	6
gone	over	9	programme	and	6	fourth	of	6
police	witnesses	9	september	nineteen	6	consider	that	6
pool	isn't	9	doesn't	it	6	car	batteries	6
pool	erm	9	chair	person	6	clothes	rationing	6
pool	because	9	years	ago	6	he'd	been	6
next	item	9	best	to	6	magazines	were	6
budget	isn't	9	during	that	6	he'd	got	6
budget	past	9	hear	about	6	green	or	6
peter	do	9	discuss	this	6	green	image	6
budget	and	9	health	authority	6	green	wedges	6
pass	them	9	buy	a	6	clearly	is	6
peter	you've	9	lady	here	6	clearly	about	6
pretty	revolting	9	hear	what	6	magazines	gave	6
peter	have	9	thinking	about	6	had	a	5
follow	this	9	tonight	and	6	has	been	5
asked	me	9	water	for	6	if	you	5

letters	we	9	central	policy	6	itself	in	5
personally	feel	9	come	to	6	recycled	cos	5
workshops	is	9	i	think	6	houses	but	5
worry	was	9	coming	to	6	hours	and	5
pretty	face	9	round	the	6	houses	had	5
pretty	depressed	9	hasn't	asked	6	val	is	5
very	much	8	let	me	6	val	well	5
services	and	8	as	a	6	houses	where	5
amnesty	does	8	got	a	6	holes	here	5
experience	where	8	longer	support	6	hours	a	5
homes	due	8	longer	being	6	propaganda	posters	5
experience	of	8	income	and	6	machine	wasn't	5
experience	it	8	income	support	6	liz	would	5
events	very	8	seventy	nine	6	include	the	5
events	which	8	bits	in	6	include	studio	5
fuel	efficiency	8	seventy	six	6	recycled	using	5
exercise	why	8	definitely	going	6	uttlesford	chaps	5
apparently	just	8	obviously	the	6	include	soap	5
exercise	within	8	college	i	6	include	clothing	5
homes	per	8	college	with	6	tropical	species	5
tom	will	8	region	and	6	better	than	4
order	a	8	street	collection	6	sounds	like	4

fuel	by	8	region	of	6	health	service	4
order	anything	8	sixty	eight	6	discuss	it	4
exercise	so	8	july	nineteen	6	happy	with	4
exercise	to	8	sixty	five	6	gonna	be	4
everybody	was	8	electricity	board	6	goes	into	4
pete	again	7	pensioners	were	6	seven	or	4

Regime 5			Regime 5		
1st word	2nd word	cost	1st word	2nd word	cost
once	a	4	speed	things	2
government	unit	4	exists	for	2
saving	fuel	3	fire	eating	2
keeping	and	3	exclude	general	2
continued	to	3	celebrating	their	2
kate	then	3	exchange	centres	2
yardage	in	3	chips	and	2
therefore	have	3	poll	tax	1
resident	i'd	3	piece	of	1
yardage	and	3	talked	to	1
kath	while	3	bishop's	stortford	1
kath	kathy	3	seems	to	1
yardage	needed	3	ought	to	1
resident	domestic	3	kind	of	1
twentieth	birthday	3	thank	you	1
involve	them	3	carry	on	1
theatres	up	3	sorts	of	1
knickers	became	3	lots	of	1
knickers	and	3	regards	waste	1
checked	with	3	regretted	the	1

knickers	from	3
knit	new	3
finding	it	3
recession	it	3
finding	out	3
inviting	and	3
knit	socks	3
beds	and	3
fifteen	years	3
realised	there	3
beds	when	3
behaviour	around	3
received	the	3
beds	round	3
pet	petticoat	3
realised	it	3
changes	of	3
subsidy	er	3
skin	refreshed	3
hadham	village	3
pet	needed	3
research	goes	3

register	an	1
imaginative	or	1
improvement	to	1
improvements	for	1
improve	matters	1
impressive	isn't	1

changed	the	3
fifteen	Tons	3
reasons	why	3
steps	of	3
feeling	well	3

APPENDIX A.2

WORD PAIR EXAMPLES WITH STORAGE COST VALUES FROM THE CHILD WORD CO-OCCURRENCE NETWORK (SECTION 6.4.3)

Regime 1			Regime 3			Regime 4		
1st word	2nd word	cost	1st word	2nd word	cost	1st word	2nd word	cost
i	splashing	720	you	wanna	110	recorder	round	60
it	juice	609	juice	yeah	110	i	fall	60
whats	toy	594	do	ahking	109	my	birdies	59
this	eggie	466	don't	fall	108	oh	okay	59
that	georgies	342	get	you	103	at	julia	58
want	play	300	want	something	100	fall	over	58
uh	daddys	290	doing	something	100	take	diaper	58
Regime 2			whats	agra	99	uh	jenko	58
down	i	285	see	milk	97	nomi	touching	57
down	on	285	mommy	sleeping	97	on	baby	57
the	piggie	259	uh	finger	97	go	out	55
i	finded	239	down	there	95	you	boo	55

this	empty	233	there	georgie	95	draw	flower	55
daddy	go	225	like	him	92	juice	somebody	55
you	standing	221	here	eyes	88	off	it	55
get	feet	206	the	moon	86	do	again	54
it	cooled	202	up	georgie	85	dont	kick	54
whats	fish	198	on	bumpy	85	eat	a	54
mommy	muddy	194	book	it	85	flower	okay	54
there	lipstick	189	go	outside	85	georgie	going	53
no	like	178	thats	nomi	80	birdie	sitting	52
that	wet	171	georgie	eat	79	coffee	animals	52
up	this	171	kitty	meow	79	this	one	52
on	get	171	yeah	too	78	jenko	finish	51
go	foo	166	to	bathroom	78	doing	to	50
georgie	bumped	159	me	love	77	its	alright	50
want	another	150	sit	by	76	want	banana	50
whats	donkey	149	and	read	73	read	book	49
uh	pour	145	more	whining	71	see	birdie	48
more	crispies	142	is	kangaroo	69	there	you	47
is	petal	139	wheres	crumb	69	broken	boat	47
wheres	toaster	137	baby	has	68	is	nomi	46
baby	smile	135	doggie	up	67	georgies	lie	46
going	do	133	going	play	66	lie	me	46

look	hand	131	yup	milk	66	wheres	thumb	46
in	kitchen	129	put	my	66	daddy	going	45
a	tacky	129	can	read	66	sleep	well	45
Regime 3			look	in	65	no	more	45
im	singinging	124	where	going	65	going	it	44
it	cries	122	hi	dad	65	ah	my	44
i	singing	120	in	mouth	65	a	go	43
my	kangaroo	119	a	bunny	65	in	dollys	43
this	mommy	117	Regime 4			peanut butter	dinner	43
at	moon	116	what	does	63	that	mommy	43
that	nomis	114	another	kitty	63	up	windowshade	43
nomi	broking	114	im	lyinging	62	open	door	42

Regime 4			Regime 4			Regime 4		
1st word	2nd word	cost	1st word	2nd word	cost	1st word	2nd word	cost
moon	sleeping	39	play	in	26	name	jackie	17
me	toast	39	a	rash	26	nomis	tired	17
cant	my	38	a	drink	26	all	done	16
fish	up	38	a	book	26	bye	baby	16
hot	bottle	38	are	tired	25	camel	chicken	16
tired	baby	38	banana	nan	25	dropped	them	16
out	way	37	gammy	talk	25	eh	jenko	16
and	broken	36	little	girl	25	truck	lying	16
cookie	daddy	36	lying	in	25	there	it	16
sun	working	36	smoke	do	24	girl	have	15
wanna	pop	36	theres	kitty	24	f	there	15
more	dinner	35	eating	grass	24	four	seven	15
find	petal	35	for	nomi	24	kittys	stand	15
green	clown	35	over	sam	24	mm	tail	15
he	a	35	kangaroo	sleepy	23	mouse	too	15
did	kangaroo	34	mouth	there	23	boy	go	14
e	ow	34	too	tight	23	babys	hands	14
it	on	34	here	bread	22	careful	nomi	14
baby	cry	34	here	jenko	22	crumb	too	14
doggie	fall	34	chair	way	22	cry	baby	14

can	stand	33	clean	my	22	dinner	okay	14
piggie	eat	33	dirty	foot	22	horsie	going	14
put	sugar	33	have	this	22	beanice	bite	13
whos	this	33	make	horse	22	because	windows	13
where	sun	32	paper	off	22	blue	thing	13
mommy	coffee	32	washing	you	22	ear	eh	13
a	basket	32	a	nap	22	hair	out	13
a	monkey	32	be	stopped	21	julia	soaking	13
agra	sleepy	32	got	toast	21	monkey	and	13
sleeping	sandy	31	how	house	21	pour	it	13
what	shredding	31	some	kangaroo	21	table	up	13
drink	cocoa	31	hat	on	20	turn	the	13
man	eyes	31	heres	melon	20	baloney	get	12
like	my	31	stars	are	20	brush	the	12
gonna	lay	30	sit	in	19	bug	fly	12
boy	you	29	toys	in	19	coco	to	12
push	that	29	jump	lolly	19	daddys	sleeve	12
cocoa	hot	28	out	mouth	19	goldie	mouse	12
blanket	bed	27	boo	georgie	18	hard	to	12
broke	here	27	change	dress	18	him	have	12
come	with	27	dish	over	18	holding	ear	12
foot	up	27	george	do	18	kiss	coco	12

okay	sit	26	water	daddy	18	lamb	doing	12
one	too	26	red	red	18	light	blue	12
kitty	nono	26	cereal	on	17	milk	on	12
good	week	26	cereal	mommy	17	ow	nany	12

Regime 5			Regime 5			Regime 5		
1st word	2nd word	cost	1st word	2nd word	cost	1st word	2nd word	
back	on	11	um	bucket	8	beef	eating	5
clean	dishes	11	an	baloney	7	birdies	eating	5
color	it	11	bear	dance	7	diaper	on	5
eight	seven	11	bottle	of	7	face	dirty	5
his	hands	11	carrot	salad	7	fell	in	5
penguin	sleep	11	drop	her	7	hes	tired	5
puppy	sleep	11	fixing	lunch	7	mom	here	5
round	and	11	gaga	bim	7	orange	spots	5
shoe	onie	11	golfball	ball	7	pick	me	5
with	dolls	11	hello	daddy	7	ribbon	see	5
boat	yup	10	jenkos	finish	7	shoes	on	5
bring	you	10	looking	donkey	7	shopping	to	5
cheese	mouth	10	meow	meow	7	stuck	them	5
horse	a	10	mine	my	7	todd	eat	5
singing	up	10	piece	paper	7	trying	beanice	5
toast	coming	10	skunk	go	7	white	as	5
eyes	hurt	9	stick	hand	7	will	light	5
under	this	9	stick	in	7	fix	babys	5
drawing	donkeys	9	stop	pushing	7	hold	that	5
dress	on	9	toaster	hot	7	nine	ten	5

fix	it	9	told	georgie	7	tail	doggie	5
foo	this	9	wait	there	7	byebye	sun	4
foot	out	9	why	in	7	feet	up	4
getting	jenkos	9	why	dad	7	coming	to	4
hands	are	9	years	old	7	get	up	4
lolly	ear	9	beanice	eat	6	ann	sleeping	4
mommys	done	9	close	it	6	being	on	4
moo	me	9	bam	bimbim	6	black	kitty	4
move	this	9	boom	mommy	6	blowing	monkey	4
peach	daddy	9	bumped	it	6	brown	cookie	4
playing	ball	9	could	i	6	candle	will	4
stand	on	9	finger	red	6	cup	green	4
ah	jumping	9	girls	dupping	6	dollys	eating	4
the	book	9	had	gum	6	fingers	red	4
cheerios	in	8	hey	sun	6	found	hen	4
gone	sun	8	ma	change	6	haha	haha	4
donkey	say	8	pajamas	feet	6	happen	that	4
drinking	bottle	8	pants	off	6	icky	bug	4
empty	plate	8	pink	catalogue	6	lipstick	under	4
feel	good	8	pretty	big	6	matzo	crumb	4
finished	cheerios	8	record	it	6	mun	munuu	4
home	tonight	8	six	seven	6	naomi	doing	4

hug	you	8	skunks	eyes	6	nice	time	4
poor	georgie	8	throwing	it	6	of	matzo	4
right	up	8	yellow	hair	6	sheep	running	4
moon	cold	8	slide	down	5	shredding	carrot	4

Regime 5			Regime 5			Regime 5		
1st word	2nd word	cost	1st word	2nd word	cost	1st word	2nd word	cost
swish	swish	4	peekaboo	daddy	3	fen	tha	2
time	is	4	piggy	box	3	food	too	2
todds	mom	4	pull	cart	3	footie	rock	2
turning	it	4	ready	nap	3	geh	geh	2
upstairs	mommy	4	tell	gammy	3	glass	of	2
was	messy	4	three	four	3	harry	toast	2
wash	it	4	throwing	ball	3	has	a	2
big	is	4	uhuh	kitty	3	having	up	2
lollipop	for	4	wawa	juice	3	hear	me	2
raining	hard	4	who	did	3	hurting	my	2
jacket	on	4	not	broken	3	kristen	crying	2
putting	my	4	talk	to	3	made	a	2
spank	you	4	sugar	on	3	mans	jacket	2
wee	down	4	uhs	nomi	3	moving	this	2
chicken	byebye	3	maybe	find	3	old	are	2
sleeper	red	3	michael	say	3	oug	oug	2
ain	ain	3	these	in	3	pants	on	2
aw	foot	3	windowshade	down	3	passey	in	2
aw	georgie	3	your	drink	3	peel	it	2
ba	bay	3	dog	barking	2	peep	peep	2

ba	ah	3	mine	going	2	pink	brush	2
bad	girl	3	them	in	2	played	toys	2
berries	on	3	youre	shuttinging	2	radio	round	2
bowl	cheerios	3	hi	daddy	2	reading	the	2
bugin	broken	3	hand	table	2	s	s	2
changing	georgie	3	clean	dish	2	salt	go	2
choochoo	horse	3	need	it	2	seeing	light	2
does	him	3	agras	tired	2	she	like	2
does	donkey	3	bee	on	2	shh	sleeping	2
eyebrows	doing	3	behind	georgie	2	sing	down	2
finish	my	3	bit	but	2	skunks	tired	2
fixed	it	3	bump	my	2	sneeze	again	2
flied	down	3	by	myself	2	standing	up	2
fly	away	3	by	the	2	swing	is	2
fooing	my	3	children	singing	2	taking	nap	2
funny	face	3	children	laugh	2	they	carrying	2
ga	throw	3	cleaned	it	2	tickle	my	2
happy	birthday	3	cleaning	floor	2	tinkertoys	away	2
knees	hurt	3	climbing	up	2	tweet	tweet	2
lab	too	3	coloring	my	2	wat	wat	2
love	boat	3	cook	my	2	way	up	2
love	that	3	cool	now	2	whoo	woo	2

mary	has	3	cups	it	2	wrap	up	2
morning	mommy	3	dee	dee	2	zip	it	2
morning	daddy	3	dey	done	2	oig	oig	2

Regime 5			Regime 5			Regime 5		
1st word	2nd word	cost	1st word	2nd word	cost	1st word	2nd word	cost
almost	fall	2	bent	over	1	golf	club	1
dat	ball	2	bimbim	bam	1	goo	georgie	1
sitting	down	2	bird	sitting	1	goose	name	1
window	open	2	birthday	to	1	grow	strong	1
throw	it	2	boots	on	1	hang	up	1
cooled	off	1	bouncing	ball	1	hanger	broken	1
had	a	1	boyge	hmm	1	hanging	up	1
mary	had	1	boys	girls	1	happened	paper	1
munuu	munuu	1	bringing	this	1	hats	floor	1
pushed	it	1	burn	me	1	horsies	name	1
riding	on	1	buttons	are	1	hugging	mommy	1
sliding	down	1	buy	at	1	hup	apple	1
so	sleep	1	called	it	1	idee	ah	1
tape	record	1	calves	going	1	julias	house	1
ten	eight	1	cannot	see	1	just	sit	1
touch	that	1	carry	a	1	kaem	kei	1
walking	around	1	co	watch	1	kaun	do	1
woof	woof	1	cocos	hat	1	kem	me	1
bay	button	1	cokey	doing	1	kimberlys	room	1
bumpy	road	1	comb	hair	1	kno	may	1

child	swimming	1	comforting	baby	1	knock	it	1
donkeys	eyes	1	covers	blanket	1	kristens	tired	1
happen	the	1	crush	flower	1	kugiz	kugu	1
shes	not	1	deres	diaper	1	kugu	du	1
vitamin	bottle	1	doed	it	1	laugh	and	1
ih	ih	1	doesnt	yell	1	laying	down	1
we	going	1	doors	open	1	leave	it	1
wearing	curtain	1	drawed	it	1	lets	have	1
look	at	1	dropdrop	broken	1	licking	the	1
climb	it	1	eated	it	1	locking	the	1
throwed	it	1	ebae	oie	1	loving	georgie	1
wake	up	1	eigh	ah	1	lyinging	down	1
lie	down	1	en	nain	1	mag	out	1
lay	down	1	enjoy	it	1	may	ma	1
6wal	no	1	er	kick	1	monkeys	at	1
ag	oig	1	falled	down	1	moth	doing	1
air	flied	1	fe	fot	1	mu	um	1
ands	going	1	feather	in	1	mum	mun	1
anek	in	1	feed	the	1	nain	ain	1
anug	oig	1	feeling	better	1	naughty	boy	1
as	snow	1	feets	cold	1	new	room	1
ate	it	1	finded	where	1	niaw	foot	1

attacking	mommy	1	finding	georgies	1	nuts	in	1
awful	lot	1	fine	boy	1	papa	horse	1

BIBLIOGRAPHY

Abbot-Smith, K., Lieven, E.V.M. and M. Tomasello. (2001). What pre-school children do and do not do with ungrammatical word orders. *Cognitive Development*, 16(2): 679-692.

Abbot-Smith, K., Lieven E.V.M. and M. Tomasello. (2008). Graded representations in the acquisition of English and German transitive constructions. *Cognitive Development,* 23(1): 48–66.

Abbot-Smith, K. and M. Tomasello. (2006). Exemplar-learning and schematization in a usage-based account of syntactic acquisition. *The Linguistic Review*, 23: 275–290.

Adamic, L.A. (2000). Zipf, Power-laws, and Pareto - a ranking tutorial. http://www.hpl.hp.com/research/idl/papers/ranking/ranking.html. Accessed 29 Sept 2009.

Adamic, L. A. and B.A. Huberman. (2002). Zipf's Law and the Internet. *Glottometrics*, 3:143-150.

Adger, D. (2003). *Core syntax: a minimalist approach*. Oxford: Oxford University Press. Akhtar, N. (1999). Acquiring basic word order: evidence for data-driven learning of syntactic structure. *Journal of Child Language*, 26(2): 339-356.

Albert, R. and A.-L. Barabási. (2002). Statisical Mechanics of Complex Networks. *Review of Modern Physics*, 74: 47-97.

Albert, R., Jeong H. and A.-L. Barabási. (1999). Diameter of the world wide web. *Nature*, 401: 130-131.

Albert, R., Jeong H. and A.-L Barabási. (2000). Error and attack tolerance of complex systems. *Nature*, 406: 378-382.

Amaral, L.A.N., M. Barthélémy and H. E. Stanley. (2000). Classes of small-world networks. *Proceeding of the National Academy of Science,* 97: 11149-11152.

Amaral, L.A.N. and J.M. Ottino. (2004). Complex networks - augmenting the framework for the study of complex systems.*European Physical Journal B*, 38: 147-162.

Amaral L. A. N., Scala A., Barthélémy M. and H.E. Stanley. (2000). Classes of small-world networks. *Proceeding of the National Academy of Science*, 97: 11149–11152.

Ambridge, B., Rowland C.F. and J.M. Pine. (2008). Is structure dependence an innate constraint? New experimental evidence from children's complex-question production. *Cognitive Science*, 32: 222-257.

Ambridge, B., Rowland C.F., Theakston A. and M. Tomasello. (2006). Comparing different accounts of children's non-inversion errors in object wh-questions: What experimental data can tell us? *Journal of Child Language*, 30(3): 519–557.

Ambridge, B., Theakston A. L., Lieven E.V.M. and M. Tomasello. (2006). The distributed learning effect for children's acquisition of an abstract syntactic construction. *Cognitive Development*, 21: 174–193.

Ariel, M. (1999). Cognitive Universals and Linguistic Conventions: The Case of Resumptive Pronouns. *Studies in Language*, 23(2): 217-269.

Artzy-Randrup, Y., Fleishman S.J., Ben-Tal N. and L. Stone. (2004). Comment on "Network Motifs: Simple Building Blocks of Complex Networks" and "Superfamilies of Evolved and Designed Networks". *Science*, 305(5687): 1107.

Aslin, R.N., Saffran J.R. and E.L. Newport. (1998). Computation of conditional probability statistics by 8-month-old infants. *Psychological Science*, 9: 321–324.

Bak, P., Tang C., and K. Wiesenfeld. (1987). Self-organized criticality: an explanation of 1 / f noise. *Physical Review Letters*, 59(4): 381–384.

Barabási, A.-L. (2003). *Linked: The New Science of Networks*. Cambridge, Massachusetts: Perseus Publishing.

Barabási, A.-L. and R. Albert. (1999). Emergence of Scaling in random Networks. *Science*, 286: 509-512.

Barabási, A.-L and Z. N. Oltvai. (2004). Network Biology: Understanding the Cell's Functional Organisation. *Nature Reviews, Genetics*, 5: 101-113.

Bassano, D. Laaha S., Maillochon I. and W.U. Dressler. (2004). Early acquisition of verb grammar and lexical development: evidence from periphrastic constructions in French and Austrian German. *First Language,* 24(1): 33–70.

Bates, E., Bretherton I. and L. Snyder. (1988). *From first words to grammar: Individual differences and dissociable mechanisms*. Cambridge, MA: Cambridge University Press.

Bates, E. and J.C. Goodman. (1997). On the inseparability of grammar and the lexicon.: Evidence from acquisition, aphasia and real-time processing. *Language and cognitive processes*, 12: 507-584.

Bates, E. and B. MacWhinney. (1982). Functionalist approaches to grammar. In E. Wanner, and L. Gleitman (Eds.), *Language acquisition: The state of the art*. New York: Cambridge University Press, 173-218.

Bell, A., Jurafsky D., Fosler-Lussier E., Girand C., Gregory M. and D. Gildea. (2003). Effects of disfluencies, predictability, and utterance position on word form variation in English conversation. *Journal of the Acoustical Society of America*, 113 (2): 1001-1024.

Berkenfield, C. (2001). The role of frequency in the realization of English that. In J. Hopper and P. Bybee (Eds.), *Frequency and the emergence of Linguistic Structure*. John Benjamins Publishing Company, 281-307.

Berko, J. (1958). The child's learning of English morphology. *Word*, 14: 150-77.

Bloom, L. (1970). *Language development: form and function in emerging grammars.* Cambridge, Mass.: M.I.T. Press.

Bloom L., Hood L. and P. Lightbown. (1974). Imitation in language development if when and why. *Cognitive Psychology*, (6): 380–420.

Bloom, L., Lifter K., and J. Hafitz. (1980). Semantics of verbs and the development of verb inflection in child language. *Language*, 56(2): 386-412.

Bloom L, Lightbown P. and L. Hood. (1975). Structure and variation in child language. *Monographs of the society for Research in Child Development*. Serial 160: 402.

Bod, R. (2001). Sentence Memory: Storage vs. Computation of Frequent Sentences. *Proceedings CUNY-(2001), Philadelphia, Pennsylvania.* http://staff.science.uva.nl/~rens/publications.html Accessed 29th September 2009.

—. (2006). Exemplar-Based Syntax: How to Get Productivity from Examples. The *Linguistic Review*, 23(3): 291–320.

—. (2009). From Exemplar to Grammar: A Probabilistic Analogy-based Model of Language Learning. *Cognitive Science*, 33(5): 752-793.

Bolinger, D. (1977). *Meaning and form. London*; New York: Longman.

Borensztajn G., Zuidema, W. and R. Bod. (2009). Children's Grammars Grow More Abstract with Age. *Topics in Cognitive Science*, 1: 175-188.

Borovsky, A. and J. Elman. (2006). Language input and semantic categories: a relation between cognition and early word learning. *Journal of Child Language*, 33(4): 759–790.

Bortfeld, H., Morgan, J., Golinkoff, R. and K. Rathbun. (2005). Mommy and me: Familiar names help launch babies into speech stream segmentation. *Psychological Science*, 16: 298-304.

Bowerman, M. (1996). Learning how to structure space for language. A cross-linguistic perspective. In P. Bloom, M. Peterson, L. Nadel, and M. Garret, (Eds.), *Language and Space*. Cambridge, MA: MIT Press, 385-436.

Bowerman, M. (1998). The "no negative evidence" problem: How do children avoid constructing an overgeneral grammar? In J.A. Hawkins (Ed.), *Explaining Language Universals*. Oxford: Basil Blackwell, 73-101.

Boyd, J.K. and A.E. Goldberg. (forthcoming). Children's failure to generalize when exposed to the same input as adults and its benefits in language learning. http://www.princeton.edu/~adele/Princeton_Construction_Site/Publications.html Accessed 19th August 2009.

Braine, M.D.S. (1963). The ontogeny of English phrase structure: the first phase. *Language*, 39: 1-13.

—. (1976). Children's first word combinations. *Monographs of the Society for Research in Child Development*, 41: 164.

Braine, M.D.S. and P. Brooks. (1995). Verb argument structure and the problem of avoiding an overgeneral grammar. In M. Tomasello and W.E. Merriman, (Eds.), *Beyond names for things: Young children's acquisition of verbs*. Hillsdale, NJ: Lawrence Erlbaum, 353-376.

Brent, M.R., and T.A. Cartwright. (1996). Distributional regularity and phonotactic constraints are useful for segmentation. *Cognition*, 61: 93–125.

British National Corpus, version 2 (BNC World). (2001). Distributed by Oxford University Computing Services on behalf of the BNC Consortium. URL: http://www.natcorp.ox.ac.uk/

Brooks, P., Tomasello, M., Lewis, L. and K. Dodson. (1999). Children's overgeneralization of fixed transitivity verbs: The entrenchment hypostudy. *Child Development*, 70: 1325-37.

Brown, J.H., West, G.B. and B.J. Enquist. (2005). Yes, West, Brown and Enquist's model of allometric scaling is both mathematically correct and biologically relevant. *Functional Ecology*, 19(4): 735 –738.

Brown, R. (1958). How shall a thing be called? *Psychological Review*, 65: 14-21.

Brown, R. and U. Bellugi. (1964). Three processes in the child's acquisition of syntax. *Harvard Educational Review* 34: 133-51.

Burlando, B. (1990). The fractal dimension of taxonomic systems. *The Journal of Theoretical Biology*, 146(1): 99-114.

—. (1993). The fractal geometry of evolution. *The Journal of Theoretical Biology*, 163(2): 161-172.

Bush, N. (2001). Frequency effects and word-boundary palatalization in English. In J. Hopper and P. Bybee (Eds.), *Frequency and the emergence of Linguistic Structure*. John Benjamins Publishing Company, 255-280.

Bybee, J.L. (1985). *Morphology: A study of the Relation between Meaning and Form*. Vol. 9: Typological Studies in Language (TSL). Amsterdam/Philadelphia: John Benjamins.

—. (1995). Regular morphology and the lexicon. *Language and Cognitive Processes*, 10: 425-455.

—. (1998). Morphology as lexical organization. In M. Hammond and M. Noonan (Eds.), *Theoretical morphology*. Academic Press, 119-141.

—. (2001). *Phonology and Language Use.* Cambridge: Cambridge University Press. Bybee, J.L. (2002a). Sequentiality as the basis of constituent structure. In T. Givón and B. Malle (Eds.), *The Evolution of Language out of Pre-Language*. Amsterdam: John Benjamins, 107-132.

—. (2002b). Phonological Evidence for Exemplar Storage of Multiword Sequences. *Studies in Second Language Acquisition,* 24(2): 215-221.

—. (2003). Mechanisms of Change in Grammaticization: The Role of Frequency. In B. D. Joseph and J. Janda (Eds), *The Handbook of Historical Linguistics*. Blackwell, 602-623.

—. (2006). From usage to grammar: the mind's response to repetition. *Language* 82(4): 711-733.

—. (2007). Review of Adele E. Goldberg, Constructions at work: the nature of generalization in language. *Journal of Child Language*, 34(3): 692-697.

Bybee, J.L., Chakraborti P., Jung D. and J. Scheibman. (1998). Prosody and segmental effect: some paths of evolution for word stress. *Studies in Language*, 22: 267-314.

Bybee, J.L. and J.L. McClelland. (2005). Alternatives to the combinatorial paradigm of linguistic theory based on domain general principles of human cognition. In N. A. Ritter (Ed.) *The Role of Linguistics in Cognitive Science. Special Issue of The Linguistic Review*, 22(2-4): 381-410.

Bybee, J.L., Perkins, R. and W. Pagliuca. (1992). *The Evolution of Grammar: Tense, Aspect, and Modality in the Languages of the World. Chicago*: University of Chicago Press.

Bybee, J. L. and J. Scheibman. (1999). The effect of usage on degrees of constituency: the reduction of don't in English. *Linguistics*, 37(4): 575-596.

Bybee, J.L. and D.I. Slobin. (1982). Rules and schemas in the development and use of the English past tense. *Language*, 58(2): 265-289.

Bybee, J. L. and S. Thompson. (2000). Three frequency effects in syntax. *Berkeley Linguistic Society*, 23: 65-85.

Cairns, P., Shillcock, R., Chater, N. and J. Levy. (1997). Bootstrapping word boundaries: A bottom-up corpus-based approach to speech segmentation. *Cognitive Psychology*, 33: 111–153.

Cameron-Faulkner, T., Lieven, E.V.M. and M. Tomasello. (2003). A construction based analysis of child directed speech. *Cognitive Science*, 27: 843–873.

Cann, R. (2000). Functional versus lexical: a cognitive dichotomy. In R.D. Borsley (Ed), *The Nature and Function of Syntactic Categories*. Academic Press, 37-78.

Carlson-Luden, V. (1979). *Causal understanding in the 10-month-old*. Unpublished doctoral dissertation. University of Colorado at Boulder.

Cartwright, T. A. and M. R. Brent. (1997). Syntactic categorization in early language acquisition: Formalizing the role of distributional analysis. *Cognition*, 63: 121–170.

Caselli, C., Casadio, P. and E. Bates. (1999). A comparison of the transition from first words to grammar in English and Italian. *Journal of Child Language*, 26(1): 69–111.

Casenhiser, D. and A. E. Goldberg. (2005).Fast Mapping of a Phrasal Form and Meaning. *Developmental Science*, 8(6): 500-508.

Caterette, E. and M. H. Jones. (1974). *Informal Speech: Alphabetic and Phonemic Texts with Statistical Analyses and Tables*. Berkeley: University of California Press.

Chemla, E., Mintz T.H., Bernal S. and A. Christophe. (2009). Categorizing Words Using 'Frequent Frames': What Cross-Linguistic Analyses Reveal About Distributional Acquisition Strategies. *Developmental Science*, 12(3): 396-406.

Cherniak, C. (1995). Neural Component Placement. Trends in Neurosciences 18(12): 522-527.

Childers, J. and M. Tomasello. (2001). The role of pronouns in young children's acquisition of the English transitive construction. *Developmental Psychology*, 37(6): 739-748.

Christiansen, M.H., Allen, J. and M.S. Seidenberg. (1998). Learning to segment speech using multiple cues: A connectionist model. *Language and Cognitive Processes*, 13: 221-268.

Christiansen, M.H. and N. Chater. (1999). Toward a connectionist model of recursion in human linguistic performance. *Cognitive Science*, 23: 157–205.

Christiansen M.H. and N. Chater. (2008). Language as shaped by the brain. *Behavioral and Brain Sciences*, 31(5):489-509.

Christiansen, M.H., Conway, C.M. and S. Curtin. (2005). Multiple-cue integration in language acquisition: A connectionist model of speech segmentation and rule-like behavior. In J. W. Minett and W. S.-Y. Wang (Eds.), *Language Acquisition, Change and Emergence: Essays in Evolutionary Linguistics*. City University of Hong Kong Press. http://cornell.academia.edu/documents/0005/5540/ccc-ACE-chapter.pdf Accessed 9th October 2009.

Clark, E.V. (1987). The principle of contrast: A constraint on language acquisition. In Brian MacWhinney (Ed.), *Mechanisms of language acquisition*. Hillsdale, N.J.; London: L. Erlbaum Associates, 1-33.

—. (1990). Speaker perspective in language acquisition. *Linguistics,* 28: 1201-1220.

—. (1996). Early verbs, event types, and inflections. In C.E. Johnson and J.H.V. Gilbert (Eds.), *Children's language, Vol 9*. Mahwah, NJ: Lawrence Erlbaum Associates, 61-73.

Clark, E.V. and J. Bernicot. (2008). Repetition as ratification: How parents and children place information in common ground. *Journal of Child Language*, 35(2): 349-7.

Clausner, T. C. and W. Croft. (1997). Productivity and Schematicity in Metaphors. *Cognitive Science*, 21(3): 247-282.

Corominas-Murtra, B. (2007). Network statistics on early English Syntax: Structural criteria. http://arxiv.org/abs/0704.3708 accessed 06/11/08.

Corominas-Murtra, B., Valverde S. and R. V. Solé (2007a). *The ontogeny of scale-free syntax networks through language acquisition*. SFI Working Paper. 07-06-012. http://www.santafe.edu/research/publications/wpabstract/200706012. Accessed 23/9/08.

Corominas-Murtra, B., Valverde S. and R. V. Solé (2007b). *Emergence of scale-free syntax networks*. http://arxiv.org/abs/0709.4344. Accessed 29/08/09.

Crain, S., and M. Nakayama. (1987). Structure dependence in grammar formation. *Language*, 63: 522-543. Reprinted in C. Otero (Ed.), *Noam Chomsky: Critical Assessments*. Routledge: London, England.

Croft, W. (1990). A conceptual framework for grammatical categories (or, a taxonomy of propositional acts). *Journal of Semantics* 7: 245-279.

—. (1991). *Syntactic Categories and grammatical relations*. Chicago, University of Chicago Press.

—. (1995). Intonation units and grammatical structure. *Linguistics*, 33: 839-82.

—. (2001). *Radical Construction Grammar: Syntactic Theory in Typological Perspective*. Oxford: Oxford University Press.

—. (2006). *Noun phrase structure in Radical Construction Grammar*. 7th Meeting of the High Desert Linguistic Society, Albuquerque. http://www.unm.edu/~wcroft/Papers/Ms. Accessed 10th July 2009.

—. (2008). On iconicity of distance. *Cognitive Linguistics*, 19(1): 49-57.

—. (2009). Ten unwarranted assumptions in syntactic argumentation. Ms. http://www.unm.edu/~wcroft/Papers/Arguments-formatted.pdf Accessed 10th July 2009.

—. (in press). Methods for finding language universals in syntax. In S. Scalise and E. Magni (Ed.), *Universals of language today*. Berlin: Springer, 145-64.

Culicover, P.W. (1999). *Syntactic Nuts: hard cases, syntactic theory, and language acquisition. Volume One of Foundations of Syntax*. Oxford University Press, Oxford.

—. (2001). Concrete minimalism, branching structure, and linear order. In P. Banski and A. Przepiorkowski (Eds.), *Proceedings of Generative Linguistics in Poland – 2.* http://www.ling.ohio-state.edu/~culicove/ Recentpublications.html. Accessed 30th September 2009.

Culicover, P.W. (2005). Linguistics, cognitive science, and all that jazz. *Special Issue of Linguistic Review*, 22(2-4): 227-248.

Culicover, P.W. and R. Jackendoff. (1999). The view from the periphery: The English comparative correlative. *Linguistic Inquiry* 30: 543-571.

Culicover, P.W. and R. Jackendoff. (2006). The simpler syntax hypostudy. *Trends in Cognitive Science*, 10(9): 413-418.

Culicover, P.W. and A. Nowak. (2003). *Dynamical Grammar. Volume Two of Foundations of Syntax*. Oxford University Press, Oxford.

Cutler, A. and D.M. Carter. (1987). The predominance of strong initial syllables in the English vocabulary. *Computer Speech and Language*, 2: 133–142.

De Groot, A.D. (1978). *Thought and Choice in Chess*. Mouton Publishers.

de Nooy, W., Mrvar A. and V. Batagelj. (2005). *Exploratory social network analysis with Pajek*. New York: Cambridge University Press.

Devescovi, A., Caselli M.C., Marchione D., Pasqualetti, P., Reilly J. and E. Bates. (2005). A crosslinguistic study of the relationship between grammar and lexical development. *Journal of Child Language*, 32(4): 759–786.

Devescovi, A., Caselli M.C., Marchione D., Reilly, J. and E. Bates. (2003). Cross-Linguistic Study of the Relationship between Grammar and Lexical Development. *Technical Report CND-0301*. Project in Cognitive and Neural Development Center for Research in Language, University of California.

Diessel, H. and M. Tomasello. (2001). The acquisition of finite complement clauses in English: A corpus-based analysis. *Cognitive Linguistics*, 12(2): 97-141.

Dodson, K. and M. Tomasello. (1998). Acquiring the transitive construction in English: The role of animacy and pronouns. *Journal of Child Language*, 25(3): 605-622.

Dorogovtsev, S.N. and J.F.F. Mendes. (2001). Language as an evolving word web. *Proceedings of the Royal Society of London B*, 268: 2603-2606.

Dorogovtsev, S.N. and J.F.F. Mendes. (2002). Accelerated growth of networks. In S. Bornholdt and H.G. Schuster (Eds.), *Handbook of Graphs and Networks: From the Genome to the Internet*, Wiley-VCH, Berlin, 318-341.

Dromi, E. (1987). *Early lexical development*. New York: Cambridge University Press.

Dryer, M. S. (1997). Are Grammatical Relations Universal? In J. Bybee, J. Haiman and S. Thompson (Eds.), *Essays on Language Function and Language Type: Dedicated to T. Givón*. Amsterdam: John Benjamins, 115 - 143.

—. (1998). Why Statistical Universals are Better Than Absolute Universals, *Chicago Linguistic Society, The Panels*, 33: 123-145.

Edelman, G.M. and J.A. Gally (2001). Degeneracy and complexity in biological systems. *Proceedings of the National Academy of Science*, 98(24): 13763–13768.

Ellis, N. C. (1996). Sequencing in SLA: Phonological memory, chunking, and points of order. *Studies in Second Language Acquisition*, 18: 91-126.

—. (2002). Frequency effects in language acquisition: A review with implications for theories of implicit and explicit language acquisition. *Studies in Second Language Acquisition*, 24: 143-188.

Engstrøm. D. and J.A.S. Kelso. (2008). Coordination dynamics of the complementary nature. *Gestalt Theory*, 30(2): 1-16.

Epps, P. (2008). From 'wood' to future tense. Nominal origins of the future construction in Hup. *Studies in Language*, 32(2): 383–404.

Erdös, P. and A. Rényi. (1959). On random graphs I. *Mathematicae Debrecen*, 6: 290-297. Erdös, P. and A. Rényi. (1960). On the Evolution of Random Graphs. *Mathematical Institute of the Hungarian Academy of Science*, 5: 17-61.

Erdös, P. and A. Rényi. (1961). On the Evolution of Random Graphs. *Bulletin of the International Statistical Institute*, 38: 343-347.

Erkelens, M.A. (submitted). Restrictions of frequent frames as cues to categories: the case of Dutch. In H. Chan, H. Jacob and E. Kapia (Eds.), *BUCLD 32 Proceedings Supplement*.

Evans, N. and S. Levinson. (to appear). The Myth of Language Universals: Language diversity and its importance for cognitive science. Behavioral and Brain Sciences.

Everett M.G. and S.P. Borgatti. (1999). The centrality of groups and classes. *Journal of Mathematical Sociology*, 23(3): 181-201.

Fellbaum, C. (1998). *Wordnet: An Electronic Lexical Database*. Bradford Books.

Ferrer i Cancho, R. (2004). The Euclidean distance between syntactically linked words. *Physical Review E 70*, 056135.

—. (2005). The structure of syntactic dependency networks: insights from recent advances in network theory. In G. Altmann, V. Levickij and V. Perebyinis (Eds.), *The problems of quantitative linguistics*, Chernivtsi: Ruta, 60-75.

—. (2006). Why do syntactic links not cross? *Europhysics Letters,* 76: 1228-1235.

Ferrer i Cancho, R., Capocci, A. and G. Caldarelli. (2007). Spectral methods cluster words of the same class in a syntactic dependency network. *International Journal of Bifurcation and Chaos*, 17(7): 2453 - 2463.

Ferrer i Cancho, R. and A. Hernández Fernández. (2008). Power laws and the golden number. In G. Altmann, I. Zadorozhna and Y. Matskulyak (Eds.), *Problems of general, germanic and slavic linguistics. Chernivtsi: Books – XXI*, 518-523.

Ferrer i Cancho, R., Lorenzo, G. and V. Longa. (2008). Long-distance dependencies are not uniquely human. In A.D.M. Smith, K. Smith and R. Ferrer i Cancho (Eds.), *The Evolution of Language: Proceedings of the 7th International Conference (EVOLANG7)*. Singapore: World Scientific Press, 115-122.

Ferrer i Cancho, R. & D. Lusseau. (2009). Efficient coding in dolphin surface behavioral patterns. *Complexity*, 14(5): 23-25.

Ferrer i Cancho, R., Riordan, O. and B. Bollobás. (2005). The consequences of Zipf's law for syntax and symbolic reference. *Proceedings of the Royal Society of London Series B*, 272 (1562): 561–565.

Ferrer i Cancho, R. and R.V. Solé. (2001a). The small world of human language. *Proceedings of the Royal Society of London B*, 268: 2261-2266.

Ferrer i Cancho, R. and R.V. Solé. (2001b). Two Regimes in the Frequency of Words and the Origins of Complex Lexicons: Zipf's Law Revisited. *Journal of Quantitative Linguistics*, 8: 165-173.

Ferrer i Cancho, R. and R.V. Solé. (2003a). Least effort and the origins of scaling in human language. *Proceedings of the National Academy of Sciences USA*, 100: 788-791.

Ferrer i Cancho, R. and R.V. Solé. (2003b). Optimization in Complex Networks. *Statistical Physics of Complex Networks, Lecture Notes in Physics, Springer (Berlin)*, 114-125 http://complex.upf.es/~ricard/optimization_Sitges.pdf. Accessed 19th Sept 2009.

Ferrer i Cancho, R., Solé, R.V. and R. Köhler. (2002). Universality in syntactic dependency networks. *Santa Fe Working Paper*, 03-06-42.

Ferrer i Cancho, R., Solé, R.V. and R. Köhler. (2004). Patterns in syntactic dependency networks. *Physical Review E*, 69: 051915.

Fillmore, C. J., Kay, P. and M.K. O'Connor. (1998). Regularity and Idiomaticity in Grammatical Constructions: The Case of Let Alone. *Language*, 64(3): 501–538.

Fiser, J. and R.N. Aslin. (2002). Statistical learning of higher-order temporal structure from visual shape sequences. *Journal of Experimental Psychology: Learning, Memory, and Cognition*, 28(3): 458–467.

Fox, B.A. and R. Jasperson. (1995). A syntactic exploration of repair in English conversation. In P. W. Davis (Ed.), *Alternative Linguistics: Descriptive and Theoretical Modes*. Amsterdam: John Benjamins, 77-134.

Fraser, N.M. (1994). Dependency grammar. In R.E. Asher (Ed.), *The Encyclopedia of Language and Linguistics*. Oxford: Pergamon Press, 860-864.

Friederici A. and J. Wessels. (1993). Phonotactic knowledge of word boundaries and its use in infant speech perception. *Perception and Psychophysics*, 54: 287-295.

Gergely, G., Bekkering H. and I. Király. (2002). Rational imitation in preverbal infants. *Brief communications, Nature*, 415: 755.

Gerken, L.A. and B.J. McIntosh. (1993). Interplay of function morphemes and prosody in early language. *Developmental Psychology*, 29(3): 448–457.

Ghazanfar, A.A. (2008). Language evolution: neural differences that make a difference. *Nature Neuroscience*, 11(4): 382-384.

Gibson, E.J. (1969). *Principles of perceptual learning and development.* Meredith Corporation, New York.

Gibson, J.J. (1966). *The senses considered as perceptual systems*. George Allen and Unwin: London.

Gleitman, L. (1990). The structural sources of verb meanings. *Language Acquisition*, 1(1): 3-55.

Gobet, F., Lane P.C.R., Croker S., Cheng, Peter C-H., Jones G., Oliver I. and J.M. Pine. (2001). Chunking mechanisms in human learning. *TRENDS in Cognitive Sciences*, 5(6): 236-43.

Goh,K.-I., Oh, E., Jeong, H., Kahng, B. and D. Kim. (2002). Classification of scale-free networks. *Proceeding of the National Academy of Science U.S.A*, 99: 12583-12588.

Goh, K.-I., E. Oh E., Kahng B. and D. Kim. (2003). Betweenness centrality correlation in social networks, *Physical Review E*, 67: 017101.

Goldberg, A.E. (1995). *Constructions: A Construction Grammar Approach to Argument Structure. Chicago and London*. University of Chicago Press.

—. (2003). Constructions: a new theoretical approach to language. *TRENDS in Cognitive Sciences*, 7(5): 219-224.

—. (2006). *Constructions at Work*. Oxford: Oxford University Press. Goldberg, A.E. (2008). Universal Grammar? Or prerequisites for natural language? *Behavioral and Brain Sciences*, 31(5): 522-3.

Goldberg, A.E., Casenhiser, D. and N. Sethuraman. (2004). Learning argument structure generalizations. *Cognitive Linguistics*, 15(3): 289-316.

Goldberg, A.E., Casenhier D. and N. Sethuraman. (2005). The Role of Prediction in Construction Learning. *Journal of Child Language*, 32(2): 407-426.

Goldin-Meadow, S. (2007). The challenge: Some properties of language can be learned without linguistic input. *The Linguistic Review*, 24(4): 417–421.

Gomes, M.A.F., Vasconcelos, G.L., Tsang, I.J. and I.R. Tsang. (1999). Scaling relations for diversity of languages. *Physica A*, 271: 489-495.

Gómez R. (2002). Variability and detection of invariant structure. *Psychological Science*, 13: 431-36.

—. (2006). Dynamically guided learning. InY. Munakata and M. Johnson (Eds.), *Processes of Change in Brain and Cognitive Development Attention and Performance XXI*. Oxford University Press, 87-110.

—. (2007). Statistical learning in infant language development. In G. Gaskell (Ed.), *Oxford Handbook of Psycholinguistics*. Oxford University Press, Part V - Language Development, chapter 36.

Gómez R. and LA. Gerken. (1998). Determining the basis of abstraction in artificial language acquisition. *Infant Behavior and Development*, 21: 434.

Gómez, R. and LA. Gerken. (1999). Artificial grammar learning by one-year-olds leads to specific and abstract knowledge. *Cognition*, 70: 109–135.

Gómez, R. and L. LaKusta. (2004). A first step in form-based category abstraction by 12- month-old infants. *Developmental Science*, 7(5): 567-580.

Gómez, R. and J. Maye. (2005). The developmental trajectory of non-adjacent dependency learning. *Infancy*, 7 (2): 183-206.

Goodman, J., Dale P. and P. Li. (2006). Does frequency count? Parental input and the acquisition of vocabulary. *Journal of Child Language*, 35(3): 515-531.

Goodsitt, J.V., Morgan J.L. and P.K. Kuhl. (1993). Perceptual Strategies in Prelingual Speech Segmentation. *Journal of Child Language*, 20(2): 229-52.

Granovetter, M. (1973). The Strength of Weak Ties. *The American Journal of Sociology*, 78: 1360-1380.

Granovetter, M. (1983). The Strength of Weak Ties: A Network Theory Revisited. *Sociological Theory*, 1: 201-233.

Green, T.R.G. (1979). The Necessity of Syntax Markers: Two Experiments with Artificial Languages. *Journal of Verbal and Verbal Behaviour*, 18: 481-496.

Guillemin, D. (2007). Definiteness and Specificity in Mauritian Creole: A syntax semantics mapping. In P. Baker and G. Fon Sing (Eds.), *The making of Mauritian Creole*. London: Battlebridge Publications, 63-91.

Gurevich, O. and A.E. Goldberg. (forthcoming). *Incidental Verbatim Memory for Language*.

Haiman, J. (1985). *Natural syntax: iconicity and erosion*. Cambridge; New York; Melbourne: Cambridge University Press.

—. (1991). Motivation, repetition and emancipation: The bureaucractisation of language. In H. C. Wolfart (Ed.), *Linguistic Studies Presented to John L. Finlay*. Winnipeg, Manitoba: Algonquian and Iroquoian Linguistics, 45-70.

—. (2008). In defence of iconicity. *Cognitive Linguistics*, 19(1): 35-48.

Haspelmath, M. (1999). Why is grammaticalization irreversible? *Linguistics*, 37(6): 1043–1068.

Haspelmath, M. (2008a). Frequency vs. iconicity in explaining grammatical asymmetries. *Cognitive Linguistics*, 19(1):1-33.

Haspelmath, M. (2008b). Reply to Haiman and Croft. *Cognitive Linguistics*, 19(1): 59-66. Haspelmath, M. (To appear). A frequentist explanation of some universals of reflexive marking. To appear in *Linguistic Discovery*.

Hauser, M.D., Newport E.L. and R.N. Aslin. (2001). Segmentation of the speech stream in a non-human primate: statistical learning in cotton-top tamarins. *Cognition*, 78(3): B53-B64.

Heine, B. (1993). *Auxiliaries: Cognitive forces and grammaticalization*. New York: Oxford University Press.

Hershberg, U. and A. Ninio. (2004). Optimal exemplar learning in cognitive systems. In G. Dalenoort and M. Pichat (Eds.), *Cognitive Systems (ESSCS) on Multidisciplinary Aspects of Learning*, 6: 181-188.

Hershberg, U. and A. Ninio. (in press). Cognitive systems and the special order of their environment. *Journal of Applied Systems Studies*.

Hoey, M. (2006). Corpus-driven Approaches to Grammar. Exploring the Lexis-Grammar Interface International Conference, 5-7 October (2006), University of Hanover, Germany.

Hoff, E. and L. Naigles (2002). How Children Use Input to Acquire a Lexicon. *Child Development*, 73(2): 418-433.

Hoffman, M.W., Grimes D.B., Shon A.P. and R.P.N. Rao. (2006). A probabilistic model of gaze imitation and shared attention. *Neural Networks*, 19: 299–310.

Höhle, B., Schmitz M., Santelmann L.M. and J. Weissenborn (2006). 19-Month-Olds: Evidence for Lexical and Structural Influences on Children's Early Processing Capacities. *Language Learning and Development*, 2(4): 277–300.

Hopper, P., and E. Closs Traugott. (2003). *Grammaticalization*. Cambridge: Cambridge University Press.

Hu H.-B. and X.-F. Wang. (2009). Disassortative mixing in online social networks. *EPL*, 86(1): 18003.

Huttenlocher, J., Vasilyeva M., Cymerman E. and S. Levine. (2002). Language input and child syntax. *Cognitive Psychology*, 45: 337-374.

Ingram, J.C.L. (1989). Connected Speech Processes in Australian English. *Australian Journal of Linguistics*, 9: 21-49.

—. (2007). *Neurolinguistics: An Introduction to Spoken Language Processing and its Disorders*. Cambridge Textbooks in Linguistics, Cambridge University Press.

Israel, M. (2002). Consistency and Creativity in Child Language. *Berkeley Linguistics Society*, 28: 123–135.

Itzkovitz, S. and U. Alon. (2005). Subgraphs and network motifs in geometric networks. *Physical Review E,* 71: 026117.

Jacob, F. (1977). Evolution and Tinkering. *Science*, New Series, 196(4295): 1161-1166. Jeong, H., Tombor B., Albert R., Oltvai Z. and A.-L. Barabási (2000).The large-scale organization of metabolic networks. *Nature*, 407: 651-655.

Jones, G., Gobet F. and J.M. Pine. (2000). A process model of children's early verb use. In L. R. Gleitman, A. K Joshi (Eds.), *Proceedings of the Twenty-second Annual Conference of the Cognitive Science Society*. Mahwah, NJ: Erlbaum, 723-728

Jurafsky, D. (2003). Probabilistic modelling in psycholinguistics: linguistic comprehension and production. In R. Bod, J. Hay and S. Jannedy (Eds.), *Probabilistic Linguistics*. Cambridge: The MIT Press, 39-95.

Jurafsky, D., Bell, A., Gregory, M. and W.D. Raymond. (2001). Probabilistic relations between words. *Frequency and the Emergence of Linguistic Structure: Typological studies in Language*: John Benjamins Publishing Company. http://www.stanford.edu/~jurafsky/cmu.pdf Accessed 9th October 2009.

Jusczyk, P.W. (1996). Investigations of the Word Segmentation Abilities of Infants. *Proceedings of ICSLP Fourth International Conference on Spoken Language Processing,* 3: 1561-1564.

Jusczyk, P.W., Cutler, A. and N. Redanz. (1993). Preference for the predominant stress patterns of English words. *Child Development*, 64: 675-687.

Jusczyk P.W., Friederici A., Wessels J., Svenkerud V. and A.M. Jusczyk. (1993). Infants' sensitivity to the sound patterns of native language words. *Journal of Memory and Language*, 32: 402-420.

Jusczyk, P.W., Luce, P.A. and J. Charles-Luce. (1994). Infants' sensitivity to phonotactic patterns in the native language. *Journal of Memory and Language*, 33(5): 630-645.

Kapustin, V. and A. Jamsen. (2007). Vertex degree distribution for the graph of word co- occurrences in Russian. In *Proceedings of the Second Workshop on TextGraphs: Graph- Based Algorithms for Natural Language Processing*, Rochester, NY, USA. Association for Computational Linguistics, 89-92.

Kashtan, N., Itzkovitz, S., Milo, R., and U. Alon,. (2004). Topological generalizations of network motifs. *Physical Review E*, 70(3): 031909.

Kashtan, N., Itzkovitz, S., Milo, R. and U. Alon (2002-2005). *Mfinder*. Rehovot, Israel. Accessed November 2005.

Kashtan, N., Itzkovitz, S., Milo, R. and U. Alon (2002-2005b). *Mfinder Tool Guide*. Accessed November 2005. Rehovot, Israel.

Kashtan, N., Itzkovitz, S., Milo, R. and U. Alon. (2002-2005c). *Mfinder Dictionary*. Rehovot, Israel. Accessed November 2005.

Kasturirangan, R. (1999). Multiple Scales in Small-World Networks. *Technical Report AIM-1663*, arixiv:cond-mat/9904055; www.lanl.gov.

Ke, J. and Y. Yao. (2008). Analysing Language Development from a Network Approach. *Journal of Quantitative Linguistics*, 15(1): 70-99.

Kemmer, S. and M. Barlow. (2000). Introduction: A Usage-Based Conception of Language. *Usage-Based Models of Language*, CSLI publications, Stanford, California, vii-xxviii.

Kibrik, A.A. (2005). On the use of 'quasi-grammeme' in Athabaskan. In J.D.Apresjan and L.L.Iomdin (Eds.), Vostok-Zapad: Vtoraja mezhdunarodnaja konferencija po modeli "Smysl ? Tekst", Moscow: JaSL, 208-218. http://www.philol.msu.ru/~otipl/new/main/people/ kibrik-aa/files/ Quasi-grammeme_Athabaskan@Smysl-Tekst_2005.pdf Accessed 9th October 2009.

Kidd, E., Lieven E.V.M. and M. Tomasello. (2004). The acquisition of complement clause constructions: A sentence repetition study. In E. V. Clark (Ed.), *Proceedings of the (2004) Stanford Child Language Research Forum: Constructions and Acquisition*, CSLI- Publications, 50-59.

Kidd, E., Lieven, E.V.M. & M. Tomasello. (2006). Examining the Role of Lexical Frequency in Children's Processing and Acquisition of sentential Complements. *Cognitive Development*, 21: 93-107.

Kilgarriff, A. (2005). Language is never, ever, ever, random. *Corpus Linguistics and Linguistic Theory*, 1(2): 263–275.

Kirkham, N.Z., Slemmer J.A. and S.P. Johnson. (2002). Visual statistical learning in infancy: evidence for a domain general learning mechanism. *Cognition*, 83(2): B35-B42.

Klarskov Mortensen, H.J. (2006). Transitional probability software. http://www.hjkm.dk/ Kleinberg, J.M. (1999). Authoritative Sources in a Hyperlinked Environment. *Journal of the ACM*, 46(5): 604–632.

Laakso, A. and L.B. Smith. (2004). Pronouns predict verb meanings in child-directed speech. In K. Forbus, D. Gentner and T. Regier (Eds.), *Proceedings of the 26th Annual Meeting of the Cognitive Science Society*, Mahwah, NJ: Lawrence Erlbaum Associates, 767-772.

Lakoff, G. (1987). *Women, Fire, and Dangerous Things*, Chicago: University of Chicago Press.

Langacker, R.W. (1987). *Foundations of Cognitive Grammar, Vol 1 and 2: Theoretical Prerequisites*. Stanford: Stanford University Press.

—. (1991). *Concept, Image, and Symbol: The Cognitive Basis of Grammar*. Berlin and New York: Mouton de Gruyter.

—. (1998). A Usage-based model. In B. Rudzka-Ostyn (Ed.), *Topics in Cognitive Linguistics (Current Issues in Linguistic Theory 50)*. Amsterdam: Benjamins, 127-61.

LaPolla, R.J. (2003). Why languages differ: Variation in the conventionalization of constraints on inference. In D. Bradley, R.J. LaPolla, B. Michailovsky & G. Thurgood (Eds.), Language Variation: *Papers on variation and change in the Sinosphere and in the Indosphere in honour of James A. Matisoff,*. Pacific Linguistics. Canberra: Australian National University, 2003, 113-144.

Lee, D. (2001). *Cognitive Linguistics: an introduction*. Oxford: Oxford University Press. Lewis, J. D. and J. L. Elman. (2002). Learnability and the Statistical Structure of Language: Poverty of Stimulus Arguments Revisited. BUCLD 26: Proceedings of the 26th annual Boston University Conference on Language Development, 359-370.

Lieven, E.V.M.., Pine J.M. and G. Baldwin. (1997). Lexically-based learning and early grammatical development. *Journal of Child Language*, 24(1): 187-219.

Lieven, E.V.M., Pine J.M. and D. Barnes. (1992) Individual differences in early vocabulary development. *Journal of Child Language*, 19(2): 287-310.

Liu, H. (2008). The complexity of Chinese dependency syntactic networks. *Physica A*, 387: 3048-3058.

Luka, B.J. and L.W. Barsalou. (2005). Structural facilitation: Mere exposure effects for grammatical acceptability as evidence for syntactic priming in comprehension. *Journal of Memory and Language*, 52: 436–459.

Macía, J. and R. V. Solé. (2008). Distributed robustness in cellular networks: insights from evolved digital circuits. *Journal of the Royal Society Interface*, 442: 259-264.

MacWhinney, B. (1989). Competition and lexical categorization. In R. Corrigan, F. Eckman, and M. Noonan (Eds.), *Linguistic categorization*. New York: Benjamins.

—. (1995). *The CHILDES Project: Tools for Analyzing Talk*, Erlbaum.

—. (2000a). *The CHILDES project: Tools for analyzing talk*. Third Edition. Mahwah, NJ: Lawrence Erlbaum Associates.

—. (2000b). Emergence from what? Comments on Sabbagh and Gelman. *Journal of Child Language*, 27(3): 727-733.

—. (2005). The emergence of linguistic form in time. *Connection Science*, 17(3-4): 191-211.

Maratsos, M. and M. A. Chalkley. (1980). The internal language of children's syntax: The ontogenesis and representation of syntactic categories. In K. Nelson (Ed.), *Children's language Vol. 2*. New York: Gardner, 127-214.

Marcus G., Vijayan S., Bandi R.S. and P. Vishton. (1999). Rule learning by seven month-old infants. *Science*, 283(5398): 77-80.

Masucci, A.P. and G. J. Rodgers. (2007). Multi-directed Eulerian growing networks. *Physica A*, 386(1): 557-563.

Mathias, N. and V. Gopal. (2001a). Small worlds: How and why, *Physical Review E (Statistical, Nonlinear, and Soft Matter Physics)*, 63(2):021117/1-12.

Mathias, N. and V. Gopal. (2001b). Small-worlds: How and why. arXiv:cond- mat/0002076v1. Accessed 29th July 2009.

Matthews, D., Lieven E.V.M., Theakston A. and M. Tomasello. (2004). The role of frequency and distributional regularity in the acquisition of word order. In E. V. Clark (Ed.), *Proceedings of the (2004) Stanford Child Language Research Forum: Constructions and Acquisition*, CSLI- Publications, 68-77.

Matthews, D., Lieven E.V.M., Theakston A. and M. Tomasello. (2005). The role of frequency in the acquisition of English word order. *Cognitive Development*, 20: 121–136.

Mattys, S.L. and P.W. Jusczyk. (2001). Do infants segment words or recurring contiguous patterns? *Journal of Experimental Psychology: Human perception and performance*. 27(3):644-55.

Mattys, S.L., Jusczyk P.W., Luce P.A. and J.L. Morgan. (1999). Phonotactic and Prosodic Effects on Word Segmentation in Infants. *Cognitive Psychology*, 38: 465–494.

McClelland, J. and J. Bybee. (2007). The gradience of gradience: a reply to Jackendoff. *The Linguistic Review*, 24(4): 437-455.

McClure, K., Pine, J.M. and E.V.M. Lieven (2006). Investigating the abstractness of children's early knowledge of argument structure. *Journal of Child Language*, 33(4): 693-720.

Miller, G.A. (1956). The magical number seven, plus or minus two: some limits on our capacity for processing information. *Psychological Review,* 63: 81–97.

—. (1990). Nouns in WordNet: a lexical inheritance system. *International Journal of Lexicography*, 3(4): 245-264.

Milo R., Itzkovitz S., Kashtan N., Levitt R. and U. Alon, (2004). Response to comment on "Network motifs: Simple building blocks of complex networks" and “Superfamilies of evolved and designed networks". *Science*, 305: 1107.

Milo, R., Itzkovitz S., Kashtan N., Levitt R., Shen-Orr S., Ayzenshtat I., Michal Sheffer M. and U. Alon. (2004). Superfamilies of Evolved and Designed Networks. *Science*, 303 (5663): 1538-1542.

Milo, R., Shen-Orr, S., Itzkovitz, S., Kashtan, N., Chklovskii, D. and U. Alon. (2002). Network Motifs: Simple Building Blocks of Complex Networks. *Science* 298: 824-827.

Mintz, T.H. (1996). *The roles of linguistic input and innate mechanisms in children's acquisition of grammatical categories.* Unpublished Doctoral Dissertation, University of Rochester.

—. (2002). Category induction from distributional cues in an artificial language. *Memory and Cognition*, 30: 678–686.

—. (2003). Frequent frames as a cue for grammatical categories in child directed speech. *Cognition*, 90: 91-117.

Mintz, T.H., Newport E.L. and T.G. Bever. (1995). Distributional regularities of grammatical categories in speech to infants. In J. Beckman (Ed.), *Proceedings of the North East Linguistics Society 25 (Vol. 2).* Amherst, MA: GLSA.

Mintz, T.H., Newport E.L. and T.G. Bever. (2002). The distributional structure of grammatical categories in speech to young children. *Cognitive Science*, 26: 393-424.

Mitchinson, G. (1991). Neural branching patterns and the economy of cortical wiring. *Proceedings of the Royal Society of London B*, 245(1313): 151-158.

Monaghan, P., Chater, N. and M.H. Christiansen. (2005). The differential contribution of phonological and distributional cues in grammatical categorisation. *Cognition*, 96: 143-182.

Monaghan, P. and M.H. Christiansen. (2004). What distributional information is useful and usable in language acquisition? *Proceedings of the 26th Annual Conference of the Cognitive Science Society.* Mahwah, NJ: Lawrence Erlbaum.

Montoya, J.M. and R.V. Solé. (2002). Small World Patterns in Food Webs. *Journal of Theoretical Biology*, 214(3): 405-412.

Morgan, J.L., Meier, R.P. and E.L. Newport. (1987). Structural packaging in the input to language learning: Contributions of prosodic and morphological marking of phrases to the acquisition of language. *Cognitive Psychology*, 19: 498–550.

Morgan, J.L., and E.L. Newport. (1981). The role of constituent structure in the induction of an artificial language. *Journal of Verbal Learning and Verbal Behavior*, 20: 67–85.

Morgan, J.L., and J.R. Saffran. (1995). Emerging integration of sequential and suprasegmental information, in preverbal speech segmentation. *Child Development*, 66, 911–936.

Morgan, J.L., Shi R., and P. Allopenna. (1996). Perceptual Bases of Rudimentary Grammatical Categories: Towards a Broader Conceptualization of Bootstrapping. In J.L. Morgan and K. Demuth (Eds.), *Signal to syntax: Bootstrapping from speech to grammar in early acquisition*. Mahwah, NJ: Lawrence Erlbaum Associates, Publishers, 263-283.

Mukherjee, A. (2009). *Self-Organization of Speech Sound Inventories in the Framework of Complex Networks*, Doctoral Dissertation, Indian Institute of Technology, Kharagpur.

Naigles, L.R. and E. Hoff-Ginsberg. (1995). Input to verb learning: Evidence for the plausibility of Syntactic Bootstrapping. *Developmental Psychology*, 31: 827-837.

Naigles, L.R. and E. Hoff-Ginsberg. (1998). Why are some verbs learned before other verbs? Effects of input frequency and structure on children's early verb use. *Journal of Child Language*, 25(1): 95-120.

Newell, A. and P.S. Rosenbloom (1981). Mechanisms of Skill Acquisition and the Law of Practice. In J. R. Anderson (Ed.), *Cognitive Skills and Their Acquisition,* Hillsdale, NJ: Lawrence Eribaum Associates, 1-55.

Newman, M.E.J. (2002). Assortative mixing in networks. *Physical Review Letters*, 89: 208701.

—. (2005). Power laws, Pareto distributions and Zipf's law. *Contemporary Physics*, 46: 323–351.

Newport, E.L. and R.N. Aslin. (2000). Innately constrained learning: Blending old and new approaches to language acquisition. In S. C. Howell, S. A. Fish, and T. Keith-Lucas (Eds.), *Proceedings of the 24th annual Boston university conference on language development.* Somerville, MA: Cascadilla Press, 1-21.

Ninio, A. (1994). Words with holes: the acquisition of the predicates. Paper presented at the Emory Conference on Cognitive and Functional Approaches to Grammatical Development, Emory, Atlanta. http://micro5.mscc.huji.ac.il/~msninio/. Accessed 9th October 2009.

—. (1996). Pathbreaking verbs in syntactic development. Paper presented at the *Seventh International Congress for the Study of Child Language*, Istanbul, Turkey, July 1996.
http://micro5.mscc.huji.ac.il/~msninio/ Accessed 9th October 2009.

—. (1999a). Pathbreaking verbs in syntactic development and the question of prototypical transitivity. *Journal of Child Language*, 26(3): 619-653.

—. (1999b). Learning from examples in syntactic development. Keynote address at the 11th Australasian Human Development Conference, Sydney University, Sydney, Australia, July 9, 1999. http://micro5.mscc.huji.ac.il/~msninio/ Accessed 9th October 2009.

—. (1999c). Model learning in syntactic development: Intransitive verbs. *International Journal of Bilingualism*, 3: 111-31.

—. (2003). No verb is an island: Negative evidence on the Verb Island hypostudy. *Psychology of Language and Communication*, 7(1): 3-21.

—. (2005a). Adult-type grammatical relations in children's early language. In B. Bokus (Ed.), *Studies in the psychology of child language: In Honor of Grace Wales Shugar*. Warsaw: Matrix, 71-83.

—. (2005b). Accelerated learning without semantic similarity: indirect objects. *Cognitive Linguistics*, 16: 531–556.

—. (2005c). Testing the role of semantic similarity in syntactic development. *Journal of Child Language,* 32(1): 1–27.

—. (2006a). *Language and the learning curve: A new theory of syntactic development*. Oxford: Oxford University Press.

—. (2006b). Kernel vocabulary and Zipf's Law in maternal input to syntactic development. In D. Bamman, T. Magnitskaia and C. Zaller (Eds.), *BUCLD 30: Proceedings of the 30th annual Boston University Conference on Language Development*. Somerville, MA: Cascadilla Press, 423-431.

—. (2008). Self-Organized Criticality and the properties of English. Paper presented at the *5th European Conference on Complex Systems (ECCS 2008),* Jerusalem, September 14-19, 2008. http://micro5.mscc.huji.ac.il/~msninio/ Accessed 9th October 2009.

Ninio, A. and T. Keren-Portnoy. (2002). Mastering the syntax of pathbreaking verbs delays learning others of the same kind. Paper presented at the *Ninth International Congress for the Study of Child Language and the Symposium on Research in Child Language Disorders (IASCL/SRCLD)*, Madison, Wisconsin, July, (2002). http://micro5.mscc.huji.ac.il/~msninio/ Accessed 9th October 2009.

Ono, T. and S.A. Thompson. (1994). Unattached NPs in English Conversation. *Proceedings of the Twentieth Annual Meeting of the Berkeley Linguistics Society, Berkeley*: University of California Press, 402-419.

Panchenko, D. (2005). Statistics for Applications. Accessed 21st July 2005. Department of Mathematics Massachusetts Institute of Technology. http://ocw.mit.edu/NR/rdonlyres/Mathematics/18-443Fall 2003/50403CA1-046E-4BCC-A634-28367938DF6F/0/18443.pdf.

Pentland, C. (2009). Lamalama verbal morphology. Unpublished Ms, University of Queensland, Australia.

Perruchet, P. and A. Vintner. (1998). PARSER: A Model for Word Segmentation. *Journal of Memory and Language*, 39(2): 246–263.

Peters, A.M. (1977). Language learning strategies: Does the whole equal the sum of the parts? *Language*, 53(3): 560-573.

—. (1993). The interdependence of social, cognitive and linguistic development: Evidence from a visually-impaired child. In H. Tager-Flusberg (Eds.), *Constraints on language acquisition: Studies of atypical children*. Hillsdale, NJ: Lawrence Erlbaum Associates, 195-219.

Peters, A.M. and L. Menn. (1993). False starts and filler syllables: Ways to learn grammatical morphemes. *Language*, 69(4): 742-777.

Pine, J.M. and E.V.M.Lieven (1993). Reanalysing rote-learned phrases: Individual differences in the transition to multi word speech. *Journal of Child Language*, 20(3): 551-571.

Pine, J.M. and E.V.M Lieven. (1997). Slot-and-frame patterns and the development of the determiner category. *Applied Psycholinguistics*, 18: 123-138.

Pine, J.M., Lieven, E.V.M. and C.F. Rowland. (1998). Comparing different models of the development of the English verb category. *Linguistics*, 36(4): 807-830.

Pinker, S. (1984). *Language Learnability and Language Development*. Cambridge, MA: Harvard University Press.

—. (1989). *Learnability and Cognition: The acquisition of Argument Structure. Cambridge*, Mass.: MIT Press.

—. (1991). Rules of language. *Science*, 253: 530-535

Pinker, S. and A. Prince. (1991). Regular and irregular morphology and the psychological status of rules of grammar. *Berkeley Linguistic Society*, 17: 230–251.

Potapov, A.P., Voss N., Sasse N. and E. Wingender. (2005). Topology of Mammalian Transcription Networks. *Genome Informatics*, 16(2): 270-278.

Press, W.H., Teukolsky, S.A., Vetterling, W.T. and B.P. Flannery. (1992). *Numerical Recipes in C: The Art of Scientific Computing: second edition*. Cambridge: Cambridge University Press.

Ravasz, E. and A.-L. Barabási. (2003). Hierarchical organization in complex networks, *Physical Review E*, 67: 026112.

Ravasz, E., Somera A.L., Mongru D.A., Oltvai Z.N. and A.-L. Barabási. (2002). Hierarchical organization of modularity in metabolic networks, *Science* 297: 1551-1555.

Reali, F. and M.H. Christiansen. (2003). Reappraising Poverty of Stimulus Argument: A Corpus Analysis Approach. Online supplement to *Proceedings of the 28th Boston University Conference on Language Development.*

Reali, F. and M.H. Christiansen. (2004). Structure dependence in language acquisition: Uncovering the richness of the stimulus. In *Proceedings of the 26th Annual Conference of the Cognitive Science Society*. Mahwah, NJ: Lawrence Erlbaum, 1131-1136.

Reali, F., and M.H. Christiansen. (2005). Uncovering the richness of the stimulus: Structure dependence and indirect statistical evidence. *Cognitive Science*, 29: 1007-1028.

Redington, M., Chater N. and S. Finch. (1998). Distributional information: A powerful cue for acquiring syntactic categories. *Cognitive Science*, 22: 435–469.

Regier, T. and S. Gahl, (2004). Learning the unlearnable: The role of missing evidence. *Cognition*, 93: 147-155.

Robinson, J.J. (1970). Dependency structures and transformational rules. *Language*, 46(2): 259-285.

Rostila, J. (2004). Lexicalization as a Way to Grammaticalization. In F. Karlsson (Ed.), *Proceedings of the 20th Scandinavian Conference of Linguistics, Helsinki, January 7-9, 2004* (University of Helsinki, Department of General Linguistics, Publications No. 36), http://www.ling.helsinki.fi/kielitiede/20scl/proceedings.shtml. Accessed 9th October 2009.

Rostila, J. (2006). Storage as a way to grammaticalization, *Constructions*. Online journal: http://www.elanguage.net/journals/index.php/constructions/issue/view/16. Accessed August 08.

Sachs, J. (1983). Talking about the there and then: The emergence of displaced reference in parent–child discourse. In K.E. Nelson (Ed.), *Children's language, Vol. 4*. Hillsdale, NJ: Lawrence Erlbaum Associates, 1-28.

Saffran, J.R. (2001). Words in a sea of sounds: The output of statistical learning. Cognition, 81, 149-169.

Saffran, J.R., Aslin, R.N. and E.L. Newport. (1996). Statistical learning by 8-month-old infants. *Science*, 274: 1926-1928.

Saffran, J.R., Hauser M., Seibel R., Kapfhamer J., Tsao F. and F. Cushman. (2008). Grammatical pattern learning by human infants and cotton-top tamarin monkeys. *Cognition*, 107: 479–500.

Saffran, J.R., Johnson, E.K., Aslin, R.N. and E.L. Newport. (1999). Statistical learning of tone sequences by human infants and adults. *Cognition*, 70: 27-52

Saffran, J.R., Newport, E. L. and R.N. Aslin. (1996). Word segmentation: The role of distributional cues. *Journal of Memory and Language*, 35: 606–621.

Saffran, J.R.., Newport E.L., Aslin R.N. Tunick R.A. and S. Barrueco. (1997). Incidental language learning: Listening (and Learning) out of the Comer of Your Ear. *Psychological Science,* 8(2): 101-105.

Saffran, J.R., Pollak S.D., Seibel R.L. and A. Shkolnik. (2007). Dog is a dog is a dog: Infant rule learning is not specific to language. *Cognition*, 105: 669–680.

Saffran J.R. and E.D. Thiessen (2003). Pattern induction by infant language learners. *Developmental Psychology*, 39: 484-94.

Saffran J.R. and D. Wilson. (2003). From syllables to syntax: Multi-level statistical learning by 12-month-old infants. *Infancy*, 4(2), 273-284.

Sandhofer, C.M., Smith L.B. and J. Luo. (2000). Counting nouns and verbs in the input: differential frequencies, different kinds of learning? *Journal of Child Language*, 27(3): 561-585.

Santelmann, L. and P. Jusczyk. (1998). Sensitivity to Discontinuous Dependencies in Language Learners: Evidence for Processing Limitations. *Cognition*, 69: 105-134.

Searle, J. (1969). *Speech Acts: An Essay in the Philosophy of Language*. New York , Cambridge University Press.

Seidenberg, M.S. (1997). Language acquisition and use: Learning and applying probabilistic constraints. *Science*, 275: 1599-1603.

Senghas, A., Kita S. and A. Ozyurek. (2004). Children Creating Core Properties of Language: Evidence from an Emerging Sign Language in Nicaragua. *Science*, 305(5691): 1779-1782.

Sethuraman, N. and J.C. Goodman. (2004a). Learning to Generalize Verbs to New Syntactic Environments. In E.V. Clark (Ed.), *Proceedings of the 2004 Stanford Child Language Research Forum: Constructions and Acquisition*. CSLI Publications, 78-87.

Sethuraman, N. and J.C. Goodman. (2004b). Children's mastery of the transitive construction. In E.V. Clark (Ed.), *Proceedings of the 2004 Stanford Child Language Research Forum: Constructions and Acquisition*. CSLI Publications, 60-67.

Sinclair, J. (1991). *Corpus, Concordance, Collocation*. Oxford: Oxford University Press.

Soares, M., Corso, G. and L.S. Lucena. (2005). The network of syllables in Portuguese. *Physica A*, 355 (2-4): 678-684.

Solé, R.V. (2005). Syntax for free? *Nature*, 434: 289.

—. (2006). Scaling laws in language evolution. In C Cioffi-Revilla (Ed.), *Power Laws in the Social Sciences: Discovering Complexity and Non-equilibrium Dynamics in the Social Universe*. Cambridge University Press. http://complex.upf.es/index.php?page=4 Accessed 9th October 2009.

—. (2008). On Networks and Monsters: The Possible and the Actual in Complex Systems. *Leonardo MIT Press*, 41(3): 253–258.

Solé, R.V., Corominas-Murta, B., Valverde, S. and L. Steels. (2005). Language Networks: their structure, function and evolution. *Santa Fe Working papers*. 05-12-042. http://www.santafe.edu/research/publications/workingpapers/05-12-042.pdf Accessed 9th October 2009.

Solé, R.V., Ferrer i Cancho R., Montoya J.M. and S. Valverde. (2003) Selection, Tinkering and Emergence in Complex Networks. *Complexity*, 8: 20-33.

Solé, R.V. and B. Goodwin. (2000). *Signs of life: How complexity pervades biology*. NY: Basic Books.

Solé, R.V. and S. Valverde. (2006). Are network motifs the spandrels of cellular complexity? *Trends in Ecology and Evolution*, 21(8): 419-422.

Solomonoff, R. and A. Rapoport. (1951). Connectivity of Random Nets. *Bulletin of Mathematical Biophysics*, 13: 107-227.

Sporns O. and R. Koetter. (2004). Motifs in brain networks. PLoS Biology, 2(11): e369. Stanley, H.E. (1971) *Introduction to Phase Transitions and Critical Phenomena.* Oxford: Oxford University Press.

Steels, L. (2005). The Role of Construction Grammar in Language Grounding. AI Communications. http://www3.isrl.illinois.edu/~junwang4/langev/localcopy/pdf/steels_constructionGrammar.p df Accessed 9th October.

Steels, L. and J. De Beule. (2006). A (very) brief Introduction to Fluid Construction Grammar. Third International Workshop on Scalable Natural Language Understanding *(ScaNaLU 2006)* June 8, 2006, following HLT/NAACL, New York City. http://arti.vub.ac.be/~joachim/acl-ny-06-3.pdf Accessed 9th October 2009.

Steyvers, M. and J. B. Tenenbaum. (2005). The large-scale structure of semantic networks: statistical analyses and a model of semantic growth. *Cognitive Science*, 29(1): 41-78. http://web.mit.edu/cocosci/Papers/03nSteyvers.pdf Accessed 9th October 2009.

Stromswold, K.J. (1990). *Learnability and the acquisition of auxiliaries.* Doctoral dissertation. Department of Brain and Cognitive Sciences, Massachusetts Institute of Technology, Cambridge, MA..

—. (1994). *The Nature and Acquisition of Lexical and Functional Categories*, manuscript, Rutgers University, New Brunswick, New Jersey.

Swingley D. and R. Aslin (2000). Spoken word recognition and lexical representation in very young children. *Cognition*, 76: 147-166.

Szagun, G. Steinbrink C., Franik M. and B. Stumper. (2006). Development of vocabulary and grammar in young German-speaking children assessed with a German language development inventory. *First Language*, 26(3): 259-280.

Theakston, A.L., Lieven E.V.M., Pine J.M. and C.F. Rowland. (2001). The role of performance limitations in the acquisition of verb-argument structure: An alternative account. *Journal of Child Language*, 28(1): 127-152.

Theakston, A.L., Lieven E.V.M, Pine J.M. and C.F. Rowland. (2004).Semantic generality, input frequency and the acquisition of syntax. *Journal of Child Language*, 31(1): 61–99.

Theakston, A.L., Lieven E.V.M, Pine J.M. and C.F. Rowland. (2006). Note of clarification on the coding of light verbs in 'Semantic generality, input frequency and the acquisition of syntax' (Journal 31(1) 61–99). *Journal of Child Language*, 33(1): 191-197.

Thiessen E. and J.R. Saffran. (2003). When cues collide: Use of stress and statistical cues to word boundaries by 7- to 9-month-old infants. *Developmental Psychology*, 39(4): 706-16.

Thompson, S.P. and E.L. Newport. (2007). Statistical Learning of Syntax: The Role of Transitional Probability. *Language Learning and Development*, 3(1): 1–42.

Tomasello, M. (1992). *First verbs: a case study of early grammatical development*. Cambridge: Cambridge University Press.

—. (2003). *Constructing a language: a usage based theory of language acquisition.* Harvard University Press.

—. (2006). Construction Grammar for kids. *Constructions. Special Volume 1 - Constructions all over: case studies and theoretical implications*, SV1-11/2006.

Tomasello, M., Akhtar, N. Dodsen, K. and L. Rekau. (1997). Differential Productivity in Young Children's Use of Nouns and Verbs. *Journal of Child Language*, 24(2): 373-87.

Tonkes, B., Blair, A. and J. Wiles. (2000). Evolving learnable languages. In S. A. Solla, T. K. Leen and K.-R.Muller (Eds.), *Advances in Neural Information Processing Systems 12*, (NIPS*99). MIT Press, 66 - 72.

Toro, J.M., Sinnett S. and S. Soto-Faraco. (2005). Speech segmentation by statistical learning depends on attention. *Cognition*, 97(2): B25-B34.

Toro, J.M. and J. Trobalón. (2005). Statistical computations over a speech stream in a rodent. *Perception and Psychophysics*. 67(5): 867-875.

Tsonis, A.A., Schultz, C. and P.A. Tsonis. (1997). Zipf's law and the structure and evolution of language. *Complexity*, 3: 12-13.

Tsonis, A.A., Schultz, C. and P.A. Tsonis. (2003). On Words and genes. *Complexity*, 8: 12-13.

Tuldava, J. (1996). The frequency spectrum of text and vocabulary. *Journal of Quantitative Linguistics,* 3: 38-50.

Tversky, A. and D. Kahneman. (1974). Judgement under uncertainty: heuristics and biases. *Science,* 185: 1124-31.

Uziel-Karl, S. (1999). Children's verb lexicon. *Proceedings of the 30th Child Language Research Forum*. Center for the Study of Language, 41-49.

Valian, V. and S. Coulsen. (1988). Anchor points in language learning: the role of marker frequency. *Journal of Memory and Language*, 27: 71-86.

Van Geert, P.L.C. and M. van Dijk. (2002), Focus on variability: New tools to study intra- individual variability in developmental data. *Infant Behavior and Development*, 25: 340-374.

Vázquez, A., Dobrin R., Sergi D., Eckmann J.-P., Oltvai, Z.N. and A.-L. Barabási. (2004). The topological relationship between the large-scale attributes and local interaction patterns of complex networks. *Proceeding of the National Academy of Science,* 101(52): 17940-17945.

Vitevitch, M.S. (2008). What can graph theory tell us about word learning and lexical retrieval? *Journal of Speech-Language-Hearing Research*, 51(2): 408-422.

Vogel Sosa, A. (2000). Phonological Reduction in Frequency-Based Constituents: The Alternation of English 'of'. In K.A. Smith and D. Nordquist (Eds.), *Proceedings of the Third Annual High Desert Linguistics Society Conference*, April 7-9, (2000), University of New Mexico, 3: 1-8.

Vogel Sosa, A. and J. MacFarlane. (2002). Evidence for frequency-based constituents in the mental lexicon: collocations involving the word of. *Brain and Language*, 83(2): 227-236.

Vouloumanos, A. (2008). Fine-grained sensitivity to statistical information in adult word learning. *Cognition*, 107: 729-742.

Wade, T. and B. Möbius. (2008). Detailed Phonetic Memory for Multi-Word and Part-Word Sequences. *Laboratory Phonology 11*. Wellington, New Zealand, 30 June -2 July (2008). http://www.ims.uni-stuttgart.de/~moebius/papers/wade_moebius_labphon11.pdf Accessed 16th August (2008).

Wagner, A. (2005). Distributed robustness versus redundancy as causes of mutational robustness. *BioEssays*, 27(2): 176–188.

Watts, D.J. (1999). Networks, dynamics and the small world phenomena. *American Journal of Sociology*, 105: 493-527.

Watts, D. J. and S.H. Strogatz (1998). Collective Dynamics of 'small world' networks. *Nature*, 393: 440-442.

Werker J, Fennell C, Corcoran K, and C. Stager. (2002). Infants' ability to learn phonetically similar words: Effects of age and vocabulary size. *Infancy*, 3(1): 1-30.

West, G.B., Brown, J.H. and B.J. Enquist. (1999). The fourth dimension of life: Fractal geometry and allometric scaling of organisms. *Science,* 284: 1677-1679.

White, J.G., Southgate, E., Thomson, J.N. and S. Brenner. (1986). The structure of the nervous system of the nematode Caenorhabditis elegans. *Philosophical Transactions of the Royal Society of London, Series B, Biological Sciences*, 314(1165): 1-340.

Wichmann, S. (2005). On the power-law distribution of language family sizes. *Journal of Linguistics*, 41:117-131.

Wikipedia. (2008). http://en.wikipedia.org/wiki/Constituent_(linguistics) accessed 21.02.09. WordNet. (2008). http://wordnet.princeton.edu/ accessed 6th October (2008).

Yang, C.D. (2004) Universal Grammar, statistics or both? *TRENDS in Cognitive Sciences*, 8(10). 451-456.

Zipf, G.K. (1935). *The Psychobiology of Language*. Houghton-Mifflin. Reprinted 1965, Cambridge, MA: MIT Press.

—. (1949). *Human Behaviour and the Principle of Least Effort. An Introduction to Human Ecology*. Cambridge, MA.: Addison-Wesley.

Zwicky, A.M. (1985). Heads. *Journal of Linguistics*, 21(1): 1-29.